BIOSYNTHESIS OF TETRAPYRROLES

New Comprehensive Biochemistry

Volume 19

General Editors

A. NEUBERGER
London

L.L.M. van DEENEN
Utrecht

ELSEVIER
Amsterdam · London · New York · Tokyo

Biosynthesis of Tetrapyrroles

Editor

P.M. JORDAN

School of Biological Sciences, University of London, London E1 4NS, U.K.

1991

ELSEVIER

Amsterdam · London · New York · Tokyo

© 1991, Elsevier Science Publishers B.V. All rights reserved.

No part of this publication may be reproduced, stored in a retrieval system, or transmitted in any form or by any means, electronic, mechanical, photocopying, recording or otherwise, without the prior written permission of the Publisher, Elsevier Science Publishers B.V., Permissions Department, P.O. Box 521, 1000 AM Amsterdam, The Netherlands.

No responsibility is assumed by the Publisher for any injury and/or damage to persons or property as a matter of products liability, negligence or otherwise, or from any use or operation of any methods, products, instructions or ideas contained in the material herein. Because of the rapid advances in the medical sciences, the Publisher recommends that independent verification of diagnoses and drug dosages should be made.

Special regulations for readers in the USA. This publication has been registered with the Copyright Clearance Center, Inc. (CCC). Salem, Massachusetts. Information can be obtained from the CCC about conditions under which the photocopying of parts of this publication may be made in the USA. All other copyright questions, including photocopying outside of the USA, should be referred to the Publisher.

ISBN 0-444-89285-0 (volume)
ISBN 0-444-80303-3 (series)

This book is printed on acid-free paper.

Published by:

Elsevier Science Publishers B.V.
P.O. Box 211
1000 AE Amsterdam
The Netherlands

Sole distributors for the USA and Canada:

Elsevier Science Publishing Company, Inc.
655 Avenue of the Americas
New York, NY 10010
USA

Library of Congress Cataloging in Publication Data

Biosynthesis of tetrapyrroles / editor, P.M. Jordan.
    p.   cm. -- (New comprehensive biochemistry)
  Includes bibliographical references.
  ISBN 0-444-89285-0
  1. Tetrapyrroles--Synthesis.   I. Jordan, P.M. (Peter M.)
II. Series.
    [DNLM: 1. Pyrroles--metabolism.   W1 NE372F / QD 441 B615]
QD415.N48
[QP671.P6]
574.19'2 s--dc20
[574.19'218]
DNLM/DLC
for Library of Congress   91-34351
                                                                                                                                                                 CIP

Printed in The Netherlands

Foreword

The study of the structure and function of tetrapyrrolic compounds has excited the interests of organic chemists, biochemists, botanists and other biologists for more than a hundred years. The structures of most naturally occurring porphyrins were first elucidated by the efforts of Hans Fischer, and great progress was made in our knowledge of chlorophyll by Stoll and Willstatter by applying the best current methods of organic chemistry.

The next major advance was made by biochemists who used newly available isotopes of carbon and nitrogen to tackle the formation of porphyrins in biological systems. This was started by the discovery by Shemin and Rittenberg that, of all the amino acids, glycine specifically supplied the nitrogen atoms of protoporphyrin IX. This led to the elucidation of one of the major pathways for the synthesis of 5-aminolaevulinic acid, and ultimately to the origin of all the atoms in protoporphyrin IX. During this period conclusive evidence was also produced that chlorophyll and bacteriochlorophyll are formed in a series of reactions in which protoporphyrin IX is an essential intermediate. At about this time Dorothy Hodgkin demonstrated the structure of vitamin B_{12} by X-ray crystallography and it also became clear that this cobalt-containing compound was derived from uroporphyrinogen III.

The present volume deals to a large extent with the progress which has been made in this field during the last ten to fifteen years. These impressive further advances have been made possible by the application of physical methods, such as NMR spectroscopy and recombinant DNA technology which has allowed enzymes to be obtained in large amounts.

A completely unexpected finding was the observation that the factor F_{430} was a tetrapyrrole nickel complex and that this metal is a necessary factor in the growth of Archaebacteria. The elucidation of the structure of this porphyrin derivative also exhibited a feature which it shares with chlorophyll, sirohaem and vitamin B_{12} and this is the methylation of carbon atoms in varying positions of the macrocyclic system.

Another surprising finding was the discovery of a second pathway for the biosynthesis of 5-aminolaevulinic acid which was not based on glycine but on glutamate. That such a pathway existed had been known for some time but the intermediate steps involving glutamyl-tRNA were established only during the last few years by Danish workers. This is one of the few reactions other than protein biosynthesis which utilizes tRNA.

A problem which has puzzled scientists for many years is the mechanism of the biosynthesis of uroporphyrinogen III. It was known that the deaminase produces a linear hydroxymethylbilane called preuroporphyrinogen which is formed by the head to tail condensation of four porphobilinogen molecules. It has now been shown by several groups of workers that this enzyme contains a novel dipyrromethane cofactor which is covalently bound to the deaminase through the sulphydryl group of cysteine. This cofactor does not participate in the catalytic process but it functions as an anchor. The product of this reaction is a hexapyrrole from which the tetrapyrrolic bilane is cleaved.

In this field there have been many more exciting and unexpected developments, which are discussed in this volume. Moreover, not all the problems have been solved. Thus we do not know the exact mechanism involved in the last step in the biosynthesis of uroporphyrinogen III.

In the biosynthesis of chlorophyll we know that the first step is the insertion of magnesium into uroporphyrinogen III. This is presumably an enzyme reaction but as far as I know it has not been demonstrated in a cell-free system. This reaction seems to be closely coupled, at least in photosynthetic bacteria, with the methylation of the carboxyl group of protoporphyrin IX. It thus appears that there is great scope for further research in this important field.

Lister Institute   Albert Neuberger
London

Contents

Foreword v

Chapter 1
The biosynthesis of 5-aminolaevulnic acid and its transformation into uroporphyrinogen III
P.M. Jordan (London, UK) 1

1. Introduction 1
2. Early isotopic studies on the origin of the tetrapyrrole ring 2
3. The biosynthesis of 5-aminolaevulinic acid from glycine 4
   3.1. Occurrence and properties of 5-aminolaevulinic acid synthase 4
   3.2. Regulation of 5-aminolaevulinic acid synthase in bacteria 5
   3.3. Regulation of 5-aminolaevulinic acid synthase in eukaryotes 6
   3.4. Substrate specificity and kinetics 9
   3.5. Mechanism of 5-aminolaevulinic acid synthase 11
   3.6. Structure of the enzyme and molecular biology 14
4. The biosynthesis of 5-aminolaevulinic acid from glutamate 15
   4.1. Discovery of the C-5 pathway 15
   4.2. The enzymes of the C-5 pathway 16
5. The biosynthesis of porphobilinogen 19
   5.1. Introduction 19
   5.2. General properties of 5-aminolaevulinic acid dehydratases 20
   5.3. Importance of sulfhydryl groups 21
   5.4. Requirement for metals 22
   5.5. Inhibition by lead 23
   5.6. Nature of the active site groups 24
   5.7. Order of binding the two substrates and the enzyme mechanism .. 26
   5.8. Catalytic groups and the nature of the active site 28
6. The biosynthesis of uroporphyrinogen III 30
   6.1. Introduction 30
   6.2. Properties of porphobilinogen and uroporphyrinogens 32
   6.3. Porphobilinogen deaminase 33
   6.4. Uroporphyrinogen III synthase (cosynthase) 34
   6.5. Early investigations on the mechanism of uroporphyrinogen biosynthesis .. 34
   6.6. Experiments with aminomethyldipyrromethanes, aminomethyltripyrranes and aminomethylbilanes 35
   6.7. The discovery of preuroporphyrinogen, the substrate for uroporphyrinogen III synthase ... 38
   6.8. Order of assembly of the four pyrrole rings of the tetrapyrrole ... 41
   6.9. Enzyme intermediate complexes between the deaminase and porphobilinogen ... 42
   6.10. Preliminary studies on the nature of the enzymic group involved in substrate covalent binding 43
   6.11. Discovery of a cofactor in all porphobilinogen deaminases – the dipyrromethane cofactor ... 43
   6.12. Further evidence for the role of the dipyrromethane cofactor from experiments with the chain termination suicide inhibitor α-bromoporphobilinogen 46
   6.13. Attachment site and mechanism of assembly of the dipyrromethane cofactor ... 47

6.14. Mechanism of action of porphobilinogen deaminase	49
6.15. Reaction of porphobilinogen deaminase with 1-aminomethylbilanes	53
6.16. Steric course of the porphobilinogen deaminase and uroporphyrinogen III synthase reactions	53
6.17. Use of synthetic analogues to investigate the uroporphyrinogen III synthase reaction	55
6.18. Mechanism of action of uroporphyrinogen III synthase	56
6.19. Molecular biology and protein structure of porphobilinogen deaminase	57
6.20. Molecular biology of uroporphyrinogen synthase	58
Acknowledgements	59
References	59

Chapter 2
Mechanism and sterochemistry of the enzymes involved in the conversion of uroporphyrinogen III into haem
M. Akhtar (Southampton, Hants., UK) 67

1.	Introduction	67
2.	Uroporphyrinogen decarboxylase (uroporphyrinogen carboxylase)	67
	2.1. Introduction	67
	2.2. Intermediates in the decarboxylation	68
	2.3. The interaction between the decarboxylase and the substrates	70
	2.4. Isolation and structural studies on uroporphyrinogen decarboxylase	72
	2.5. The number of active sites	73
	2.6. Mechanistic and stereochemical studies	74
	2.7. The mechanism	75
3.	Coproporphyrinogen III oxidase	77
	3.1. Introduction	77
	3.2. Isolation of, and structural studies on, coproporphyrinogen oxidase	78
	3.3. The order in which the propionic side-chains are decarboxylated	79
	3.4. Stereochemical and mechanistic studies on coproporphyrinogen III oxidase	79
	3.5. The mechanism of the aerobic coproporphyrinogen III oxidase catalysed reaction	81
	3.6. Concerning the stereochemistry and mechanism of the anaerobic coproporphyrinogen III oxidase	83
4.	Protoporphyrinogen IX oxidase	84
	4.1. Introduction	84
	4.2. Isolation of, and structural studies on, protoporphyrinogen IX oxidase	85
	4.3. Stereochemical and mechanistic studies on protoporphyrinogen IX oxidase	86
	4.4. The mechanism of the protoporphyrinogen IX oxidase catalysed reaction	90
5.	Ferrochelatase	92
	5.1. Occurrence and isolation of ferrochelatases	92
	5.2. Properties of ferrochelatases	92
	5.3. Substrate specificity	93
	5.4. Metal requirements	94
	5.5. Inhibition of ferrochelatases by N-alkylporphyrins	94
	5.6. Kinetics, active site and mechanism of action of ferrochelatases	95
Acknowledgements		96
References		96

Chapter 3
The biosynthesis of vitamin B_{12}
   A.I. Scott and P.J. Santander (College Station, TX, USA)   101

1. Introduction   101
2. The carbon balance   104
3. Stereochemistry of methyl group insertion in corrinoid biosynthesis   104
4. Concerning the fate of the methyl group protons   106
5. Uro'gen III is a precursor of vitamin B_{12}   106
6. Characterization and intermediacy of the isobacteriochlorins of *P. shermanii*   107
7. The methylation sequence: Pulse experiments   112
8. Timing of the decarboxylation step   116
9. The protein balance of vitamin B_{12} biosynthesis   120
10. Factors S_1-S_4, isomeric, tetramethylated corphinoids derived from uro'gen I   123
11. The methyl transferases   128
12. Biosynthesis of the neucleotide loop and coenzyme B_{12}   132
13. Evolutionary aspects of B_{12} biosynthesis   133
    13.1. The C_5 pathway   133
    13.2. Chemical methods   134
    13.3. Why type III?   134
References   135

Chapter 4
Biochemistry of coenzyme F430, a nickel porphinoid involved in methanogenesis
   H.C. Friedmann, A. Klein and R.K. Thauer (Marburg, FRG and Chicago, IL, USA)   139

1. Introduction   139
2. Structural relations to other tetrapyrroles   140
3. Biosynthesis from glutamate via uroporphyrinogen III and dihydrosirohydrochlorin   143
4. Properties of free coenzyme F430 including its redox behaviour   144
5. Function of coenzyme F430 as prosthetic group of methyl coenzyme M reductase in methanogenesis   147
6. Comparative analysis of genes encoding methyl coenzyme M reductase   149
Acknowledgements   152
References   152

Chapter 5
Biochemistry and regulation of photosynthetic pigment formation in plants and algae
   S.I. Beale and J.D. Weinstein (Providence, RI and Clemson, SC, USA) . .   155

1. The variety and functions of plant and algal tetrapyrroles   155
    1.1. Introduction to the branched tetrapyrrole biosynthetic pathway   155
    1.2. Chlorophylls and bacteriochlorophylls   155

1.3.	Hemes		156
1.4.	Phycobilins and the phytochrome chromophore		156
1.5.	Other tetrapyrroles found in phototrophic organisms		158
2. The biosynthetic route			158
2.1.	ALA formation		158
	2.1.1.	ALA biosynthesis from glycine and succinyl-CoA	158
	2.1.2.	ALA biosynthesis from glutamate	161
	2.1.2.1.	*In vivo* evidence for ALA and tetrapyrrole formation from glutamate	161
	2.1.2.2.	Mechanism of ALA formation from glutamate	162
	2.1.2.2.1.	tRNA^{Glu}	162
	2.1.2.2.2.	Glutamyl-tRNA	163
	2.1.2.2.3.	Glutamyl-tRNA synthetase	164
	2.1.2.2.4.	Glutamate-1-semialdehyde	164
	2.1.2.2.5.	Dehydrogenase	165
	2.2.2.2.6.	Aminotransferase	165
	2.1.3.	Phylogenetic distribution of the two ALA-forming pathways	166
2.2.	The Pathway from ALA to protoporphyrin IX		167
	2.2.1.	ALA dehydratase	167
	2.2.2.	PBG deaminase	168
	2.2.3.	Uroporphyrinogen III synthase	169
	2.2.4.	Uroporphyrinogen decarboxylase	169
	2.2.5.	Coproporphyrinogen oxidase	170
	2.2.6.	Protoporphyrinogen oxidase	170
2.3.	The Fe branch		170
	2.3.1.	Ferrochelatase and protoheme formation	170
	2.3.2.	Bilin formation	172
	2.3.2.1.	Heme as a phycobilin precursor	172
	2.3.2.2.	Biliverdin as a phycobilin precursor	172
	2.3.2.3.	Algal heme oxygenase	173
	2.3.2.4.	Biliverdin reduction to phycocyanobilin	174
	2.3.2.5.	Biosynthesis of other phycobilins	177
	2.3.2.6.	Biosynthetic relationship of the phytochrome chromophore to phycobilins	177
	2.3.3.	Ligation of phycobilins to apoproteins	178
2.4.	The Mg branch		179
	2.4.1.	Chlorophyll *a* formation	179
	2.4.1.1.	Chelation of Mg	179
	2.4.1.2.	Mg-protoporphyrin methyl transferase	182
	2.4.1.3.	Isocyclic ring formation	183
	2.4.1.4.	Vinyl group reduction	187
	2.4.1.5.	Protochlorophyllide reduction	188
	2.4.1.5.1.	Light-requiring reduction catalyzed by protochlorophyllide oxidoreductase	188
	2.4.1.5.2.	Dark protochlorophyllide reduction	192
	2.4.1.6.	Esterification of the ring-D propionate	193
	2.4.2.	Chlorophyll *b* formation	195
	2.4.2.1.	*In vivo* studies	195
	2.4.2.2.	*In vitro* studies	198
	2.4.3.	Possible existence of multiple forms of chlorophylls *a* and *b*	199
	2.4.4.	Chlorophylls *c*	200
	2.4.5.	Reaction center chlorophylls	201
	2.4.6.	Reaction center pheophytins	201

3. Regulation 202
   3.1. Environmental and physiological modulation of pigment content 202
      3.1.1. Physiology of greening in plants and algae 202
      3.1.2. Effects of Fe chelators and anaerobiosis on greening in intact tissue and chloroplasts 205
      3.1.3. Chlorophyll and heme turnover 206
      3.1.4. Nonplastid heme synthesis and turnover 208
   3.2. Regulation of biosynthetic steps 209
      3.2.1. Subcellular compartmentation of tetrapyrrole biosynthesis 209
      3.2.2. Regulations of metabolic activity 210
         3.2.2.1. Effector-mediated regulation of ALA-forming activity 211
         3.2.2.2. Expression and turnover of ALA-forming enzymes 213
         3.2.2.3. Regulation of ALA dehydratase activity 215
         3.2.2.4. Branch point regulation 215
         3.2.2.5. Regulation of protochlorophyllide photoreduction 216
   3.3. Regulation of phycobilin content by light and nutritional status 218
      3.3.1. Light effects 218
      3.3.2. Responses to nitrogen status 221
Acknowledgments 222
References 223

Chapter 6
The structure and biosynthesis of bacteriochlorophylls
Kevin M. Smith (Davis, CA, USA) 237

1. Nomenclature 237
2. Occurrence and structures 238
   2.1. Bacteriochlorophylls *a, b* 238
   2.2. Bacteriochlorophylls *c, d, e, f* 239
   2.3. Bacteriochlorophyll *g* 239
3. Biosynthesis 240
   3.1. Bacteriochlorophylls *a, b, g* 240
   3.2. Bacteriochlorophylls *c, d, e, f* 246
Acknowledgment 253
References 253

Chapter 7
The genes of tetrapyrrole biosynthesis
P.M. Jordan and B.I.A. Mgbeje (London, UK) 257

1. Genes of haem biosynthesis 257
   1.1. Mapping of haem biosynthesis genes 257
      1.1.1. The *hem* genes in *Escherichia coli* 258
      1.1.2. The *hem* genes in *Salmonella typhimurium* 259
      1.1.3. The *hem* genes in *Staphylococcus aureus* 260
      1.1.4. The *hem* genes in *Bacillus subtilis* 260
      1.1.5. The *hem* genes in *Rhodobacter sphaeroides* 261
      1.1.6. The *hem* genes in yeast 261

		1.1.7.	The *hem* genes in mammalian species	262
	1.2.	Molecular biology of haem biosynthesis genes		263
		1.2.1.	5-Aminolaevulinic acid synthesis	263
		1.2.1.1.	5-Aminolaevulinic acid synthase (the glycine pathway)	263
		1.2.1.2.	The genes of the glutamate (C_5) pathway	266
		1.2.2.	5-Aminolaevulinic acid dehydratase	268
		1.2.3.	Porphobilinogen deaminase	269
		1.2.4.	Uroporphyrinogen III synthase (cosynthase)	272
		1.2.5.	Uroporphyrinogen decarboxylase	273
		1.2.6.	Coproporphyrinogen oxidase	274
		1.2.7.	Protoporphyrinogen oxidase	275
		1.2.8.	Ferrochelatase	275
2.	Genes of cobalamin (vitamin B_{12}) biosynthesis			276
3.	Genes of bacteriochlorophyll and chlorophyll biosynthesis			282
	3.1.	Bacteriochlorophyll biosynthesis genes		282
	3.2.	Chlorophyll biosynthesis genes		287
References				288

Subject Index 295

CHAPTER 1

The biosynthesis of 5-aminolaevulinic acid and its transformation into uroporphyrinogen III

PETER M. JORDAN

School of Biological Sciences, Queen Mary and Westfield College, University of London, Mile End Road, London, E1 4NS, U.K.

1. Introduction

The tetrapyrrole biosynthesis pathway is broadly similar in all living systems, the first common intermediate being the highly reactive aminoketone, 5-aminolaevulinic acid. There are two totally distinct routes by which 5-aminolaevulinic acid is produced, one utilising the carbon skeleton of glutamate and the other involving glycine and succinyl-CoA. The glutamate, or C_5 pathway is found in many anaerobic bacteria and in plants, whereas the pathway using glycine and succinyl-CoA, although occurring in some bacteria, is mainly confined to animals and other eukaryotes. Once formed, 5-aminolaevulinic acid is transformed into the tetrapyrrole macrocycle in only three enzymic stages. In the first of these, two molecules of 5-aminolaevulinic acid condense with one another forming the basic pyrrole building block, porphobilinogen. Next, four molecules of porphobilinogen polymerize together in a chain to generate a highly unstable 1-hydroxymethylbilane called preuroporphyrinogen. Lastly, preuroporphyrinogen is cyclised, with rearrangement of one of the pyrrole rings, the D ring, to yield the key intermediate uroporphyrinogen III. Uroporphyrinogen III is the universal precursor from which porphyrins, haems, chlorophylls, corrins and all other tetrapyrroles are derived. Uroporphyrinogen III is at a major branch point in the pathway and may either be converted into protoporphyrin IX en route to haem, chlorophylls and bacteriochlorophylls or, alternatively, it may be methylated for transformation into sirohaem, vitamin B_{12} or the nickel cofactor F_{430}. Although no single organism elaborates all these final tetrapyrrole products a global summary of these pathways is shown in Scheme 1.

This chapter deals with the enzymic synthesis of 5-aminolaevulinic acid and the subsequent three enzymic steps to porphobilinogen, preuroporphyrinogen and uroporphyrinogen III. These three stages are common to all living systems which synthesise tetrapyrroles. Additional information on the biosynthesis of uroporphyrinogen III in plants is covered in Chapter 5. The enzymic stages from uroporphyrinogen III

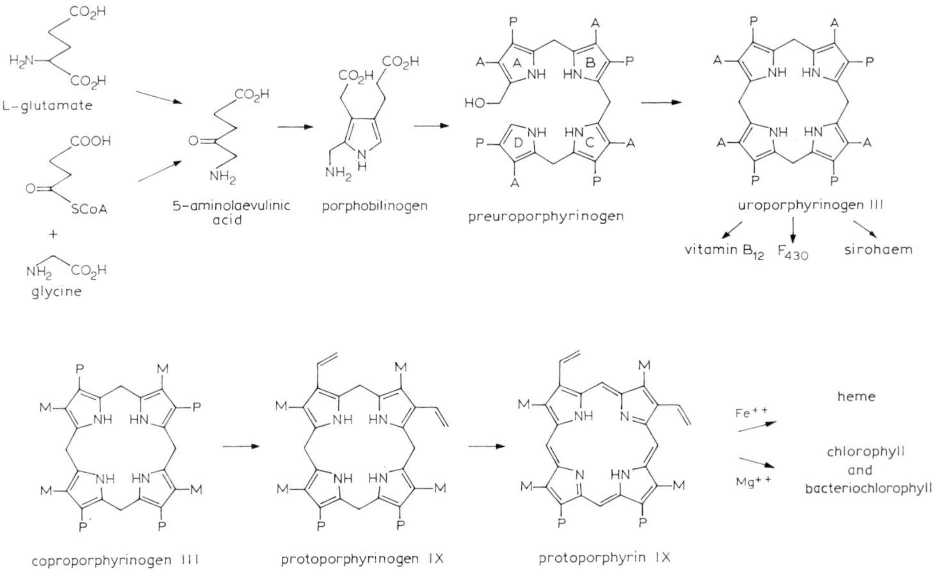

Scheme 1. The biosynthesis of tetrapyrroles

to haem are described in detail in Chapter 2 and the molecular biology of the pathway is reviewed in Chapter 7. The transformation of uroporphyrinogen III into vitamin B_{12} and F_{430} is discussed in Chapters 3 and 4, respectively. Comprehensive accounts of chlorophyll biosynthesis and bacteriochlorophyll biosynthesis are given in Chapters 5 and 6.

2. Early isotopic studies on the origin of the tetrapyrrole ring

The elucidation of the early stages in the tetrapyrrole biosynthesis pathway was largely complete by the end of the 1950s and all the intermediates, with the exception of preuroporphyrinogen, had been characterized. The major contributions stemmed from the investigations carried out by David Shemin and his co-workers in the USA and the group of Albert Neuberger in the UK. Central to their methodology was the use of isotopically labelled substrates, which had just become available, in order to follow the incorporation of potential precursors into haem. Shemin, using himself initially as an experimental subject, orally administered [^{15}N] glycine and established that this amino acid was the most efficient precursor of the haem nitrogen atoms [1]. Subsequently, experiments carried out with [^{14}C] labelled glycine using enzyme extracts from avian erythrocytes established that the C-2 carbon atom, but not the carboxyl-carbon atom, of glycine was incorporated into 8 of the positions in the haem macrocyclic ring [2,3]. This was deduced by degrading, painstakingly and systematically, the labelled haem by a procedure which yielded ethylmethylmaleimide and haematinic acid. The maleimides were then further degraded by methods which permitted the isolation and positional assignment of each of the carbon atoms present in the haem macrocycle. Additional labelling experiments using 1-[^{14}C] and 2-[^{14}C] acetate established that the remaining 26 carbon atoms of haem were derived from an unsymmetrical 4-carbon compound arising from the tricarboxylic acid cycle [4] and a succinyl-derivative such as succinyl-CoA was suggested as a possible candidate. Subsequently, experimental proof for the involvement of succinyl-CoA was obtained from in vitro experiments using avian erythrocyte preparations [5].

From these pioneering studies the origin of all the carbon and nitrogen atoms of haeme were established. The findings are summarized in Scheme 2.

This labelling pattern applies to tetrapyrroles biosynthesized in mammalian, avian and some bacterial systems but not to those formed in higher plants, algae and anaerobic bacteria where glutamic acid provides the carbon and nitrogen atoms. This latter aspect is only discussed briefly below but is covered in detail in Chapter 5.

The labelling data obtained from the incorporation experiments with radioactive glycine and acetate, and especially the observation that the carboxyl-carbon atom of glycine was not incorporated into haem, led to the hypothesis that glycine and succinyl-CoA condense together to form 2-amino 3-ketoadipic acid which yields 5-aminolaevulinic acid on decarboxylation [6]. This was confirmed when 5-amino[5-^{14}C]laevulinic acid was chemically synthesised and shown to be incorporated into haem in high yield with a labelling pattern identical to that given by [2-^{14}C]glycine [7]. The discovery of 5-aminolaevulinic acid ranks highly in the history of biochemical investigation. It is most educational to read the first-hand accounts from the laboratories of Shemin and Neuberger describing this early phase in the elucidation of the tetrapyrrole pathway.

At the time that 5-aminolaevulinic acid was first described, the existence of the next intermediate in the porphyrin biosynthetic pathway, porphobilinogen, had been

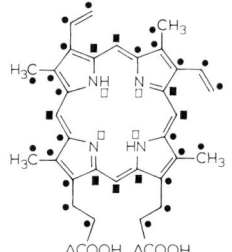

● = CH$_3$ carbon of acetate
△ = COOH carbon of acetate
■ = CH$_2$ carbon of glycine
□ = N of glycine

Scheme 2. Origin of carbon and nitrogen atoms in protoporphyrin IX known for several years, having been isolated from the urine of patients suffering from acute intermittent porphyria [8]. The significance of porphobilinogen as an intermediate in the haem biosynthetic pathway was not appreciated, however, until its structure had been elucidated by X-ray crystallography [9] and found to be strikingly similar to that postulated for the haem precursor pyrrole. It was then quite evident that two molecules of 5-aminolaevulinic acid could account for all the carbon and nitrogen atoms of porphobilinogen thus explaining the results from the radioactive labelling experiments. Porphobilinogen was shown subsequently to act as an excellent precursor for porphyrins [10] and its role as an intermediate was finally confirmed when it was shown to be formed enzymatically from 5-aminolevulinic acid [11,12]. Three years later Bogorad [13,14] described enzyme systems that would transform porphobilinogen into uroporphyrinogen III and although the complexity of this system was not fully appreciated, the molecular basis for the early stages of the tetrapyrrole biosynthesis pathway was largely complete. Comprehensive accounts on haem biosynthesis by Dolphin [15], Granick and Beale [16], Battersby and McDonald [17], Akhtar and Jordan [18], Leeper [19,20] and Dailey [21] are available. The reader should also refer to Chapters 2, 5 and 7 in this volume.

3. The biosynthesis of 5-aminolaevulinic acid from glycine

3.1. Occurrence and properties of 5-aminolaevulinic acid synthase

5-Aminolaevulinic acid synthase catalyses the condensation between glycine and succinyl-CoA to yield 5-aminolaevulinic acid, CoA and carbon dioxide as shown in Scheme 3.

5-Aminolaevulinic acid synthase (E.C. 2.3.1.37) was described simultaneously by Shemin and his co-workers [22] in bacterial extracts and by Neuberger and his group in avian preparations [5]. Since that time the enzyme has been purified from many

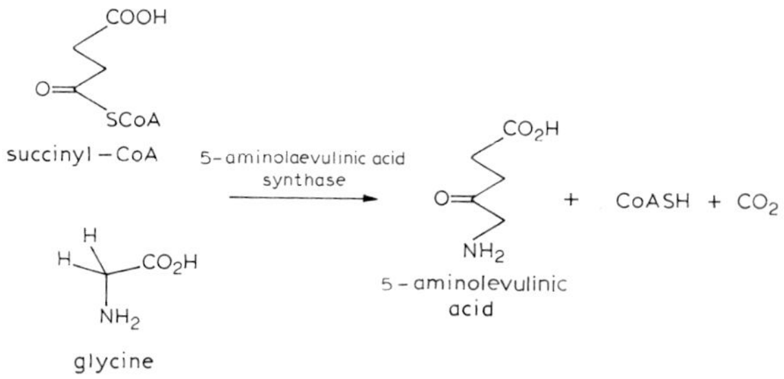

Scheme 3. The 5-aminolaevulinic acid synthase reaction

sources including *Rhodobacter sphaeroides* [23,24,25], rat liver [26,27], chicken liver [28], *Euglena* [29] and yeast [30,31].

In eukaryotes the enzyme is located in mitochondria, reflecting the requirement for succinyl-CoA as one of the substrates. Although succinyl-CoA is generated by other enzyme systems such as methylmalonyl-CoA mutase, succinate thiokinase and acetoacetyl-CoA:succinate transferase, the major source of succinyl-CoA for haem synthesis is from the tricarboxylic acid cycle. It is worth commenting on the fact that the synthesis of 5-aminolaevulinic acid is the only stage in haem biosynthesis which requires an activated substrate, succinyl-CoA, and that the remaining steps of the porphyrin biosynthesis pathway are energetically 'downhill' as a result of aromatic ring formation, deaminations, decarboxylations or oxidations

Because of the mitochondrial location of the 5-aminolaevulinic acid synthase in eukaryotes a considerable amount of investigation has been focussed on the mechanism by which the enzyme is first synthesised in the cytosol and then transported across the mitochondrial membrane to its ultimate location. Experiments with chick embryo liver [28,32] and yeast [31] have established that the enzyme is first synthesised in the cytosol as a pre-enzyme which is rapidly imported into the mitochondria and processed to give the mature form. The chick embryo liver enzyme appears to be synthesised initially as a precursor molecule of M_r 74,000 Da with an amphipathic N-terminus which is recognised by the mitochondrial import system. After its import across the mitochondrial membrane the first 56 N-terminal amino acids are removed to yield the mature enzyme, M_r 68,000 Da. The rat liver 5-aminolaevulinic acid synthase precursor also has a M_r of 75,000 Da [27]. A similar situation exists in yeast where the pre-enzyme, which has a M_r of 59,000 Da, is rapidly converted into the mature enzyme after import into mitochondria. The N-terminal portion of the yeast pre-enzyme is quite basic and, in addition, contains several threonine and serine residues representing a good recognition sequence for mitochondrial targetting. The mature yeast 5-aminolaevulinic acid synthase subunit has a M_r of 55,000 Da. The human liver 5-aminolaevulinic acid synthase cDNA also shows the presence of an N-terminal sequence of 56 amino acids [33].

3.2. Regulation of 5-aminolaevulinic acid synthase in bacteria

As the first enzyme of the pathway, 5-aminolaevulinic acid synthase has been a major focus of attention for the study of the regulation of tetrapyrrole biosynthesis. In bacteria such as *R. sphaeroides*, which can rapidly adapt from non-photosynthetic aerobic growth to photosynthetic anaerobic growth, a large number of proteins are induced, many of which are required for the generation of the photosynthetic apparatus. Lascelles has established that the demand for bacteriochlorophyll biosynthesis increases by as much as 100 fold [34] and 5-aminolaevulinic acid synthase is induced to high levels. Haem is an important regulator acting as both a feed-back inhibitor of the enzyme directly and also as a regulator at the nucleic acid level.

There is evidence from enzymic studies that two forms of 5-aminolaevulinic acid synthase are present in *R. sphaeroides*. The two enzyme forms have been purified [35] although it is not known whether these represent a precursor and mature forms of the same enzyme or completely distinct enzymes. The proportion of the enzymic forms and the enzyme activity vary according to the conditions under which the bacteria are grown. Recently, the identification by Kaplan and associates of two genes, *hemA* and *hemT*, encoding 5-aminolaevulinic acid synthase [36], one on each chromosome, has provided unambiguous evidence for the existence of two 5-aminolaevulinic acid synthases, however, it has not yet been demonstrated whether they have specific metabolic roles or specify the two enzymes identified previously by Tuboi et al. [35]. *R. sphaeroides* is unusual in being able to biosynthesise haeme, bacteriochlorophyll and corrin which necessitates the regulation of tetrapyrrole precursors along three branches. It is tempting to speculate that one of the 5-aminolaevulinic acid synthase genes subserves the background levels of 5-aminolaevulinic acid required for haeme biosynthesis under aerobic conditions and that the second gene may encode an enzyme dedicated to the synthesis of 5-aminolaevulinic acid destined for bacteriochlorophyll and corrin biosynthesis under anaerobic conditions. There is little direct evidence for this proposal although the presence of 'aerobic' and 'anaerobic' coproporphyrinogen oxidase enzymes [37] indicates that at least one other stage of the pathway may require dual gene and enzyme systems.

Another regulatory mechanism which may be present in photosynthetic bacteria relates to the regulation of pre-existing levels of 5-aminolaevulinic acid synthase by a mechanism linked to the oxygen status of the cells. Evidence for the presence of both an inhibitor and an activator of the enzyme has been obtained which may be regulated through sulphur-containing metabolites such as the trisulphides of cysteine and glutathione [38]. These compounds appear to be able to activate the 5-aminolaevulinate synthase enzyme in vitro by converting a low activity form to a high activity form. Such observations deserve further investigation at the enzyme level both in vitro and in vivo. Under constant semi-anaerobic conditions, 5-aminolaevulinic acid levels are several times higher in dim light compared to intense light suggesting that a regulatory mechanism triggered by light is present. Details of regulation of the bacteriochlorophyll pathway are included in chapter 7.

3.3. Regulation of 5-aminolaevulinic acid synthase in eukaryotes

5-Aminolaevulinic acid synthase from eukaryotes has also been, predictably, the focus of attention with respect to the regulation of the flux of intermediates through the haem biosynthesis pathway. As with bacteria, such as *R. sphaeroides*, there are also two 5-aminolaevulinic acid synthase genes in mammals, one of which is expressed in erythropoietic tissues and the other which is expressed as the 'housekeeping' protein in all tissues (see also chapter 7).

Most investigations on the regulation of 5-aminolaevulinic acid synthase have been focussed on the rat liver and chicken embryo liver systems. However the first indication that 5-aminolaevulinic acid synthase is a key regulatory step came from observations on patients suffering from acute intermittent porphyria in which 5-aminolaevulinic acid was found to be elevated during an acute attack [39,40]. The demonstration by Granick that 5-aminolaevulinic acid synthase is induced by drugs in experimental animals [41] in order to generate haem for de novo cytochrome P_{450} formation also greatly stimulated research on the role of haem in the feed-back regulation of its own synthesis.

In hepatic tissues, the regulation of 5-aminolaevulinic acid synthase by haem is accomplished by several mechanisms which extend from the regulation of the transcription of 5-aminolaevulinic acid-specific mRNA to a direct effect on the activity of the enzyme itself. One of the first experiments which suggested that low intracellular haem levels lead to the induction of 5-aminolaevulinic acid synthase stemmed from the observation that enzyme which had been induced to high levels by drugs can be suppressed by the presence of haem [42]. More recently, haem depletion using the 5-aminolaevulinic acid dehydratase inhibitor, succinylacetone, has been shown to result in 5-aminolaevulinic acid synthase induction in murine erythroleukaemia (MEL) cells [43]. These experiments have been extended to the study of dimethylsulphoxide-promoted 5-aminolaevulinic acid synthase induction by succinylacetone which is prevented by haem [44]. Using avian embryo liver and rat liver systems the mechanism of suppression has been shown to be through the inhibition of 5-aminolaevulinic acid-specific mRNA synthesis which is reduced in the presence of haem [45,46].

The fact that the mRNA specifying 5-aminolaevulinic acid synthase has an unusually short half life of about 1 hour in the rat [39], 3 hours in mouse liver [47] and 5 hours in chick embryo liver [48] indicates an additional regulatory device. Thus, in the presence of high haem concentrations, when 5-aminolaevulinic acid synthase-specific mRNA synthesis is suppressed, the capacity for 5-aminolaevulinic acid synthase translation decreases rapidly allowing a quick response to the haem pool level.

The process of translation itself is also thought to be under the influence of haem. Granick et al. [49] suggested a model for the interaction of haem at the polysome level. Evidence for interaction of haem with a specific regulatory protein has been suggested as a mechanism for translational control [48,50]. Once synthesised in the cytoplasm, the 5-aminolaevulinic acid synthase pre-enzyme needs to traverse the mitochondrial membrane and it has been proposed that haem plays a significant role in the rate of the translocation process. Convincing evidence suggests that haem may prevent mitochondrial import of the protein in rats [51] and in chicken embryo liver [52,53].

Having been imported, the regulation of the 5-aminolaevulinic acid synthase level is a function of the enzyme stability. Several groups of investigators have shown that the half life of the enzyme is as little as 1 to 2 hours [39,47,48]. Evidence for intrinsic instability of rat liver 5-aminolaevulinate synthase indicates that this is also a most important regulatory mechanism [51]. Similar findings have been forthcoming from studies on both mitochondrial and cytosolic forms of the enzyme from induced chicken liver cells [54]. Thus under conditions where the formation of new enzyme is suppressed by any of the above mechanisms, the existing enzyme will be depleted within a few hours. Clearly alterations in the balance between rapid synthesis and rapid breakdown of the enzyme permits a fast response to the cellular requirements for haem. An additional regulatory mechanism may involve the direct interaction of haem with the 5-aminolaevulinic acid synthase enzyme. Studies using rat liver have shown that haem at relatively low concentrations ($K_i=20\ \mu M$) inhibits the activity of 5-aminolaevulinic acid [55]. This mechanism is not thought to be operative generally in mammalian systems in the same way that haem inhibits directly the 5-aminolaevulinic acid synthase from *R. sphaeroides*.

Extensive investigations on the effects of many of the porphyrinogenic drugs have established that induction of 5-aminolaevulinic acid synthase occurs as a result of the lowering of the intracellular haem pool due, in almost all cases, to the increased requirement for haem for the biosynthesis of cytochrome P_{450} enzymes. Drugs such as phenobarbitone cause a dramatic induction of 5-aminolaevulinic acid synthase in rat liver [56]. In the case of the 5β-steroid metabolites there is also strong evidence for the induction of 5-aminolaevulinic acid synthase through a mechanism involving the depletion of the free haem pool for cytochrome P_{450} synthesis since the induction of the enzyme is reversed when depletion of the intracellular haem pool is prevented [57,58]. The haem pool has been estimated as $10^{-7}–10^{-9}M$ by Granick et al [49]. Other compounds, such as 3,5-dicarbethoxy-1,4-dihydrocollidine (DDC) lead to the induction of 5-aminolaevulinic acid synthase [59] via a different mechanism of haem depletion. This compound causes the breakdown of cytochrome P_{450} haem by a 'suicide' reaction which involves the alkylation of the tetrapyrrole ring to give green pigments identified as N-alkylporphyrins. The N-alkylporphyrins are powerful competitive inhibitors of the terminal enzyme of the pathway, ferrochelatase [60] with values for K_i in the low micromolar range. The combined effect of haem destruction and ferrochelatase inhibition lowers the intracellular haem pool [61,62]. In all cases described above, the lowered level of cellular haem leads to the induction of 5-aminolaevulinic acid synthase which may reach greatly elevated levels. It may be concluded, at least in general terms, that all compounds which are metabolised by cytochrome P_{450} enzymes have the ability to induce 5-aminolaevulinic acid synthase in this way [63]. Since strong correlations between drug levels and elevated 5-aminolaevulinic acid synthase-specific mRNA have consistently been shown to depend on intracellular haem levels it is now generally accepted that the prime regulatory device for activating the 5-aminolaevulinic acid synthase gene is a reduced haem level. The precise details of the mechanism of induction, however, remain still somewhat unclear. The various mechanisms discussed above by which 5-aminolaevulinic acid synthase is regulated are summarised below.

The mRNA specifying erythroid 5-aminolaevulinic acid synthases during differen-

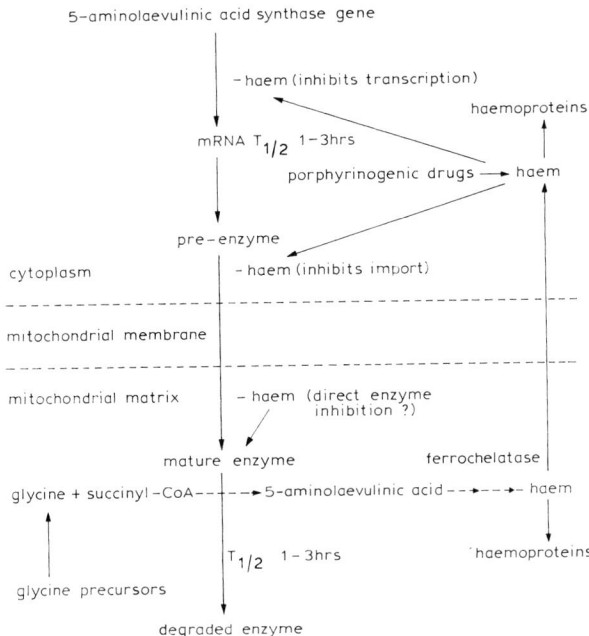

Figure 1. The mechanisms of 5-aminolaevulinic acid synthase regulation by haem pool size.

tiation of precursor cells into reticulocytes is regulated by different mechanisms compared to those outlined for the non-erythroid systems. Most of the studies on the erythroid system have been confined to MEL cells which show some properties similar to reticulocyte precursor cells. This mechanism involves the coordinated synthesis of haem and globins for haemoglobin formation. The 5-aminolaevulinic acid synthase mRNA specifying the erythroid enzyme may be induced to levels approaching 100 times the non induced levels in MEL cells [64]. The activity of the enzyme appears to respond to haem levels [44]. Induction of MEL cells has shown that other haem synthesis enzymes also become raised [44,65]. The regulation of 5-aminolaevulinate synthase levels, and the levels of other haem synthesis enzymes, in differentiating erythroid cells is complex and outside the scope of this chapter. The reader is directed towards detailed texts elsewhere [66,67].

3.4. Substrate specificity and kinetics

The elucidation of the mechanism by which 5-aminolaevulinic acid synthetase catalyses the synthesis of 5-aminolaevulinic acid has been approached from several directions including kinetic analysis, studies with exchange reactions and the use of stereospecifically tritiated glycine. Most of the mechanistic details have been deduced from investigations with the enzyme isolated from *R. sphaeroides* and there is no reason to expect that other 5-aminolaevulinic acid synthases follow a different mechanistic course.

All the 5-aminolaevulinic acid synthase enzymes purified to date appear to exist as homodimers with subunit M_rs varying from 40,000 Da to 60,000 Da. None of the enzymes appear to accept any other amino acid substrate except glycine. Even glycine is bound to the enzyme with a relatively low affinity and K_m values in the mM range are found for most of the enzymes whose kinetics have been investigated [68]. Glycine analogues such as methylamine, β-alanine and aminoethanol are only very poor inhibitors. Some amino acids such as alanine and serine also act as weak inhibitors, the L-amino acids showing preference over the D-amino acids. The most spectacular inhibition is shown by aminomalonate, with a K_i of 6 μM for the enzyme from R. sphaeroides [69]. Aminomalonate does not act as a substrate and is not, perhaps surprisingly, decarboxylated to glycine.

The rigid substrate specificity seen for glycine is not adhered to when one considers the acyl-CoA substrate. Although succinyl-CoA is by far the most favoured substrate with K_m values in the low μM range, succinyl-CoA monomethyl ester, α-glutamyl-CoA and acetyl-CoA are all accepted as poor substrates by the enzyme yielding the corresponding aminoketone products in yields which relate closely to their chemical reactivity [68].

Analysis of the 5-aminolaevulinic acid synthase from R. sphaeroides Y using steady state kinetics [70] has revealed a reaction course in which there is an ordered binding of glycine followed by succinyl-CoA. The CoASH is released from the enzyme before the product 5-aminolaevulinic acid.

All 5-aminolaevulinic acid synthases require pyridoxal 5'-phosphate for activity but unlike many other pyridoxal 5'-phosphate dependent enzymes the cofactor is very loosely bound and may be removed by dialysis. The activity of the R. sphaeroides enzyme is maximal at 50 mM pyridoxal 5'-phosphate [23] but the nature of the interaction of the coenzyme with the enzyme is complex such that it is not possible to determine the binding constant with any precision. Spectroscopic studies [70] have shown that the holoenzyme does not exhibit the characteristic peak of a Schiff base at 420nm but rather indicate the presence of a carbinolamine species absorbing at about 333nm. Treatment of the holoenzyme with sodium borohydride leads to inactivation of the enzyme but only efficiently at pHs either side of pH 7 (Jordan and Laghai – unpublished observations). At neutral pH the enzyme is not inactivated to any great extent by the reagent. This suggests that under neutral conditions the pyridoxal 5'-phosphate is bound in a less reactive form possibly as a carbinolamine substituted by an enzyme thiol rather than as a fully formed Schiff base as has been suggested previously [70].

3.5. Mechanism of 5-aminolaevulinic acid synthase

The mechanistic and steric course of the 5-aminolaevulinic acid synthase reaction has received comprehensive investigation by the author working with Professor Akhtar and our co-workers. Because the condensation between glycine and succinyl-CoA occurs at the α-carbon of glycine, a detailed study was undertaken to study the behavior of the α-hydrogen atoms of glycine during the transformation into 5-aminolaevulinic acid. In common with other pyridoxal 5'-phosphate-dependent enzyme reactions [71] the first event in the synthesis of 5-aminolaevulinic acid involves the binding of glycine to the pyridoxal 5'-phosphate enzyme complex (Scheme 4(i)) to form an enzyme-pyridoxal 5'-phosphate-glycine Schiff base complex (Scheme 4(ii)). The further reaction of this complex requires the generation of a stabilised carbanion at the glycine α-carbon atom which could, in principle, occur either by loss of one of the two α-hydrogen atoms or by decarboxylation. When 2RS-[^{3}H$_2$]glycine was incubated with succinyl-CoA and highly purified 5-aminolaevulinic acid synthase only half of the tritium label was incorporated into 5-aminolaevulinic acid suggesting that the carbanion species which reacts with succinyl-CoA is generated by the loss of a proton (Scheme 4(iii)) [72]. Had the alternative reaction course occurred, namely decarboxylation followed by condensation, then all of the tritium originally present in the glycine would have been incorporated into the 5-aminolaevulinic acid.

Having established the broad mechanistic features of the 5-aminolaevulinic acid

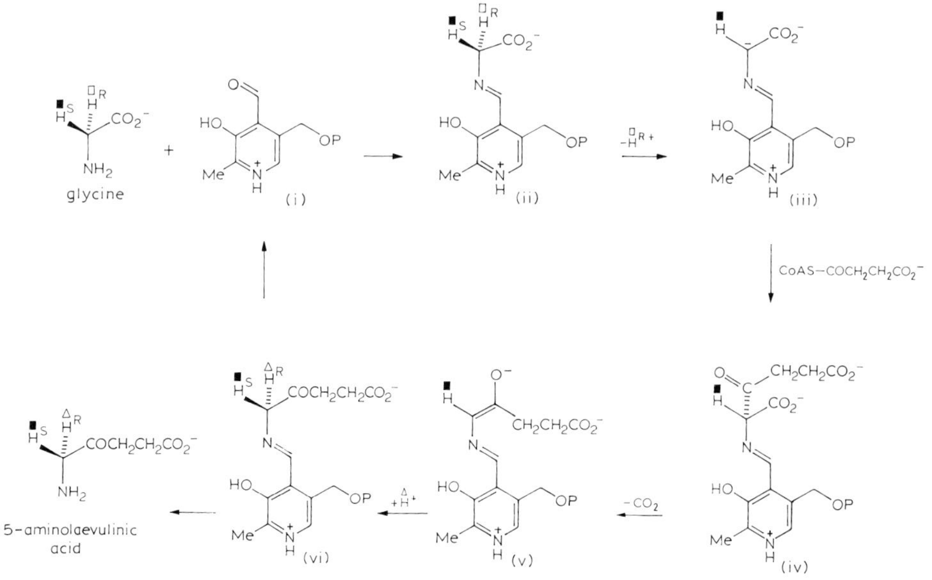

Scheme 4. Mechanism of 5-aminolaevulinic acid synthase

synthase reaction, the steric course was investigated by following the incorporation of the two tritiated enantiomers of glycine into 5-aminolaevulinic acid. 2R-[^{3}H]Glycine and 2S-[^{3}H]glycine were synthesised enzymatically using the enzyme serine hydroxymethyltransferase which stereospecifically exchanges the *proS*-hydrogen atom of glycine [73]. Incorporation of 2R-[^{3}H]glycine into 5-aminolaevulinic acid proceeded with loss of the tritium label whereas the tritium from the 2S-[^{3}H]glycine was largely incorporated into 5-aminolaevulinic acid. These experiments established that it is the *proR*-hydrogen atom of glycine which is removed in the overall enzymic transformation [74]. Exchange reactions carried out with highly purified 5-aminolaevulinic acid synthase from *R. sphaeroides* [75] confirmed these findings. In the absence of the second substrate, succinyl-CoA, the enzyme catalyses a partial reaction in which the *proR*-hydrogen atom of glycine is exchanged with the protons of the solvent. The *proR*-hydrogen atom of glycine occupies the same orientation in space as the α-hydrogen atom of L-amino acids and probably accounts for the fact that L-amino acids are better inhibitors of the enzyme than D-amino acids [68].

Since the reaction of succinyl-CoA with the enzyme-pyridoxal 5'-phosphate-glycine carbanion intermediate (Scheme 4 (iii)) precedes decarboxylation, it follows that the condensation will yield an enzyme-pyridoxal 5'-phosphate-2-amino-3-ketoadipic acid complex (Scheme 4 (iv)) The further transformation of this intermediate into 5-aminolaevulinic acid could, in principle, occur by one of two routes – either by hydrolysis of the Schiff base in this intermediate to yield free 2-amino-3-ketoadipic acid which could decarboxylate non-enzymically to 5-aminolaevulinic acid, or by an alternative mechanism in which the 5-aminolaevulinic acid synthase also acts as a decarboxylase. The latter mechanism requires that the enzyme-bound intermediate *enol* (Scheme 4 (v)) generated by the decarboxylation would be protonated stereospecifically to generate the enzyme-pyridoxal 5'-phosphate-Schiff base of 5-aminolaevulinic acid (Scheme 4 (vi)), hydrolysis of which would yield 5-aminolaevulinic acid. These two possibilities were resolved by incorporating 2RS-[^{3}H$_2$]glycine into 5-aminolevulinic acid and determining whether the remaining tritium is stereospecifically located in the product. Experimental evidence [76] showed that the tritium originally present in the *proS*-configuration is in fact incorporated into the *proS*-configuration at the C-5 position in the product thus establishing that 5-aminolaevulinic acid synthase acts as a decarboxylase as well as a condensing enzyme. This conclusion is strengthened by the finding that the enzyme catalyses the stereospecific exchange of the *proR*-hydrogen atom at the C-5 position of 5-aminolaevulinic acid [77]. These observations would not have been expected had 2-amino-3-ketoadipic acid been the final enzymic product. Further powerful evidence for the enzyme catalysed decarboxylation has stemmed from studies of the 5-aminolaevulinic acid synthase reaction working in reverse. Incubation of 5-aminolaevulinic acid with ^{14}C carbon dioxide in the presence of enzyme led to the formation of radioactive glycine, but only in the presence of CoASH [78].

The overall reaction of 5-aminolaevulinic acid synthase follows a cryptic steric course since the reaction occurs with overall inversion of configuration yet the tritium label originally in the *proS*-configuration of the glycine is found finally in the *proS*-configuration of the product. This stereochemistry may be explained by the fact that the succinyl group (-COCH$_2$CH$_2$CO$_2$H) of 5-aminolaevulinic acid occupies the same position in space as the original glycine carboxyl group but, because of the nature of the mechanism, it replaces the proton lost from the *proR*-position of glycine and adopts its final configuration only after the decarboxylation and reprotonation have occurred.

A survey of the literature on the mechanism of pyridoxal 5'-phosphate-requiring enzymes [79] reveals an almost unerring uniformity in the steric course of substitution reactions in which the new substituting group occupies the same position in space as the outgoing proton. The stereoelectronic reasons for this have been discussed at some length by Dunathan [80]. There is thus a precedent for the addition of the succinyl-moiety with retention of configuration. By the same argument, however, the reverse of the reaction involving the carboxylation of 5-aminolaevulinic acid [78], would also be expected to proceed with retention of configuration. One must therefore look for additional factors which could lead to inversion at one of these two stages. One distinct possibility is that the driving force for decarboxylation comes not from the positively charged nitrogen of the pyridine ring but from the newly formed carbonyl group (or its equivalent) of the enzyme-bound 2-amino-3-ketoadipic acid. Decarboxylation of the *β*-ketoacid in this way would generate the *enol* species 4(v) which, being planar, could easily be protonated from the opposite face and lead to the observed overall inversion. This is even more attractive when one considers that the enzymic group responsible for the original deprotonation of the *proR*-hydrogen atom of the glycine Schiff base complex would be perfectly placed to carry out the final reprotonation to give the desired stereochemistry in the 5-aminolaevulinic acid. Consistent with this view is the observation that the enzyme catalyses a partial reaction in which the *proR*-hydrogen atom at C-5 of 5-aminolaevulinic acid is stereospecifically exchanged with the protons of the medium. Thus both the glycine [75] and the 5-aminolaevulinic acid [77] can participate in proton exchange reactions with what is almost certainly the same catalytic group at the enzyme active site.

A survey of the literature on decarboxylase enzymes which do not require pyridoxal-5'-phosphate as a coenzyme reveals examples of both retention and inversion mechanisms. Uroporphyrinogen III decarboxylase, which is discussed in detail in Chapter 2 of this volume, proceeds by a retention mechanism [81] although it may be argued that the stereoelectronic considerations said to apply to pyridoxal-5'-phosphate dependent enzymes [80] apply equally well to this decarboxylase in view of the perceived participation of the pyrrole ring as a pyridine analogue. Several examples of *β*-keto acid decarboxylations which are known to proceed by inversion [82] set a precedent, at least, for the decarboxylation catalysed by 5-aminolaevulinic acid synthase. To determine the mechanistic course of 5-aminolaevulinic acid synthase, additional experiments will need to be carried out with intermediate analogues since the synthesis and resolution of the two 2-amino-3-ketoadipic acid (succinylglycine) isomers may not be possible for reasons of their extreme instability [83].

The effect of inhibitors of the enzyme on the exchange reactions has revealed additional information about the *R. sphaeroides* enzyme. The glycine exchange reaction was inhibited by 5-aminolaevulinic acid and the 5-aminolaevulinic acid exchange reaction was, in turn, inhibited by glycine [75,77]. Both exchange reactions were strongly inhibited by 0.125 mM aminomalonate consistent with the inhibition of the overall reaction by this glycine analogue. Most interestingly haem, at a concentration of 2 μM, strongly inhibited the 5-aminolaevulinic acid exchange reaction but had no effect on the glycine exchange reaction. This suggests that the binding site at which haem exerts its inhibitory effect may overlap with the area of the active site required for binding the succinyl-moiety (-COCH$_2$CH$_2$CO$_2$H). Alternatively the interaction of the haem with the enzyme could induce a conformational change which perturbs selectively the protein structure in the vicinity of the succinyl-CoA binding site. The latter is more likely since the inhibition depended upon preincubation before the effect was maximized [77].

3.6. Structure of the enzyme and molecular biology

The structure of the enzyme and the nature of the enzyme groups involved with catalysis have not yet been identified since the amounts of 5-aminolaevulinic acid synthase available for study have been very limited. The enzymes from all sources are sensitive to sulphydryl reagents such as iodoacetic acid and p-chloromercuribenzoate and cysteine has been implicated as a binding group for pyridoxal 5'-phosphate [70].

Recently, important structural information has become available as a result of the application of molecular biology to the 5-aminolaevulinic acid synthase enzyme which has led to the cloning and sequencing of the cDNA for the chick embryo liver pre-enzyme [32,84], the yeast *HEM1* gene [31], the human liver cDNA [33], the mouse erythropoietic cDNA [85] and the *Bradyrhizobium japonicum hemA* gene [86]. Partial sequence information has been obtained from *Rhizobium meliloti* [87]. Sequence information for the two 5-aminolaevulinic acid synthase genes in R. sphaeroides (*hemA* and *hemT*) has also been obtained (Neidle and Kaplan – personal communication). The nucleotide sequences from both yeast, chicken and human have provided information about the primary structure of the 5-aminolevulinic acid synthase precursor protein and the nature of the sequences required for mitochondrial targetting and protein maturation. A cDNA specifying the erythroid form of 5-aminolaevulinic acid synthase has been isolated [88] and sequenced [89].

The nucleotide sequence data have also revealed interesting similarities between the primary structures which are obvious targets for further investigation. For instance the predicted protein primary structures of the 5-aminolaevulinic acid synthases from *R. meliloti* [87], *B. japonicum* [86], chicken embryo liver [84] and human liver [33] all show highly conserved sequences suggesting that they have similar tertiary structures and arise from a common evolutionary gene. Highly conserved sequences may assist in the identification of key catalytic residues. Overall the comparison between the derived protein sequences of yeast and chick embryo liver revealed, after optimal alignment, about 41% similarity. Comparison between the sequences from *Bradyrhizobium* and chicken embryo liver showed about 49% similarity. Comparison of the human sequence with the rat and chicken sequences shows 83% and 78% similarity respectively. There is, however, only 43% similarity between human and the mouse erythropoietic cDNA-derived amino acid sequence.

A comparison of the derived primary sequences of the 5-aminolaevulinic acid synthases indicates a possible basic 5-aminolaevulinic acid synthase motif which is found in all species. This motif extends from residue 192 in the chicken enzyme to residue 515 near the carboxyl-terminus. The information from molecular biology has also permitted a more accurate determination of the molecular sizes of the pre-protein and mature protein. The pre-proteins from chicken and human comprise 635 and 642 amino acids whereas the mature enzymes have 579 and 586 residues respectively. Both rat pre-protein and mature protein are of similar length to their human counterparts. Thus both precursors lose a peptide of 56 amino acids large parts of which are remarkable for their high similarity. The loss of this presequence changes the M_r from 70,000 Da to 64,000 Da. The molecular weight of the mouse erythropoietic enzyme is much smaller consisting of 518 amino acids with M_r of 55,000. The availability of the genes for 5-aminolaevulinic acid synthase opens up new possibilities for the study of both regulatory and mechanistic aspects as well as mitochondrial targeting of the enzyme. A detailed account of the molecular biology of 5-aminolaevulinic acid synthase and other enzymes of the pathway is given in Chapter 7.

4. The biosynthesis of 5-aminolaevulinic acid from glutamate

4.1. Discovery of the C-5 pathway

Investigators were puzzled for many years by their inability to detect 5-aminolaevulinic acid synthase in higher plants and, except for a few isolated reports, there was little convincing evidence to suggest that glycine and succinyl-CoA were the immediate precursors of tetrapyrroles. The puzzle was first solved by Beale and Castelfranco in the mid 1970s [90] when, in what are now classical experiments, they demonstrated that 5-aminolaevulinic acid was derived from the carbon skeleton of glutamate in greening cucumber cotyledons and the 'C-5 pathway' was born. Further experiments with [^{14}C]-glutamate in greening barley showed that it was incorporated into 5-aminolaevulinic acid with the carbon-skeleton intact, and with the C-1 of glutamate giving rise to the C-5 of 5-aminolaevulinic acid [91]. The existence of the C-5 pathway was further confirmed in barley [92]. ^{13}C N.m.r experiments in which [^{13}C]-labelled glutamate was incorporated into chlorophyll in *Scenedesmus obliquus* [93] and higher plants [94] also confirmed conclusively that the glutamic acid carbon skeleton was incorporated into 5-aminolaevulinic acid intact and that the C-1 of glutamate occupies those positions in the resulting chlorophyll which would have arisen from the C-5 position of 5-aminolaevulinic acid. The presence of the C-5 pathway was shown subsequently in *Cyanidium caldarium* [95], *Chlamydomonas reinhardtii* [96] and in *Chlorella vulgaris* [97]. As well as occurring in plants and algae, the C-5 pathway has been demonstrated in *Euglena gracilis* [98,99], Cyanobacteriaceae like *Spirulina platensis* [100], Archaebacteriaceae such as *Methanobacterium thermoautotrophicum* [101] and the Eubacteriaceae such as *Clostridium tetanomorphum* [102]. The pathway also exists in *E. coli* [103] and *Salmonella typhimurium* [104]. The list of plants, algae and bacteria in which the operation of the C-5 pathway has been demonstrated is growing rapidly and it is apparent that it is more widespread than the glycine pathway in the biosphere. For a comprehensive account of the C-5 pathway the reader is referred to Chapter 5 of this volume.

4.2. The enzymes of the C-5 pathway

The precise pathway by which glutamate is transformed into 5-aminolaevulinic acid was the subject of intensive research during the 1980s and during this time the essential details of the enzymic pathway were established by Kannangara and his associates in a series of pioneering experiments using chloroplasts from greening barley. Altogether, three enzyme fractions were isolated from chloroplast extracts [105] but the most extraordinary discovery was that of the involvement of $tRNA_{glu}$ as an essential component of the system [106–109]. The $tRNA_{glu}$ from barley has been sequenced and shown to have the anticodon UUC [107], as have $tRNA_{glu}$ isolated from several plants and algae [110]. The barley chloroplast $tRNA_{glu}$ contains several unusual bases and appears to play a role in protein synthesis as well as in chlorophyll formation [111]. The chloroplast $tRNA_{glu}$ is specified by the chloroplast DNA. In recent years it has emerged that tRNA-amino acid species have many other roles in biochemical reactions and are not exclusively reserved for protein synthesis [112].

At least three enzymes are thought to be involved with the transformation of glutamate into 5-aminolaevulinic acid. The first enzyme, glutamate-$tRNA_{glu}$ synthase (ligase) has been isolated from barley chloroplasts [113], *Chlorella* [114] and from *C. reinhardtii* [115]. The barley enzyme, M_r 54,000 Da, catalyses the coupling of glutamate to $tRNA_{glu}$ in the presence of ATP and magnesium ions. The second enzyme, glutamyl-$tRNA_{glu}$ reductase (dehydrogenase) catalyses the reduction of glutamyl-$tRNA_{glu}$ to the aldehyde oxidation level in a NADPH requiring reaction [96,105,108,109,113,116,117].

In mechanistic and energetic terms glutamate in the form of glutamyl-tRNA, or its enzyme-bound equivalent, is an obligatory activated system necessary for the reduction of the glutamate α-carboxyl group. The requirement for such an activated system has many mechanistic precedents. For instance, in the Calvin cycle the transformation of 3-phosphoglyceric acid into 3-phosphoglyceraldehyde occurs by initial activation of the acid with phosphoglycerate kinase to give 1,3-diphosphoglyceric acid. The 1,3-diphosphoglyceric acid reacts with a sulphydryl group at the active site of glyceraldehyde 3-phosphate dehydrogenase and it is the resulting thioester which is reduced by NADPH to the thiohemiacetal at the aldehyde oxidation level. The thioester in this case is analogous to the glutamyl-tRNA ester linkage, or its enzyme-bound equivalent. There is little evidence to suggest that glutamate 1-phosphate is involved although this was originally suggested as a plausible intermediate [91] – before the involvement of tRNA was established.

There has been some debate about the precise nature of the product of the reductase (dehydrogenase) enzyme. Kannangara and his colleagues proposed the α-aminoaldehyde, glutamate 1-semialdehyde, as the product of the reductase and substrate for a specific aminotransferase (aminomutase), the final enzymic step in the formation of 5-aminolaevulinic acid. Their conclusions were based on the development of a chemical synthesis for glutamate 1-semialdehyde [118,119]. On chemical grounds this seemed astonishing since α-aminoaldehydes related to the α-amino acids are extremely labile compounds and, despite many attempts, have never been isolated. Furthermore the n.m.r. spectrum of the synthetic material revealed no aldehyde ^{1}H resonance. It was even more remarkable when it was reported that glutamate 1-semialdehyde accumulated in vivo and could be isolated in crystalline form under conditions when eliolated barley leaves had been treated with the inhibitor gabaculine (3-amino-2,3-dihydrobenzoic acid) [120]. The material extracted from this experiment had the same properties as the synthetic compound prepared by Houen et al. [119] and acted as the substrate for the final aminotransferase stage to yield 5-aminolaevulinic acid.

Extensive investigations in our laboratory on the material synthesised by the method of Kannangara and Gough [118] and by the ozonolysis of 4-vinyl-4-aminobutyric acid [121] established its structure as a six-membered ring. This was based on several pieces of evidence from ^{1}H and ^{13}C n.m.r. spectra; i) the two hydrogen atoms adjacent to the assymetric centre have very different chemical shifts typical of the conformational restriction expected from a cyclic structure. Similar chemical shifts are found in the related cyclic compound glutamic anhydride; ii) the J-values arising from the coupling between the hydrogen atoms at positions 5 and 6 are entirely consistent with a rigid stucture expected from a ring system; iii) no chemical shift can be assigned to an aldehyde proton but a resonance at $\delta=5.2$ p.p.m. characteristic of a hemiacetal as found in glucosamine is present; iv) the ^{13}C n.m.r. spectrum indicates a carbon resonance with a chemical shift $\delta=91$ p.p.m. indicative of a carbon atom substituted by two oxygen atoms as in glucosamine; v) the ^{1}H n.m.r. NOESY spectrum is consistent with a six-membered ring structure. We propose therefore that glutamate 1-semialdehyde does not exist either as the free aldehyde or as a hydrated aldehyde, as proposed [122], but that it has a cyclic structure as shown in Scheme 5. This structure is formally named 5-amino-3,4,5,6-tetrahydro-6-hydroxy-2*H*-pyran-

2-one, but 'abbreviated' to hydroxyaminotetrahydropyranone (HAT) [123]. We are of opinion that the synthetic route used initially by Kannangara and Gough [118] and by Kannangara et al. [121] yielded fortuitously the stable cyclic form. By analogy with a related cyclic structure, glucoseamine, HAT is reasonably stable in acid but highly unstable in base in which the ring opens to unmask the reactive aldehyde group. The observed properties of the intermediate, synthesised both chemically and enzymically, are consistent with this cyclic structure. From the above discussion it follows that the reductase enzyme must also catalyse the ring closure and that the aminotransferase which utilises HAT as its substrate must catalyse ring opening before the amino group can be transferred.

The aminotransferase enzyme was first described by Kannangara and Gough [118] in barley chloroplasts and shown to convert the glutamate 1-semialdehyde into 5-aminolaevulinic acid. The enzyme was purified subsequently from barley and shown to have a subunit M_r of 80,000 Da [124]. The aminotransferase appears to function without the necessity for any amino acid amino-donor such as glutamate and the substrate itself supplies the amino group. The mechanism has been studied using glutamate doubly labelled with $^{13}C, ^{15}N$ by Mau and Wang [125] who have suggested that intermolecular transfer of the amino group takes place. These results, however, need to be interpreted with caution because of the possibility of exchange.

One of the most interesting properties of the aminotransferase is its remarkable sensitivity to gabaculine [122] suggesting that pyridoxal 5'-phosphate is involved in the enzyme reaction. More direct evidence for the participation of pyridoxal 5'-phosphate has been provided by other investigators [116]. A plausible mechanism for the aminotransferase reaction has been proposed by Hoober et al. [122], in which the glutamate 1-semialdehyde first binds to the enzyme with the cofactor in the pyridoxamine 5'-phosphate form at the active site. The amino group from the pyridoxamine 5'-phosphate is transferred to the substrate to yield an enzyme-bound 4,5-diaminovaleric acid intermediate. This is transformed into 5-aminolaevulinic acid by the transfer of the amino group at the C-4 position to the pyridoxal 5'-phosphate to regenerate the pyridoxamine 5'-phosphate-enzyme complex. This mechanism explains why no amino donor other than the substrate is required. The likely involvement of the cyclic glutamate semialdehyde intermediate (HAT) as a substrate need not affect the above mechanism since the transaminase would catalyse ring opening as the first step in the mechanism. Such ring opening (mutarotation) occurs routinely in enzymic transformations involving glucose and other sugars. Most recently the aminotransferase cDNA has been isolated from barley [126] and the derived protein sequence has been shown to be similar to that of *Salmonella typhimurium* [104].

The aminotransferase reaction in which glutamate 1-semialdehyde is transformed into 5-aminolaevulinic acid by a single enzyme, glutamate 1-semialdehyde aminotransferase has been challenged by Dornemann [127] as a result of experiments carried out with *Scenedesmus*. In this organism 4,5-dioxovalerate appears to be a better precursor for 5-aminolaevulinic acid than glutamate 1-semialdehyde. The presence of

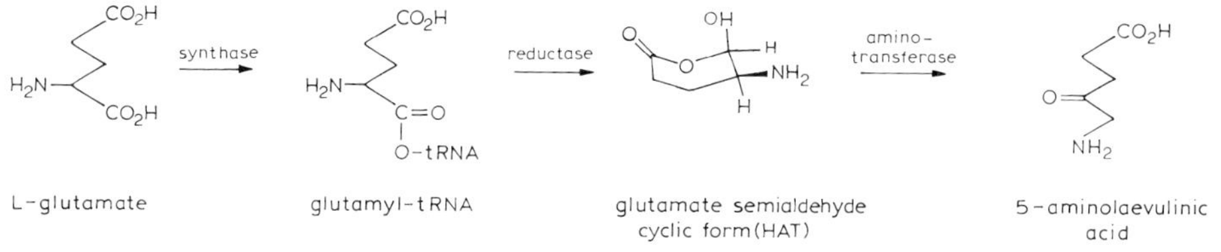

Scheme 5. The glutamate pathway for 5-aminolaevulinic acid synthesis

the enzyme 4,5-dioxovalerate transaminase has also been presented as evidence for a two stage transformation of glutamate 1-semialdehyde into 5-aminolaevulinic acid via 4,5-dioxovalerate. As yet these points of contention are unresolved. It is possible that variations on the pathway exist in different organisms with 4,5-dioxovalerate as an intermediate between glutamate 1-semialdehyde and 5-aminolaevulinic acid in *Scenedesmus* but not in barley [127].

There is a growing amount of evidence to suggest that some photosynthetic eukaryotes may operate both glycine and glutamate pathways, the former being employed for the synthesis of haem and the latter for chlorophyll production. In *E. gracilis* [^{14}C] glycine is preferentially incorporated into haem whereas radioactivity from [^{14}C] glutamate is found predominantly in chlorophyll [98]. In higher plants it is possible that the glycine path plays a transient role, for instance, during germination when mitochondrial respiration is the predominant energy-yielding metabolic process. Aspects of the C-5 pathway have been expertly reviewed elsewhere [128,129] and a detailed account is presented in Chapter 5 of this volume. The involvement of tRNA$_{glu}$ has also been reviewed in a scholarly account [112].

5. The biosynthesis of porphobilinogen

5.1. Introduction

After the discovery of 5-aminolaevulinic acid and the elucidation of the structure of porphobilinogen, the enzyme 5-aminolaevulinic acid dehydratase (E.C.4.2.1.24), also called porphobilinogen synthase, was soon described in ox liver [12] and avian erythrocytes [11]. The enzyme has since been shown to exist in virtually all living systems and has been purified to homogeneity from a variety of sources including human erythrocytes [130,131], bovine liver [132,133] and *R. sphaeroides* [134]. The enzyme has also been purified from yeast [135], *Nicotiana tabacum* [136], radish [137] and spinach [138]. The reaction catalysed by the enzyme involves the dimerization of two molecules of 5-aminolaevulinic acid and the elimination of two molecules of water in what is essentially a Knorr reaction as shown in Scheme 6.

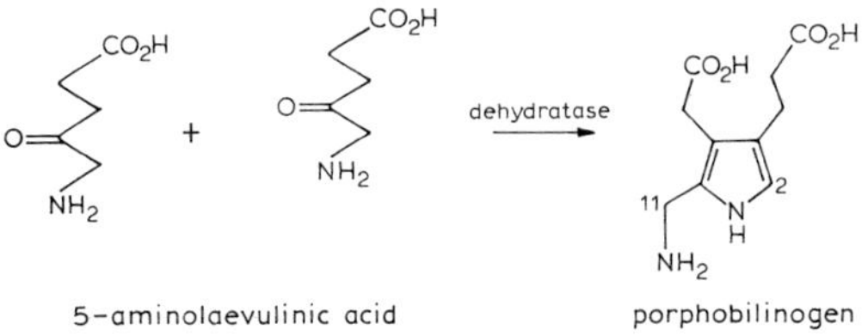

Scheme 6. The formation of porphobilinogen from 5-aminolaevulinic acid

5.2. General properties of 5-aminolaevulinic acid dehydratases

Comprehensive reviews on the properties of 5-aminolaevulinic acid dehydratases by Shemin [139] and Cheh and Neilands [140] have been written. The dehydratase enzymes have been grouped into two main classes according to their metal requirements, susceptibility to inhibition with EDTA and pH optima and the possibility that there are metallo- and Schiff base dehydratases, as in the case of aldolases, has even been proposed. It is not altogether clear, however, if such a grouping is useful since in recent years the availability of cDNA and gene-derived primary protein sequence data tend to suggest that the dehydratases may all be quite closely related in structure and mechanism despite their different metal requirements (see below and Chapter 7).

The mammalian and avian enzymes are zinc metalloenzymes with pH optima in the range 6.3–7.0 and are inhibited strongly by EDTA. During purification, a small proportion of the zinc may be lost from the enzyme but full enzymic activity can be restored by adding exogenous zinc, as long as the enzyme is maintained under reducing conditions by the presence of an exogenous thiol such as dithioerythritol [131,133,140,141]. These eukaryotic dehydratases have a M_r of about 280,000 Da and are composed of eight identical subunits of M_r 35,000 Da.

The bacterial enzyme isolated from *R. sphaeroides* [134] has very different properties from the mammalian enzymes with a pH optimum from 8 to 8.5 and a requirement for monovalent cations such as potassium ions for maximum activity. Although this enzyme does not appear to be inhibited by EDTA and is not activated by divalent metal ions this does not necessarily preclude the presence of a very tight metal binding site. The enzyme from *R. sphaeroides* has a M_r of 240,000 Da and early reports suggested that it is made up of 6 identical subunits each of M_r 39,500 Da [142]. More recent experiments have shown that this enzyme may in fact exist as an octamer [143] like the mammalian dehydratases. In the presence of potassium ions the *R. sphaeriodes* enzyme appears to aggregate to 'octameric' dimers, trimers and tetramers with an accompanying increase in the specific activity of the enzyme [144]. The dehydratase from *R. sphaeroides* has a specific activity substantially higher than that of the mammalian enzymes.

5-Aminolaevulinic acid dehydratase isolated from spinach [138] has a M_r of 300,000 Da and appears to require magnesium ions for maximal activity. The enzyme exhibits a pH optimum midway between the mammalian and *R. sphaeroides* enzymes. The yeast dehydratase [135], like the mammalian enzymes, appears to be a zinc metalloenzyme and is inhibited by EDTA.

In contrast, the bacterial 5-aminolaevulinic acid dehydratase from *E. coli* has been purified to homogeneity and shown to require zinc ions for activity [145]. Removal of the zinc with EDTA results in the loss of catalytic activity which can be recovered fully on the addition of zinc. Thus two classes of 5-aminolaevulinic acid dehydratases appear to exist.

Electron microscopy of the bovine 5-aminolaevulinic acid dehydratase has revealed that the eight subunits are arranged as if they were at the corners of a cube with dihedral symmetry [146]. A model of the quarternary structure of bovine 5-aminolaevulinic acid dehydratase has also been deduced from small-angle X-ray scattering and indicates that the octamer exists as a quadratic arrangement of four dimeric stacks [147]. The spinach enzyme, however, exists as a hexamer [138].

The interaction between the subunits has been studied by treatment of the enzyme with increasing concentrations of urea (Jordan and Chaudhry – unpublished data). The octameric enzyme dissociates into tetramers at 4M urea and a dimer at 5.5M urea, both of which can be separated by gel filtration. Both forms exhibit some catalytic activity, however increasing the urea concentration further leads to the precipitation of the enzyme and complete loss of activity, as the dimeric structure collapses to give free subunits. These observations suggest that the octamer is a dimer of tetramers and that each tetramer is a dimer of dimers. The conclusions from these studies are consistent with the enzyme structure being composed of four functional dimers. In this connection there is a considerable amount of evidence to suggest that the bovine enzyme, at least, exhibits half site reactivity [140] and this facet will be discussed below. It has been observed that 5-aminolaevulinic acid dehydratase from *R. sphaeroides* also dissociates into two functional tetramers in the absence of potassium ions. This has been shown by an elegant experiment [143] in which the bacterial enzyme was first immobilized in a chromatography column. The column was subsequently treated with buffers lacking potassium ions which released half of the enzyme activity, the other half remaining bound to the column as a functional tetramer. The immobilized enzyme has also been successfully employed for the synthesis of porphobilinogen [143].

5.3. Importance of sulfhydryl groups

One common feature of many dehydratases is the remarkable sensitivity of their sulfhydryl groups to oxygen or to thiol reagents. On exposure of the native bovine enzyme to oxygen there is a rapid loss of enzyme activity which is associated with the oxidation of two highly reactive SH groups to form a S-S bond [141,148,149]. Full catalytic activity can be restored on treatment of the enzyme with an exogenous thiol. Addition of 5,5'-dithiobis(2-nitrobenzoic acid) to the active enzyme also results in the formation of the S-S bond between these two sulfhydryl groups. Two other less reactive sulfhydryl groups react with 5,5'-dithiobis(2-nitrobenzoic acid) [149] and a further three or four become available when the enzyme is denatured with sodium dodecylsulfate [141]. The importance of the reactive SH groups in the binding of the metal ion has been demonstrated by labelling the human 5-aminolaevulinic acid dehydratase with ^{65}zinc ions [150]. Exposure of the labelled enzyme to oxygen or treatment of the enzyme with 5,5'-dithiobis(2-nitrobenzoic acid) causes the immediate displacement of the labelled ^{65}zinc ion and a concomitant loss of enzyme activity. These studies suggest that sulphydryl groups are involved with the binding of the zinc ion.

5.4. Requirement for metals

The metal requirements for the dehydratase enzymes have been investigated largely with the mammalian enzymes, especially those isolated from human erythrocytes and bovine liver. It has been established that one zinc ion is bound per subunit, eight per octamer, on the basis of experiments with both nonradioactive zinc [141] and with ^{65}zinc [150].

Investigations by Tsukamoto et al. [141] suggest that zinc plays a conformational role and is not essential for activity, whereas others [132] have suggested that only four zinc ions per octamer may be required for full catalytic activity and that only half of the subunits may be catalytically active at one time. Using ^{65}zinc ions Gibbs and Jordan [150] have demonstrated that eight zinc ions are present in the human octameric enzyme. When four zinc ions are bound per octamer, the enzyme appears to be half active and eight zinc ions are necessary for full catalytic activity [151]. Zinc has also been shown to activate the enzyme in vivo in human [152]. Cadmium displaces ^{65}zinc ions from the enzyme more rapidly than unlabelled zinc [150] in keeping with the higher affinity of the enzyme for cadmium ions [140].

The availability of the cDNA sequences specifying human [153], rat [154a] and yeast [154b] 5-aminolaevulinic acid dehydratases and the *hemB* gene sequence from *E. coli* [155,156] has provided the opportunity to compare the derived amino acid sequences and to explore the possible location of the sulfhydryl groups involved in zinc ion binding. An amino acid sequence is present in the enzymes (residues 119–132 in the human enzyme [153]) which is very similar to the concensus sequence for a zinc-finger motif found in many zinc containing proteins [157a]. In contrast, the spinach and pea enzymes have three aspartic acid substitutions which may constitute the magnesium site [157b, c].

human enzyme	C	D	V	C	L	C	P	Y	T	S	H	G	H	C	G
	*			*		*		*	*	*	*	*	*	*	
rat enzyme	C	D	V	C	L	C	P	Y	T	S	H	G	H	C	G
	*			*		*		*	*	*	*	*	*	*	
E coli enzyme	S	D	T	C	F	C	E	Y	T	S	H	G			
	*			*		*		*	*	*	*	*	*	*	
Yeast enzyme	C	D	V	C	L	C	E	Y	T	S	H	G	H	C	G
Spinach enzyme	T	D	V	A	L	D	P	Y	Y	Y	D	G	H	D	G
Pea enzyme	T	D	V	A	L	D	P	Y	S	S	D	G	H	D	G

In addition to the conserved cysteines in the zinc enzyme sequences there are two conserved histidine residues separated by a conserved glycine and one, or both, of these two histidines may interact with the zinc ion. Evidence for the involvement of histidine has been provided from inactivation studies [141] however whether the putative metal binding site comprises one of these histidines has not been confirmed experimentally. The human and rat sequences are identical and very close to that of the yeast and *E. coli* enzymes with all but one of the cysteine residues and both histidine residues conserved (*). The importance of a zinc ion binding site in the *E. coli* sequence, which has since been confirmed experimentally [145] throws some doubt on the previous attempts to classify the 5-aminolaevulinic acid dehydratases on a prokaryotic and eukaryotic basis. Unless the *E. coli* enzyme is atypical of bacterial enzymes, small differences in amino acid sequences could make a large difference in the tightness of the metal binding and could account for the lack of EDTA inhibition seen in the enzyme from *R. sphaeroides*, although there is some evidence that the latter enzyme may fall into the magnesium class found in plants.

An interesting approach has been employed for the study of zinc binding to bovine 5-aminolaevulinic acid dehydratase using extended X-ray absorption fine structure (EXAFS) spectroscopy [158]. Although the original EXAFS data from this study suggested that zinc is ligated to three sulfur atoms, a more detailed investigation [159] has revealed two distinct types of zinc binding coordinations in the octamer termed the $zinc_A$ and $zinc_B$ sites. There are reported to be four $zinc_A$ sites and four $zinc_B$ sites. Occupancy of the four $zinc_A$ sites occurs on addition of zinc to the apodehydratase and is essential for the formation of the catalytically active enzyme. It is suggested that the $zinc_A$ ligands comprise two/three histidine nitrogens, two oxygen ligands (tyrosine, aspartate or water) and one cysteine sulphur. The $zinc_B$ coordination comprises four sulphur ligands. The interesting possibility that the putative metal-binding amino acid sequence can interact with the zinc in two ways has been proposed to explain the EXAFS data. The differences in the folding of the metal-binding sequence are attributed to induced asymmetry in the interaction between monomers in the dimers [159]. In the absence of an X-ray structure, the precise positions of the zinc ligands in the primary structure of the enzyme are not known.

5.5. Inhibition by lead

One of the most interesting properties of the mammalian 5-aminolaevulinic acid dehydratases is their reaction with, and inactivation by, lead ions. The activities of human [152] and rat [160] erythrocyte 5-aminolaevulinic acid dehydratases are inhibited by lead, both in vivo and in vitro, however zinc is able to overcome this inhibition. Such is the sensitivity of the dehydratase to lead in vivo that its activity has been used to monitor plasma lead levels in humans with lead poisoning [161]. One of the effects of lead poisoning which directly relates to the inhibition of the 5-aminolaevulinic acid dehydratase enzyme is the sharp rise in 5-aminolaevulinic acid in the blood and the severe accompanying neurological symptoms caused by this compound. 5-Aminolaevulinic acid has been shown to act as a 4-aminobutyric acid agonist and found to inhibit the activity of mammalian spinal motoneurones probably via 4-aminobutyric acid inhibitory synapses [162]. Similar symptoms have also been observed in patients with 5-aminolevulinic acid dehydratase porphyria [163], acute intermittent porphyria and variegate porphyria [164].

The most obvious mechanism by which lead inhibits the mammalian dehydratases is by displacement of the zinc. However studies in our laboratory have shown that lead completely inactivates the $^{65}Zn^{++}$-labelled human dehydratase even though half of the ^{65}zinc radioactivity remains bound to the enzyme [151]. These findings suggest that either zinc and lead may interact with the human enzyme at more than one site [165] or that half site reactivity is involved. Alternatively a regulatory mechanism may be involved in which the binding of lead to four of the subunits in the octamer may induce a conformational change in the other four subunits which results in the total loss of activity. The EXAFS data discussed above [159] may be consistent with these observations.

5.6. Nature of the active site groups

The first detailed studies on the mechanism of action of 5-aminolaevulinic acid dehydratase were those carried out by Shemin and his colleagues with the enzyme from *R. sphaeroides*. One of the most interesting facets of this work was the evidence for the involvement of a Schiff base linkage between the keto-group of 5-aminolaevulinic acid and the dehydratase enzyme [166]. Incubation of the substrate, 5-aminolaevulinic acid, with dehydratase in the presence of sodium borohydride led to irreversible inactivation of the enzyme. When 5-amino[^{14}C]laevulinic acid was used, radioactive label was incorporated into the inactivated enzyme protein. Further investigations established that substrate analogues like laevulinic acid (4-ketopentanoic acid) are also able to form a Schiff base with the enzyme. The main structural requirement for the recognition of a ligand by the enzyme appeared to be a carboxylic acid with a keto-function in the 4-position. A neutral or positively charged group is thought to be required at the C-5 position since 2-ketoglutarate, although a 4-ketoacid, was found to be a very poor inhibitor whereas laevulinic acid bound very effectively to the enzyme. The inhibitory property of laevulinic acid has been exploited effectively in plants to investigate the biosynthesis of 5-aminolaevulinic acid. More recently succinylacetone (4,6-dioxoheptanoic acid) has been found to act as an extremely potent inhibitor of the dehydratases [167] and has been most successfully used to study the regulation of haem synthesis by preventing the formation of porphobilinogen and haem [43,44].

The amino acid involved with the binding of 5-aminolaevulinic acid at the catalytic site has been confirmed as a lysine residue in the *Rhodopseudomonas sphaeroides* enzyme [168]. Similarly, lysine has been shown to be involved in the human and bovine enzymes [169] by the isolation of the reduced adduct between the ε-amino group of lysine and 5-aminolaevulinic acid. The precise location of the active site lysine in 5-aminolaevulinic acid dehydratase was determined by Gibbs and Jordan [169] using a combination of protein chemistry and molecular biology. The active site lysine was permanently labelled in both the human and bovine 5-aminolaevulinic acid dehydratases by reduction of the enzymes with borohydride in the presence of 5-amino[^{14}C]laevulinic acid. Radioactive cyanogen bromide peptides from both enzymes were then isolated and sequenced [169]. Although the first five amino acids in the vicinity of the active site lysine, in the peptide MVKPG, are present in both human and bovine dehydratase the sixth amino acid is methionine in the human enzyme [153] but arginine in the bovine dehydratase – a change which can be accounted for by the difference of a single base in the codons – ATG for methionine to CTG for arginine. The cyanogen bromide peptide isolated from the human enzyme is thus a hexapeptide in contrast to the peptide isolated from the bovine enzyme.

The cloning and sequencing of the full-length cDNA for the 5-aminolaevulinic acid dehydratases from both human [153] and rat [154] have resulted in derived protein sequences for both enzymes which also reveal a high degree (88%) of conservation. The bacterial protein sequence derived from the *hemB* gene from *E.coli* [155,156] also shows a substantial similarity with the mammalian sequences. A comparison of the peptide sequences and derived seqences in the vicinity of the active-site lysine are shown below with the similarities indicated (*).

cyanogen bromide peptides

```
                    * * * * *
human               M V K P G M
                    * * * * *   * * * *   *   *
bovine              M V K P G R P Y L D L V R E
```

cDNA

```
                    * * *   * * * * *   * * * *   *   *
human               G A D M L M V K P G M P Y L D I V R E
                    * * *   * * * * *   * * *     *   *
rat                 G A D I L M V K P G L P Y L D M V Q E
E. coli             G A D C L M V K P A G A Y L D I V R E
Spinach and pea     G A D I L L V K P G L P Y L D I I R L
```

On the basis of the active site peptide sequences [169] it has been possible to define the exact position of the active site lysine in the derived protein sequences which are at position-252 in human and, by analogy, at positions-252 and 258 in the rat and *E. coli* enzymes respectively. The polar active-site lysine is flanked by sequences of hydrophobic amino acids. The highly significant similarity between the prokaryote enzyme, the plant enzymes and the mammalian enzymes, especially in the vicinity of the active-site lysine suggest they may have very similar three-dimensional structures and mechanisms of action.

Protein chemistry has also been used to investigate the protein sequences derived from the *E. coli hemB* gene. The sequence of Echelard et al. [155] shows 9 cysteine and 2 tryptophan residues but the sequence of Li et al. [156] shows 6 cysteine and 1 tryptophan. Amino acid-specific cleavages of the purified *E. coli* 5-aminolaevulinic acid dehydratase at cysteine and tryptophan residues [145] gave peptides which were consistent both in size and number with the latter sequence.

5.7. Order of binding the two substrates and the enzyme mechanism

5-Aminolaevulinic acid dehydratases are unusual in that they catalyse the condensation between two molecules of the same substrate and hence the order of substrate binding cannot be determined by classical steady state kinetics. Since a knowledge of the order of substrate binding is crucial for the elucidation of any enzyme mechanism, alternative approaches are required. In initial studies on the *R. sphaeroides* enzyme, a mixture of the substrate and a substrate analogue (laevulinic acid) was used in an attempt to distinguish between the two substrate binding sites, since the bacterial enzyme will catalyse the formation of a 'mixed' or heterologous pyrrole between 5-aminolaevulinic acid and laevulinic acid. In the 'mixed' pyrrole (Scheme 7) the acetic acid ('A') side of the product arises from laevulinic acid and the propionic acid side ('P') from 5-aminolaevulinic acid [166].

The observations that laevulinic acid also acts as a competitive inhibitor and forms a covalent Schiff base with the enzyme was used to propose a mechanism in which the

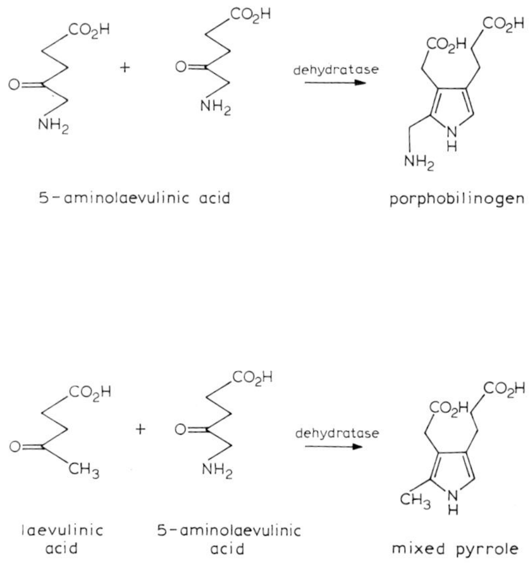

Scheme 7. Formation of a mixed pyrrole between 5-aminolaevulinic acid and laevulinic acid

substrate binding as the Schiff base gives rise to the 'A' side of the product porphobilinogen [166]. This mechanism was accepted until relatively recently when single turnover experiments were used by Jordan and his colleagues to reinvestigate the precise order of binding of the two substrates. Instead of using substrate analogues, isotopically labelled substrate was used with the bovine and bacterial dehydratases [170,171] and with the human enzyme [172]. In these experiments stoichiometric amounts of labelled substrate (either ^{13}C or ^{14}C) were reacted rapidly with the enzyme to occupy preferentially the substrate binding site with the higher affinity. Unlabelled substrate was then added to complete the 'turnover' and to drive the labelled substrate through into the product porphobilinogen. It was anticipated that the labelled 5-aminolaevulinic acid initially bound to the enzyme would occupy only one 'side' of the porphobilinogen and that a knowledge of the position of the label would reveal which of the two substrate molecules had originally bound to the enzyme. The position of the label in the porphobilinogen was determined either by n.m.r or by chemical degredation.

The results from these experiments [170–172] showed clearly that the 5-aminolaevulinic acid molecule initially bound to the enzyme gives rise to the 'P' side of the substrate and not the 'A' side as had been suggested previously [166] by Nandi and Shemin. These workers had assumed incorrectly that laevulinic acid was acting as a substrate and an inhibitor at the same site – the 'A' site. However laevulinic acid is clearly acting as a substrate at the 'A' site but acting as a good inhibitor at the 'P' site. Since laevulinic acid has no amino function it cannot supply the nitrogen atom necessary to complete the pyrrole ring and thus cannot act as a substrate at the 'P' site. Most importantly, the substrate bound initially to the 'P' site was shown experimentally to do so through a Schiff base and could be trapped with sodium borohydride under the conditions of the single turnover experiments. It thus appears that the mechanism of human, bovine and bacterial enzymes all proceed with the initial binding of a molecule of 5-aminolaevulinic acid to form a Schiff base and that this substrate gives rise to the 'P' side of the porphobilinogen.

On the basis of the order in which the two substrates molecules bind, new mechanisms may be proposed. The formation of porphobilinogen from two molecules of 5-aminolaevulinic acid requires an aldol condensation, the formation of a C-N bond, the loss of two molecules of water and a tautomeric shift. It is proposed that the initial event is the binding of the first substrate molecule to the 'P' site in the form of a Schiff base. Two broad mechanistic courses are then possible – either the aldol condensation followed by Schiff base formation or the converse. In the former mechanism the 5-aminolaevulinic acid-enzyme Schiff base provides a good electrophile for the aldol reaction with C-3 of the second substrate molecule. An enzymic base is required to deprotonate the 3-position (twice) and a candidate for this group may be one of the conserved histidine residues. Elimination of the enzyme lysine nitrogen completes the aldol condensation. The formation of a Schiff base to furnish the five-membered ring followed by enzyme catalysed tautomerisation involving the loss of one of the hydrogen atoms at the 2-position completes the reaction. In an alternative mechanism, the initial reaction between the substrates involves the formation of a Schiff base between the amino group of the substrate molecule covalently attached to the enzyme and the carbonyl carbon of the second substrate at the 'A' site. The presence of this Schiff base could serve to labilise the hydrogen atoms at the 3-position via an eneimine and to facilitate the aldol reaction. The final elimination of the ε-nitrogen of lysine and the tautomerisation would follow as before. This latter mechanistic derivative is less attractive for stereoelectronic reasons.

The elimination of one of the two hydrogen atoms from the 5-position of the 5-aminolaevulinic acid which becomes C-2 in porphobilinogen is not left to chance and is catalysed stereospecifically by the enzyme. Evidence for this has come from the use of 5-aminolaevulinic acid tritiated in the 5-position [173,174]. Incubation of 5-amino[5RS-^{3}H$_2$]laevulinic acid with dehydratases from bacterial or mammalian sources leads to the removal of exactly half of the label at the 2-position of porphobilinogen suggesting that this is an enzyme catalysed event. The use of 5-aminolaevulinic acid, stereospecifically tritiated in the *proS*-position, gives rise to porphobilinogen with complete retention of the label. These observations therefore establish that the *proR* hydrogen atom originally present in 5-aminolaevulinic acid is stereoselectively removed and must therefore occur whilst the intermediate is bound to the active site. These mechanistic details are included in Scheme 8.

5.8. Catalytic groups and the nature of the active site

The enzyme groups responsible for catalysis have not been identified with any certainty with the exception of the active site lysine [169] K-252 in the human and rat

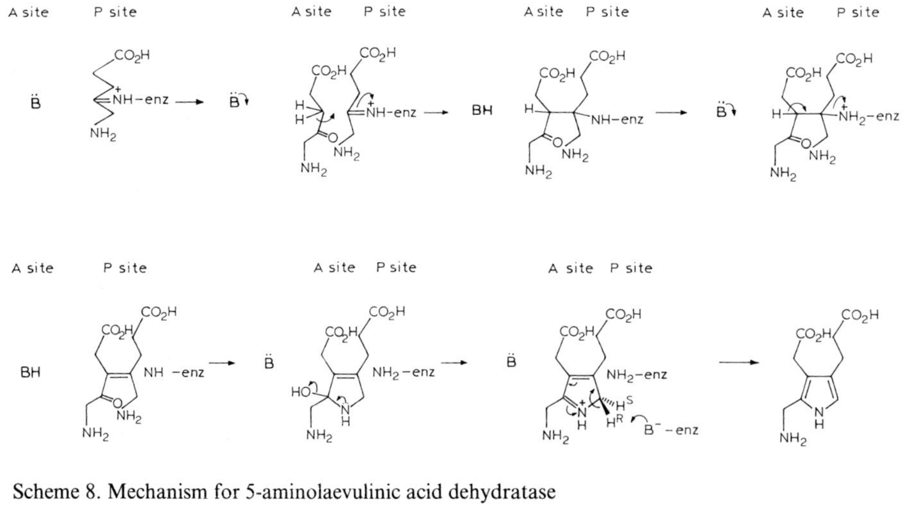

Scheme 8. Mechanism for 5-aminolaevulinic acid dehydratase

enzymes and K-258 in the *E. coli* enzyme. The involvement of cysteine residues in the enzyme mechanism has been suggested from inactivation studies with iodoacetic acid and iodoacetamide which modify two highly reactive groups [148]. A survey of the literature, however, reveals that cysteine is not favoured by enzymes for acid-base catalysis. Other evidence suggests the role of these active cysteines is in zinc binding. Two histidine residues have been implicated from photoinactivation studies [141] but whether these are catalytic groups or zinc ligands is not known. Interestingly, there are two conserved histidines in the putative zinc binding site which may correspond to one or both residues which are modified in this study [175]. A comparison of the available derived protein sequences from the gene and cDNA sequencing reveals several potential candidates for catalytic residues in other areas of high conservation.

Despite a considerable amount of information, the precise role of zinc and whether it plays a structural or mechanistic function – or both – is still an unanswered question. Apart from a purely structural function, other zinc could participate in at least three broad roles; i) as a divalent cation to orient and bind the negatively charged carboxylates of the two substrates; ii) in a mechanistic capacity to act as a Lewis acid to promote the polarisation of the carbonyl group and enolisation of the 5-aminolaevulinic acid bound to the 'A' site; iii) to activate a bound water molecule in the form of a zinc hydroxide group to catalyse the deprotonations at the C-3 of the substrate molecule bound to the 'A' site. Although the weight of evidence points to zinc being essential for activity, the results of some experiments have been interpreted to suggest its non-involvement in substrate binding. For instance, replacement of ^{113}cadmium ions at the zinc binding site yields an active enzyme with a single ^{113}cadmium n.m.r. resonance at $\delta=79$ ppm. This signal is not affected by the addition of 5-aminolaevulinic acid suggesting that the substrate does not ligate directly with the metal ion [176]. The finding that the zinc site is modified to a magnesium site in the plant enzymes suggests a structural or regulatory role is more likely for the metal ion although i) and ii) are possible.

Since 5-aminolaevulinic acid dehydratases possess two substrate binding sites, any approach which allows a distinction between them is potentially useful in the study of the mechanism. Experiments by Jordan et al. [177] with the bovine dehydratase were designed to determine whether the enzyme which had been inactivated by alkylation with iodoacetic acid is still capable of interacting with substrate. Alkylation of the most reactive thiol with iodoacetic acid, although causing the loss of zinc from the enzyme, did not affect the ability of the enzyme to form a Schiff base with the substrate. In fact when 5-amino[^{14}C]laevulinic acid and borohydride were used, twice the amount of radioactivity was incorporated into the alkylated protein compared with the native enzyme. At this time however the order of substrate addition was not known. These findings have been confirmed and extended by Jaffe and Hanes [178] with observations on the bovine enzyme temporarily alkylated with methylmethane-thiosulphonate, a reagent which reacts reversibly with SH groups. These experiments establish that an intact zinc binding site is not essential for the initial formation of the Schiff base intermediate but may be involved in some way with the ability of the enzyme to accept the second molecule of substrate. It is interesting to note that enzyme, temporarily incapacitated by the methylmethanethiosulphonate reagent, is twice as sensitive to inactivation with borohydride in the presence of substrate and loses almost all its recoverable activity, whereas native enzyme loses only half of its activity under similar conditions. Thus modification with iodoacetic acid or methylmethanethiosulphonate whilst totally preventing the overall formation of porphobilinogen maximises the binding of substrate at the Schiff base ('P') site.

The property of half-site reactivity has been documented extensively with respect to the dehydratase enzymes. Half of the subunits of the bovine dehydratase [178,179] are modified on reduction with borohydride in the presence of 5-amino[5-^{14}C]laevulinic acid. These findings in particular have been used to propose that half site reactivity may be operative. Studies with the active site directed reagents, 3-chlorolaevulinic acid and 5-chlorolaevulinic acid, have also revealed an element of half site reactivity under some conditions [180]. Aspects of half-site reactivity of 5-aminolaevulinic acid dehydratases have been discussed in detail elsewhere [140] (Cheh and Neilands, 1976).

An alternative explanation for these findings is assisted by the observations of Nandi and Shemin [166] who found that enzyme treated with laevulinic acid and borohydride is totally inactivated whereas just over half (55–65%) of the activity is lost in equivalent experiments with the substrate. It must be concluded that if turnover is prevented, the enzyme binds more efficiently to either the substrate, or an analogue such as laevulinic acid in the form of a reducible Schiff base intermediate at the 'P' site. In the native enzyme where turnover is occurring, catalytic intermediates and porphobilinogen occupy the active site at the expense of the Schiff base intermediate with an accompanying protection against inactivation. Porphobilinogen, the product is in fact a respectable competitive inhibitor, inhibiting the bacterial enzyme 50% at a concentration of 2 mM (Jordan – unpublished data) and has been shown to occupy the active site using n.m.r. spectroscopy [181]. The other explanation for the half site binding by the substrate is that the enzyme exhibits true half-site activity.

6. The biosynthesis of uroporphyrinogen III

6.1. Introduction

Uroporphyrinogen III is the universal cyclic tetrapyrrole from which all haems, chlorophylls, bacteriochlorophylls, corrin, F_{430} and all other cyclic and linear tetrapyrroles and related compounds are derived.

The construction of the first cyclic tetrapyrrole intermediate, uroporphyrinogen III from four molecules of the monopyrrolic precursor porphobilinogen requires the participation of two enzymes. The first of these enzymes is called porphobilinogen deaminase (EC 4.3.1.8) although it was originally known as uroporphyrinogen I synthase –a name still used in the medical sphere. More recently the use of another name, hydroxymethylbilane synthase, has made matters even more confusing to the reader outside the field. Since the name porphobilinogen deaminase is the least ambiguous and is the one most widely used by investigators in the field it will be used throughout this section. The second enzyme required for the biosynthesis of uroporphyrinogen III is called uroporphyrinogen III synthase (EC 4.2.1.75). This enzyme was first called uroporphyrinogen III cosynthase (or cosynthetase) since originally it was thought to function together with the deaminase in a complex rather than to catalyse its own individual reaction. It has also been referred to as uroporphyrinogen isomerase in some early literature. Whilst on the subject of nomenclature, it should be pointed out that uroporphyrinogen III is strictly a hexahydroporphyrin or more precisely a 5,10,15,20,22,24-hexahydroporphyrin with 20 macrocyclic carbon atoms (positions 1–20) and four nitrogens (21–24) in the macrocyclic ring. The reader is encouraged to read an introduction to tetrapyrrole nomenclature by Bonnett [182] and then proceed to a full treatise on the subject [183]. See also Chapter 6.

The role of porphobilinogen deaminase is to tetrapolymerise the pyrrole porphobilinogen into preuroporphyrinogen, a highly labile 1-hydroxymethylbilane. This involves the stepwise addition of four porphobilinogen molecules to the enzyme with loss of each amino group as ammonia. The enzyme-bound linear tetrapyrrole is then released into solution as preuroporphyrinogen which acts as the substrate for uroporphyrinogen III synthase. The uroporphyrinogen III synthase catalyses an astonishing reaction in which the fourth ring (ring D) of preuroporphyrinogen is rearranged and the molecule is cyclized to give the uroporphyrinogen III isomer. The precise mechanism of these two events is not yet known. In the absence of uroporphyrinogen III synthase, the highly reactive preuroporphyrinogen cyclises without rearrangement, in a non enzymic reaction, to give uroporphyrinogen I. Since many of the studies on the deaminase and synthase (cosynthase) have been closely interwoven, the two enzymes will be considered in this section together. Their combined reactions are shown below in Scheme 9.

The enzymic production of uroporphyrinogen III from porphobilinogen was first demonstrated by Bogorad and Granick [184] using enzyme extracts from spinach. Most significantly, heat treatment of these extracts prior to incubation with porphobilinogen led to the formation of the uroporphyrinogen I isomer providing the first indication that a labile component was responsible for the isomerization reaction. Porphobilinogen deaminase and uroporphyrinogen III synthase were subsequently isolated by Bogorad [13,14], although at that time the uroporphyrinogen III synthase was termed uroporphyrinogen isomerase. Bogorad found that extracts of acetone powder from spinach leaves containing the porphobilinogen deaminase catalysed the formation of uroporphyrinogen I from four molecules of porphobilinogen with the liberation of four molecules of ammonia. When semi-purified 'isomerase' from wheat germ was added to the spinach enzyme preparation, uroporphyrinogen III was pro-

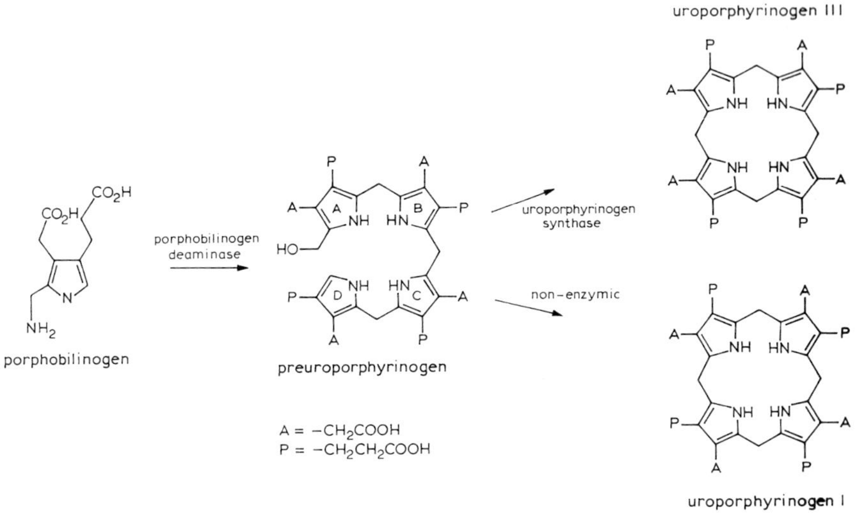

Scheme 9. The synthesis of uroporphyrinogens I and III from porphobilinogen

duced. Bogorad concluded that the role of the deaminase was to catalyse the formation of a 'polypyrromethane' intermediate and that the 'isomerase' acted on this intermediate, in the presence of porphobilinogen, to produce the uroporphyrinogen III isomer. Most importantly it was established that the 'isomerase' alone neither catalysed the isomerization of uroporphyrinogen I into uroporphyrinogen III nor would it act solely on porphobilinogen as a substrate.

6.2. Properties of porphobilinogen and uroporphyrinogens

Since Bogorad's classical experiments the pursuance of the mechanism by which the two enzymes catalyse the biosynthesis of uroporphyrinogen III has, until relatively recently, baffled consistently all investigators and has left a trail of incorrect theories and erroneous conclusions. Of all the problems in the tetrapyrrole field it is still one of the most fascinating, the most enigmatic aspect being the mechanism by which the ring D in uroporphyrinogen III becomes rearranged during the overall transformation. Interestingly, the intrinsic chemistry of porphobilinogen and its polymerization products appear to favour the uroporphyrinogen III isomer since investigations by Mauzerall into the non-enzymic polymerization of porphobilinogen revealed that in acid uroporphyrinogen III is the major product. Analysis of the uroporphyrinogen isomers produced showed a statistical distribution of uroporphyrinogens III:IV:II:I (Scheme 10) in a ratio of 4:2:1:1 [185,186]. Furthermore it was shown that any one of the uroporphyrinogen isomers, when heated in acid, yielded the same ratio of isomers indicating that the uroporphyrinogens are in free equilibrium with one another.

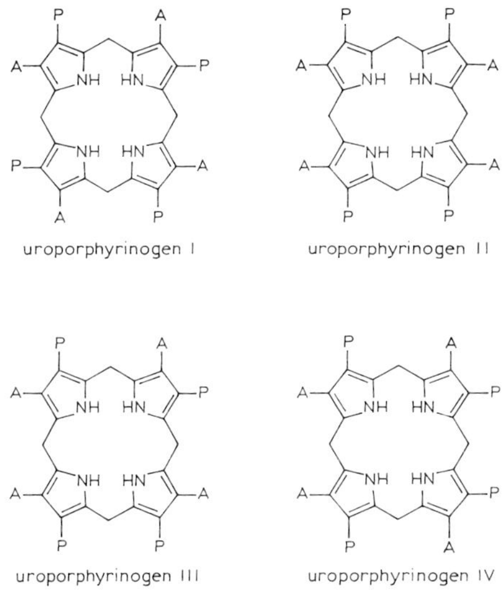

Scheme 10. The structure of uroporphyrinogen isomers

These experiments established that in chemical terms at least, it was possible for the -CH_2NH_2 side chain of porphobilinogen to react, after elimination of ammonia, with either a free or a substituted α-position of another porphobilinogen unit. The involvement of hydroxymethyl intermediates were suggested to explain the mechanism by which any one of the uroporphyrinogen isomers could interconvert to the same equilibrium mixture of all the four isomers – although other alternatives are possible. Subsequent chemical studies in Frydman's laboratory showed that the α-position of porphobilinogen reacts more favourably at the substituted α-position of another molecule of porphobilinogen. Furthermore he showed that the 'polypyrroles' formed on self condensation of porphobilinogen are unusually stable in solution [187]. Nature has thus accepted that uroporphyrinogen III is the chemically favoured product and has exploited the asymmetry of this isomer later in the pathway. For instance, coproporphyrinogen oxidase catalyses only the decarboxylation of the type III isomer and hence the final product of the pathway, haem, is also asymmetric (see chapter 2). This asymmetry or 'sidedness' of haem is perfectly suited to its role as a prosthetic group in proteins since the hydrophobic portion interacts with the non-polar amino acid side chains in the interior of the protein whilst the paired propionic acid side chains occupy positions near the more polar surface of the protein.

6.3. Porphobilinogen deaminase

The isolation of porphobilinogen deaminase and uroporphyrinogen synthase have been accomplished from a wide variety of sources. The two enzymes have been purified either together as 'porphobilinogenase' [188,189,190] or as separate enzymes. Porphobilinogen deaminases have been purified to homogeneity from *R. spheroides* [191,192], spinach [193], human erythrocytes [194], *Chlorella regularis* [195], *E. gracilis* [196] and rat liver [197]. More recently, the porphobilinogen deaminase has been isolated from strains of *E. coli* harbouring the *hemC* gene [198,199] and has been crystallised [200]. All the porphobilinogen deaminases exist as monomeric proteins of M_r from 34,000 Da to 44,000 Da and most have similar properties with optimal activities at pH 8.0–8.5 and isoelectric points between pH 4 and 4.5. The K_m values for porphobilinogen are all in the low μM range. One unifying property of the deaminases is their remarkable heat stability. This contrasts sharply with the uroporphyrinogen III synthases which are, in general, extremely heat labile. Porphobilinogen deaminases are rather 'slow' enzymes with turnover numbers of 0.5 sec^{-1} for the tetrapolymerization of porphobilinogen.

6.4. Uroporphyrinogen III synthase (cosynthase)

Far fewer uroporphyrinogen III synthases (EC.4.2.1.75) have been isolated in homogeneous form due to their extreme instability and the lack, until recently, of a convenient assay method. These problems have largely been overcome by the advent of fast protein liquid chromatography (f.p.l.c) purification methods and the development by Jordan of a rapid assay method for the enzyme [201]. The uroporphyrinogen synthases have been isolated in homogeneous form from human erythrocytes [202] and from a genetically engineered strain of *E. coli* [203,204]. The enzyme has also been obtained in high purity from spinach [193], rat liver [205,206] and *E. gracilis* [207]. The synthases are generally smaller enzymes than the deaminases, the M_rs of the human and *E. coli* enzymes being 29,500 Da and 28,000 Da respectively. The *E. gracilis* enzyme is slightly larger, M_r 31,000 Da. All the synthases, like the deaminases, appear to be monomeric enzymes with isoelectric points around pH 5. The K_m values for preuroporphyrinogen vary according to source but are in the range 10–25 μM. Uroporphyrinogen synthases have turn-over numbers of at least 200 sec^{-1}, much higher than the deaminases. Because of this higher turn-over number preuroporphyrinogen never accumulates and the transformation of porphobilinogen into uroporphyrinogen III is assured with very little formation of uroporphyrinogen I.

6.5. Early investigations on the mechanism of uroporphyrinogen biosynthesis

The mechanism by which the porphobilinogen deaminase and uroporphyrinogen III synthase catalyse the transformation of four porphobilinogen molecules into the tetrapyrrole uroporphyrinogen III has attracted the attention of numerous researchers for well over a quarter of a century and as many as thirty mechanisms have been proposed to explain the reaction. Several of these have been covered in reviews which were published in the 1970s [17,18,208]. The elucidation of the mechanism by which the deaminase and synthase catalyse the transformation of porphobilinogen into uroporphyrinogen III, has tantalized even the finest minds and few workers in this field have escaped without making incorrect conclusions at some point. Included in the catalogue of errors is the early rearrangement mechanism [209], the headless dipyrrole intermediate [210], the aminomethylbilane intermediate [211,212,213] (preuroporphyrinogen with an amino group instead of a hydroxyl group) and N-alkyl porphyrinogen [214].

The first significant pointer to the mechanism by which porphobilinogen deaminase catalyses its reaction came from observations made in the laboratories of Bogorad [215] and Neuberger [192] who found that incubation of the deaminase with porphobilinogen in the presence of high concentrations of the bases NH_3, NH_2OH or NH_2OCH_3 led to the apparent interception of enzyme-bound species which were liberated into the medium in the form of monopyrroles, dipyrromethanes, tripyrranes and bilanes linked to the inhibitory base [215]. Incubation of the deaminase with porphobilinogen in the presence of tritiated methoxyamine ($NH_2OC^3H_3$) led to the formation of tritiated monopyrrole and bilane (linear tetrapyrrole) [192]. The results suggested that a covalent linkage exists between the enzyme and the bound substrates and pointed to a mechanism in which the four porphobilinogen molecules are incorporated into the tetrapyrrole in a stepwise fashion.

6.6. Experiments with aminomethyldipyrromethanes, aminomethyltripyrranes and aminomethylbilanes

A period of investigation followed in which various synthetic aminomethyldipyrromethanes and aminomethyltripyrranes were prepared chemically to determine whether they were able to act as tetrapyrrole precursors. A great deal of controversy arose as to whether the rearrangement of the D ring occurred early [209] or late in the deaminase/synthase reaction [216]. However, when any of these compounds were incubated with either deaminase or with the combined deaminase/synthase system in the presence of porphobilinogen the results were disappointing since the deaminase preferred to use porphobilinogen rather than the aminomethyldipyrromethanes and aminomethyltripyrranes which instead acted as enzyme inhibitors [209,216,217].

An important breakthrough was made by Batlle in 1977 [218] and her colleagues when it was found that two polypyrroles (termed 'D' and 'P'), accumulated in solution when extracts of *Euglena* containing porphobilinogen deaminase or the combined deaminase/synthase system respectively were incubated with porphobilinogen. It was suggested that these compounds were natural intermediates in the biosynthesis of uroporphyrinogens and 'D' was proposed as NH_2CH_2-AP-AP-AP-AP and 'P' as NH_2CH_2-PA-PA-PA-AP (Scheme 11). The small amount of material available, however, prevented the determination of their structures. When polypyrrole 'D' was

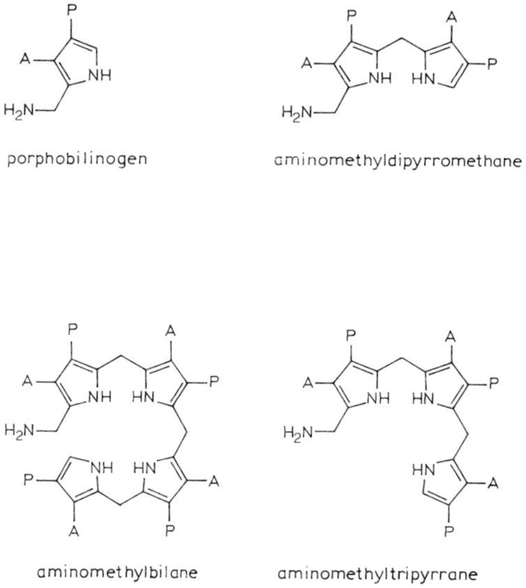

Scheme 11. Structure of the aminomethyldipyrromethane, aminomethyltripyrrane and aminomethylbilane related to porphobilinogen

incubated with the deaminase/synthase enzymes some uroporphyrinogen III was formed whereas with the deaminase alone uroporphyrinogen I was the major product. In the case of the compound 'P', deaminase alone yielded equal amounts of the two uroporphyrinogen isomers whereas incubation with the two enzymes resulted in the exclusive production of uroporphyrinogen III. Although there was insufficient definitive structural information about the nature of 'P' and 'D', their possible involvement in the biosynthetic mechanism was a great stimulus to further studies on aminomethylbilanes. The difficult chemical synthesis of the aminomethylbilane NH_2CH_2-AP-AP-AP-AP soon followed in the laboratory of Battersby [219] and its incubation with the combined deaminase/synthase system produced good yields of uroporphyrinogen III.

Although it had been well established by early investigators that the D ring is inverted during the biosynthesis of uroporphyrinogen III, elegant ^{13}C nmr studies by Battersby and his colleagues, using [2,11-$^{13}C_2$] porphobilinogen, which was diluted with non-labelled porphobilinogen, left no doubt that a single intramolecular rearrangement occurred [216]. This was further confirmed at the tetrapyrrole level by synthesising the aminomethylbilane NH_2CH_2-AP-AP-AP-AP, (Scheme 11) labelled regiospecifically with ^{13}C in either positions 16 and 20 (■) or at positions 15 and 19 (●) [211]. The uroporphyrinogen III biosynthesised from either of these labelled materials showed direct ^{13}C-^{13}C coupling since the label originally at position 16 now occupied the adjacent position 19 and in the other case, the label once at position 19, now coupled to C-15 as a result of its new location at position 16 (Scheme 12).

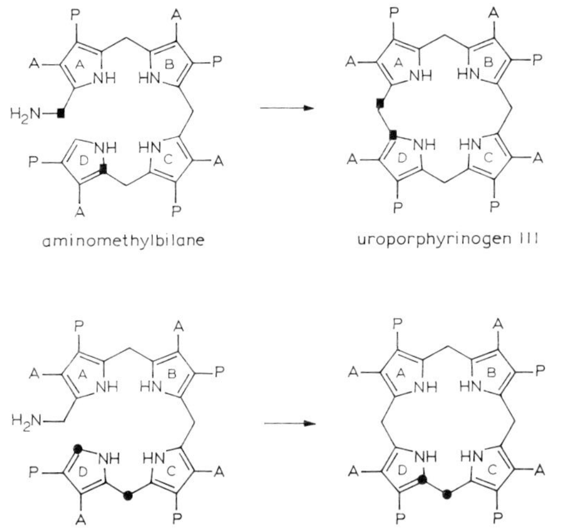

Scheme 12. Intramolecular rearrangement of regiospecifically ^{13}C labelled aminomethylbilane NH$_2$CH$_2$-AP-AP-AP-AP using ^{13}C n.m.r.

These experiments also established that the terminal pyrrole ring bearing the aminomethyl group in the aminomethylbilane NH$_2$CH$_2$-AP-AP-AP-AP gave rise to the ring A in uroporphyrinogen III. On the basis of these observations it was proposed that the aminomethylbilane NH$_2$-AP-AP-AP-AP was the key intermediate which was produced by the deaminase and which subsequently acted as the substrate for the synthase [211,212,213].

In order to explore further the role of aminomethylbilanes in the enzymic formation of uroporphyrinogens, a range of aminomethylbilanes were chemically synthesised, each of which had a single ring rearranged, and their ability to act as substrates with the combined deaminase/synthase system from *E. gracilis* was investigated [212].

aminomethybilane	rate
NH$_2$CH$_2$-AP-AP-AP-AP	+++++++
NH$_2$CH$_2$-PA-AP-AP-AP	−
NH$_2$CH$_2$-AP-PA-AP-AP	++++
NH$_2$CH$_2$-AP-AP-PA-AP	+
NH$_2$CH$_2$-AP-AP-AP-PA	++
NH$_2$CH$_2$-PA-PA-PCA-AP	+

The compounds which contained the acetic acid (A) and propionic acid (P) substituents reversed in the A ring (i.e. the ring with the aminomethyl substituent) were not utilized by the enzymes, although, as will be seen later, this was almost certainly a reflection of the inability of the porphobilinogen deaminase to interact with such aminomethylbilanes and not a property of the synthase. Although Battersby and his colleagues favoured the linear aminomethylbilane NH$_2$CH$_2$-AP-AP-AP-AP as the key enzymic intermediate in the biosynthesis of uroporphyrinogen III, in the absence of purified, deaminase-free uroporphyrinogen III synthase it was not possible to test the proposal unambiguously. It was observed, however, that deaminase alone accelerated the formation of uroporphyrinogen I from the aminomethylbilane NH$_2$-AP-AP-AP-AP but the significance of this was yet to be appreciated. Related studies with several aminomethylbilanes were also carried out by Sburlati et al. [220] using the enzyme from *R. sphaeroides*.

6.7. The discovery of preuroporphyrinogen, the substrate for uroporphyrinogen III synthase

The discovery of preuroporphyrinogen by Jordan and Burton in Scott's laboratory at Texas A & M University in 1978 was the major breakthrough in our understanding of the mechanism by which porphobilinogen deaminase and uroporphyrinogen III synthase participate together in the biosynthesis of uroporphyrinogen III and will be given special coverage here. As often in scientific discoveries, serendipity played a large part since the formation of preuroporphyrinogen was first noticed in incubations of porphobilinogen deaminase with 11-^{13}C porphobilinogen in the n.m.r. tube during attempts to observe the formation of enzyme-bound intermediate complexes. In addition to the n.m.r. signal at δ=22 p.p.m. arising from positions 5,10,15 and 20 of uroporphyrinogen I, we observed an additional complex signal at about δ=23 p.p.m. which appeared transiently over a period of several minutes. Further investigation with larger amounts of enzyme revealed that in addition to the complex signal at δ=23 p.p.m. there was a further resonance at δ=57 p.p.m., approximately 20 p.p.m. from the aminomethyl resonance of the regiospecifically labelled substrate (δ=36 p.p.m.). Most importantly the integration of the signals showed an almost perfect 3:1 distribution for the resonances at δ=23 p.p.m. and δ=57 p.p.m. respectively [221]. Clearly these were not signals due to uroporphyrinogen I but to some tetrapyrrole intermediate which contained three very similar, though not identical, labelled carbon atoms together with a fourth carbon atom in a very differnt environment. The aminomethylbilane NH$_2$-AP-AP-AP-AP previously advocated [213] in Battersby's laboratory would have given n.m.r. signals at about δ=23 p.p.m. and δ=36 p.p.m. so this was eliminated as a candidate immediately. A hydroxymethylbilane was proposed as one of the most likely structures for the intermediate [221] since substitution of -OH for -NH$_2$ characteristically produces a shift of about 20 ppm, similar to that observed. In the absence of an authentic standard other possibilities were also considered [221]. The most important question, however, still needed to be answered, namely, what was the significance of the transient species and had we discovered the elusive substrate for the uroporphyrinogen III synthase enzyme? The answer to this question was partially obtained when a similar n.m.r. experiment was performed but in the presence of uroporphyrinogen III synthase. Under these conditions the transient n.m.r. signal was not formed and uroporphyrinogen III was the product. Thus the transient species appeared to act as the substrate for the synthase [221,222].

In a series of enzymic experiments Jordan and Burton isolated the labile intermediate by incubating porphobilinogen with a large amount of deaminase enzyme so that the consumption of porphobilinogen was complete in about 10 minutes. The incubation solution was cooled to 0 °C to stabilise the intermediate which was totally separated from the deaminase by ultrafiltration. The next stage was to incubate the intermediate with highly purified uroporphyrinogen III synthase to determine the extent of uroporphyrinogen III production. As we had predicted from the previous n.m.r. experiments, the intermediate was transformed quantitatively and at remarkable speed, even at 0 °C, into uroporphyrinogen III [222]. Interestingly, when the intermediate was incubated with buffer alone uroporphyrinogen I was generated with a half life of 4.5 minutes at 37 °C. The overall conclusions from these experiments are as follows [222].

1) The deaminase is responsible for assembling a tetrapyrrole intermediate from four molecules of porphobilinogen in the absence of uroporphyrinogen III synthase.

2) The intermediate is not an aminomethylbilane of the type NH_2CH_2-AP-AP-AP-AP previously proposed [213].

3) The intermediate is highly unstable with a half life of only 4.5 minutes at pH 8.5 and is converted into uroporphyrinogen I in a non-enzymic reaction.

4) The intermediate acts as the substrate for uroporphyrinogen III synthase and is rapidly and quantitatively transformed into uroporphyrinogen III.

5) The transformation of the intermediate into uroporphyrinogen III does not require the participation of porphobilinogen deaminase or additional porphobilinogen units.

6) The two enzymes function independently and sequentially in the overall transformation of porphobilinogen into uroporphyrinogen III.

These enzymic experiments thus firmly established the transient species as the substrate for the uroporphyrinogen III synthase [222] and at a stroke eliminated virtually all previous mechanistic postulates.

Since the intermediate was the enzymic precursor for uroporphyrinogen III and the non-enzymic precursor for uroporphyrinogen I it was named preuroporphyrinogen [221,222]. Further investigations by Jordan and Berry [223] with uroporphyrinogen III synthases from several sources firmly established that preuroporphyrinogen was a substrate for the synthases from human erythrocytes, spinach leaves and yeast suggesting that animals, plants and microorganisms utilise preuroporphyrinogen as a universal intermediate. Furthermore, the fact that preuroporphyrinogen generated by *R. sphaeriodes* was consumed by uroporphyrinogen III synthases from different sources suggested that the two enzymes were working completely independently and not in a functional complex as had been previously suggested [193]. Further experiments revealed that the non-physiological aminomethylbilane NH_2CH_2-AP-AP-AP-AP is accepted as a poor substrate for porphobilinogen deaminase and is deaminated slowly to preuroporphyrinogen which then acts as the substrate for uroporphyrinogen III synthase explaining why it had previously been considered as an intermediate in the enzyme reaction [212,219]. The conclusions from these investigations are summarized in Scheme 13.

The n.m.r. experiments described above were soon confirmed by Battersby and his colleagues with similar results. Most importantly preuroporphyrinogen was synthesised chemically and shown unambiguously to be a 1-hydroxymethylbilane with a structure $HOCH_2$-AP-AP-AP-AP [224]. The synthesis of this compound was most challenging because of its instability, nevertheless it was shown to have identical n.m.r. signals and chemical and enzymic properties to the intermediate which had been isolated originally by Burton, Jordan and Scott in the enzymic experiments [221,222]. The work from Battersby's laboratory has been comprehensively reviewed [19].

Further experiments carried out in Texas A & M University in which porphobilinogen deaminase was incubated with [1-^{15}N,11-^{13}C] porphobilinogen yielded an n.m.r. spectrum which appeared to exhibit $^{13}C-^{15}N$ single bond coupling. This was incorrectly interpreted as evidence for an N-alkylporphyrinogen type of compound as being the structure of preuroporphyrinogen [214] but this was soon dismissed. The coupling observed was in fact due to long range coupling between the ^{15}N in ring A with the ^{13}C atom at position 20 of the hydroxymethylbilane which gave rise to the

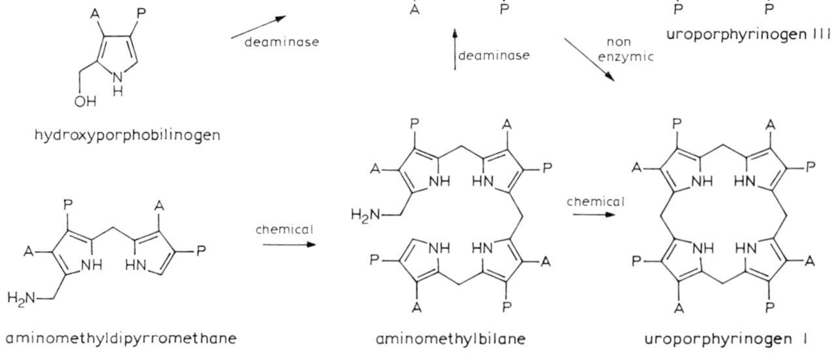

Scheme 13. Formation of preuroporphyrinogen and its conversion to uroporphyrinogens

splitting of the signal at δ=57 p.p.m. actually providing additional evidence for the true structure of preuroporphyrinogen.

Having solved the essential features of the enzymology of the porphobilinogen deaminase/uroporphyrinogen synthase reaction it was possible, for the first time, to design a rapid assay method for the synthase enzyme [201]. Previously this had involved laborious analyses of the uroporphyrinogen isomers by oxidation and decarboxylation to the corresponding coproporphyrins and identification of the esters by t.l.c. The rapid assay principle was based on the fact that uroporphyrinogens are formed more slowly from porphobilinogen with deaminase alone than when the deaminase and synthase enzymes are present together. Thus the synthase-catalysed reaction 'outstrips' the chemical cyclisation which, in the absence of the synthase, generates the unrearranged uroporphyrinogen I. It is therefore possible to assay the uroporphyrinogen III synthase directly by subtracting the amount of uroporphyrinogen I formed in a control containing deaminase alone from the uroporphyrinogens formed in the presence of the two enzymes. With minor corrections, this gives the amount of uroporphyrinogen III produced [201]. The ability to be able to assay the synthase in minutes rather than hours has enabled what is a very unstable enzyme, to be isolated far more rapidly than had been possible previously.

6.8. Order of assembly of the four pyrrole rings of the tetrapyrrole

The recognition that the role of the deaminase was to construct a linear tetrapyrrole prompted investigations to determine the order in which the four pyrrole rings are incorporated into preuroporphyrinogen by the deaminase enzyme. Two realistic possibilities existed – either the four porphobilinogen units could be assembled by the deaminase in the order A,B,C and D or alternatively in the order D,C,B, and A. The problem was solved by two different approaches from two independent groups.

Using a radiochemical approach Jordan and Seehra [225,226] exposed purified porphobilinogen deaminase to a stoichiometric amount of $[3,5^{14}C_2]$ pophobilinogen with the aim of labelling specifically the first enzyme site to bind substrate. Addition of nonlabelled porphobilinogen to complete the turn-over of the enzyme-bound label was followed by conversion of the resulting preuroporphyrinogen into protporphyrin with enzymes from *R. sphaeroides*. Chemical degradation with chromic oxide [3] of the protoporphyrin gave ethylmethylmaleimide from rings A and B and hematinic acid from rings C and D. Analysis showed that essentially all the label was present in the ethylmethylmaleimide with almost none in the hematinic acid suggesting that the A and B rings were first incorporated into the tetrapyrrole [225]. Further regiospecific degradation of the protoporphyrin established that ring A was incorporated initially [226].

This problem was also addressed by Battersby and his associates using a most elegant ^{13}C n.m.r. method [227]. The deaminase enzyme was first exposed to a limiting amount of unlabelled porphobilinogen in order to fill the active sites which first bind the substrate. This was then followed by [11-^{13}C] porphobilinogen to occupy the remaining binding sites. The resulting tetrapyrrole product was transformed into uroporphyrinogen III and thence to coproporphyrin III which was analysed by n.m.r. as the tetramethylester. The spectra showed clearly that the ^{12}C content of the *meso*-bridge atoms decreased in the order C20, C5, C10 and C15. The conclusions from this independent approach was the same as that above, namely, that preuroporphyrinogen is built up by the deaminase with ring A first binding to the enzyme followed by rings B, C and finally D.

6.9. Enzyme intermediate complexes between the deaminase and porphobilinogen

Shortly after these studies Anderson and Desnick [194] made the important observation that incubation of porphobilinogen deaminase with porphobilinogen led to the formation of several additional enzyme species. These species were more negatively charged than the native enzyme and could be separated from one another either by ion exchange chromatography or by electrophoresis. Using tritiated porphobilinogen the enzyme species were shown to be enzyme-intermediate complexes with one, two, three and four pyrrole units linked to the enzyme. These will be referred to as ES, ES_2, ES_3 and ES_4 respectively. Similar behavior was observed on adding substrate to the purified deaminases isolated and characterised from *R. spheroides* [228] and *E.coli* [229] although the ES_4 complexes from the bacterial enzymes were too unstable to be isolated. Importantly, sodium dodecylsulphate treatment of the complexes failed to release the bound substrate providing strong evidence for the existence of a covalent link between the enzyme and its substrates [230]. The overall conclusion from these observations is that the deaminase catalyses the synthesis of

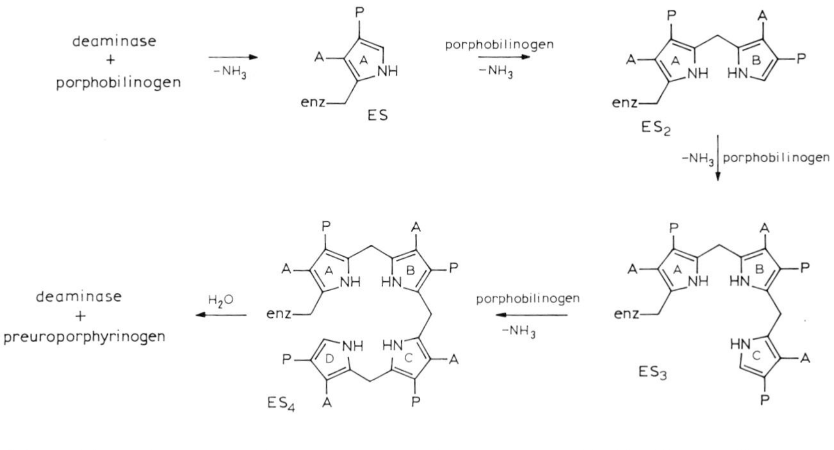

Scheme 14. Porphobilinogen deaminase intermediate complexes ES, ES_2, ES_3 and ES_4

the tetrapyrrole in a stepwise mechanism with a covalent enzyme-intermediate complex at each step (Scheme 14).

6.10. Preliminary studies on the nature of the enzymic group involved in substrate covalent binding

The evidence from the experiments described above [225–227,230], the observations from experiments by Davies and Neuberger [192] in which the amino group of porphobilinogen could be exchanged for an inhibitory base such as NH_2OCH_3 and the studies of Pluscec and Bogorad [215] all suggested that the substrate and intermediates of the reaction are linked covalently to the deaminase. The initial search for the enzymic group involved in the covalent linkage centered on n.m.r. experiments and protein chemistry. On the basis of ^{13}C n.m.r. evidence, a covalent link between the substrate and a lysine residue was initially proposed by Battersby's laboratory for the *E. gracilis* enzyme [231]. Inhibition studies with pyridoxal 5'-phosphate and borohydride consolidated the case for lysine [232] and a reduced pyridoxyl-lysine adduct was isolated from the inactivated enzyme [233]. Studies using ^3H n.m.r. with the *R. spheroides* enzyme pointed to the involvement of a cysteine residue as the amino acid responsible for the covalent link with the substrate [234]. In fact neither of these amino acids are involved in any covalent linkage with the substrate although as will be indicated later these residues may play some role in non covalent interaction with the substrate.

6.11. Discovery of a cofactor in all porphobilinogen deaminases – the dipyrromethane cofactor

Initial attempts in our laboratory to isolate the amino acid to which the substrate is covalently attached met with little success using the *R. sphaeroides* deaminase since the link between the enzyme and bound substrate is rather labile under analytical conditions. However whilst purifying peptides at acid pH from a proteolytic digest of deaminase enzyme-intermediate complexes prepared from [^{14}C]-porphobilinogen a pink, non-radioactive chromophore was observed which had spectroscopic properties identical to those of an oxidised and protonated dipyrromethane (Jordan and Berry 1983 – unpublished data). The study of the deaminase protein chemistry was greatly stimulated by the identification, cloning and sequencing of the *hemC* gene encoding the porphobilinogen deaminase from *E. coli* by Thomas and Jordan 1986 which permitted the construction of strains producing milligrammes of the deaminase enzyme [198,200]. Treatment of the native *E. coli* deaminase with formic acid also resulted in the formation of the pink chromophore and prolonged treatment with acid generated a highly fluorescent material which was identified as uroporphyrin I. Since this work had been carried out on the native enzyme rather than on an enzyme intermediate complex we concluded that the native *E. coli* deaminase also contains resident pyrrole residues which, on acid treatment, give rise to uroporphyrin. The treatment of the native deaminase with Ehrlich's reagent confirmed our suspicions [235] and showed a reaction typical of a dipyrromethane [215]. It was concluded that the pink chromophore observed at acid pH is the oxidised, protonated form of a resident dipyrromethane system thus corroborating our earlier observations with the deaminase from *R. sphaeroides*.

The next task was to determine the significance and role of the enzyme-bound dipyrromethane in the functioning of the deaminase. When the *E. coli* enzyme was treated with substrate to give the enzyme-intermediate ES_2, this complex exhibited a very rapid color reaction with Ehrlich's reagent indicative of a linear tetrapyrrole (bilane) [236]. This result could only mean that the two substrate molecules were interacting directly with the enzyme-bound dipyrromethane to form an enzyme-bound linear tetrapyrrole (bilane). Prolonged treatment of the ES_2 complex with formic acid yielded twice as much uroporphyrin compared to the reaction with the enzyme alone. In algebraic terms if the enzyme, E, gives rise to an amount of uroporphyrin x and the ES_2 complex generates an amount of uroporphyrin 2x it follows that E must have contained x=2 pyrrole ring equivalents. These results pointed to a most important conclusion, namely, the existence of a dipyrromethane which was resident at the active site of the enzyme and which provided the attachment site for the covalent binding of substrate [235].

One crucial experiment needed to be carried out to prove beyond doubt that the dipyrromethane was not merely an artifact due to bound substrate in some abortive complex or to inactive enzyme. If the dipyrromethane cofactor was able to generate uroporphyrin in acid it followed that it must be made up of pyrrole units similar to the substrate porphobilinogen. In that case it should be possible to label the dipyrromethane cofactor specifically with ^{14}C radioactivity by growing a deaminase overproducing strain of *E. coli* in the presence of 5-amino[5-^{14}C]laevulinic acid. The labelling was made even more effective by using a *hemA$^-$* strain so that no endogenous 5-aminolaevulinic acid would dilute the label. This approach generated [^{14}C]-porphobilinogen deaminase of high specific activity [235,237]. When the ^{14}C deaminase was isolated and incubated with non-radioactive porphobilinogen no radioactivity was incorporated into the enzymic product and, most significantly, the enzyme retained all the original label [235,237]. Only prolonged formic acid treatment released the radioactivity which was then found in uroporphyrin. These investigations proved without doubt that the deaminase from *E. coli* contains a resident dipyrromethane which is not subject to catalytic turnover and which is responsible for the covalent binding of the substrate molecules during the assembly of the tetrapyrrole. Jordan and Warren therefore introduced the name dipyrromethane cofactor [235]. Further investigations in our laboratory have revealed that the dipyrromethane cofactor is not just confined to the *E. coli* enzyme but that it is present in the deaminases from animals (human), dicotyledonous and monocotyledonous plants as well as in *R.*

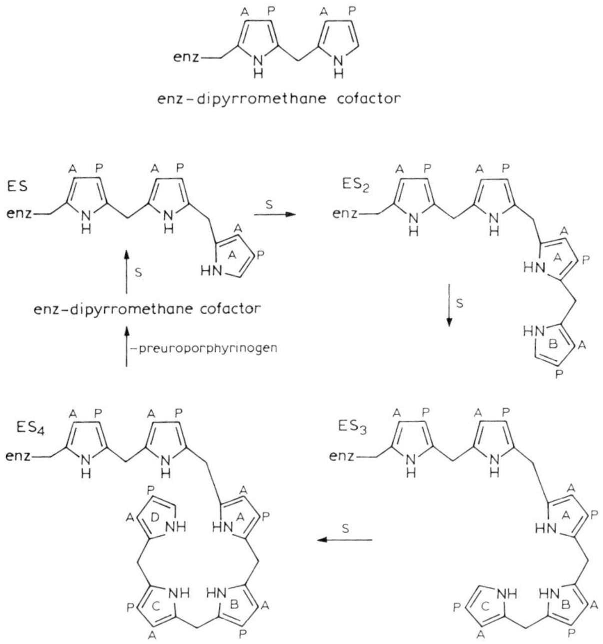

Scheme 15. Structure of the dipyrromethane cofactor and its role in the tetrapolymerization of porphobilinogen

sphaeroides [237]. Scheme 15 illustrates the polymerization process attached to the dipyrromethane cofactor.

Quite independently, other investigators also came to the conclusion that a pyrromethane system is present as the substrate binding group of porphobilinogen deaminase from *E. coli* [238]. This conclusion was deduced from the fact that a ^{13}C n.m.r. signal at $\delta=24.6$ was exhibited by the ES complex prepared from *E. coli* deaminase and 11-[^{13}C]porphobilinogen. This chemical shift was assigned to a methylene (-CH$_2$) group between two pyrrole rings and is about 20 p.p.m. upfield of a signal expected from a pyrrole α-methylene attached to the ε-amino group of lysine. The Ehrlich positive reaction, typical of a dipyrromethane was also noted by these workers.

Additional evidence in support of the dipyrromethane cofactor came from the fact that a *hemA$^-$* mutant of *E. coli* was also a 'mutant' for porphobilinogen deaminase. However, growth of this strain in medium containing 5-aminolaevulinic acid restored the activity of the deaminase to normal wild-type levels [229,239]. Similarly, it had also been noticed that *hemB* mutants of *E. coli* were also 'mutants' for porphobilinogen deaminase [240]. Thus the inability to biosynthesise porphobilinogen prevents the assembly of the dipyrromethane cofactor necessary for the generation of the active mature deaminase even though the capacity to form the apoprotein is unimpaired.

6.12. Further evidence for the role of the dipyrromethane cofactor from experiments with the chain termination suicide inhibitor α-bromoporphobilinogen

The evidence that the substrate molecules are added stepwise to the dipyrromethane cofactor prompted investigations with the porphobilinogen analogue α-bromoporphobilinogen [229]. α-Bromoporphobilinogen has the reactive α-position blocked with a bromine atom and should therefore act as a chain terminator of the polymerisation reaction. When α-bromoporphobilinogen was incubated with the native deaminase not only was the enzyme inactivated but the Ehrlich's reaction was inhibited showing unambiguously that the inhibitor had reacted directly with the dipyrromethane cofactor. Clearly the inhibitor is recognized by the deaminase, it is deaminated like a substrate and covalently linked to the dipyrromethane cofactor. However, the presence of the bromine atom blocks the reaction of Ehrlich's reagent and prevents the reaction with substrate. α-Bromoporphobilinogen thus acts as a true suicide inhibitor. Further experiments showed that E, ES, ES_2 and ES_3 all reacted with α-bromoporphobilinogen B to yield EB, ESB, ES_2B and ES_3B where B is the α-bromoporphobilinogen [229]. Interestingly, these termination complexes could all be isolated except for ES_3B which rapidly decomposed to S_3B, bromopreuroporphyrinogen, and free enzyme which then re-reacted with the inhibitor to form the termination complex EB.

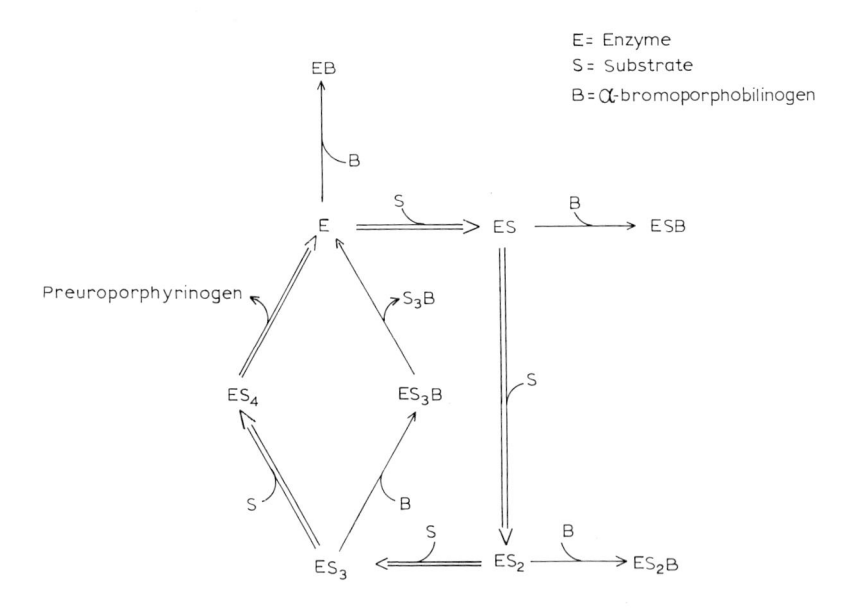

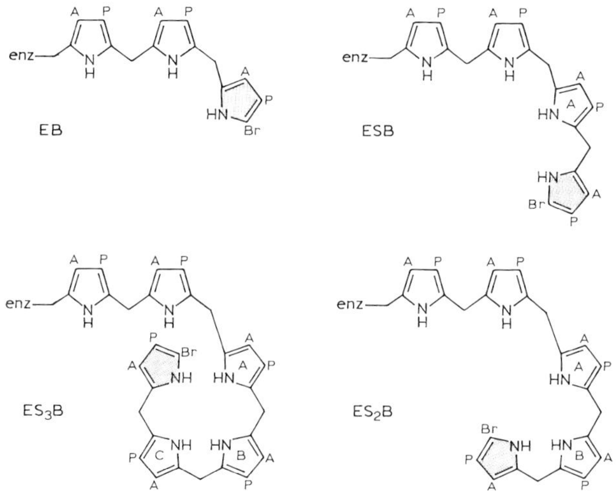

Scheme 16. Structure of termination complexes with α-bromoporphobilinogen

Subsequent experiments showed that the bromopreuroporphyrinogen released from the deaminase was an inhibitor of the uroporphyrinogen III synthase at low concentrations (Warren, Alwan and Jordan – unpublished observations). The structure of the termination complexes are shown in Scheme 16 (inhibitor shaded).

6.13. Attachment site and mechanism of assembly of the dipyrromethane cofactor

The discovery that the substrate binding site of porphobilinogen deaminases contains a covalently bound dipyrromethane cofactor posed important questions, namely, how the cofactor is attached to the deaminase and what is the mechanism of its assembly? The answer to the former question has come from the adaptation of the ^{14}C experiments described above for the labelling of the dipyrromethane [235]. Thus a recombinant strain of *E. coli* was grown in the presence of amino[^{13}C]laevulinic acid to generate deaminase containing the ^{13}C labelled cofactor [241]. The dipyrrole cofactor would be expected to exhibit two aromatic signals and two aliphatic signals,

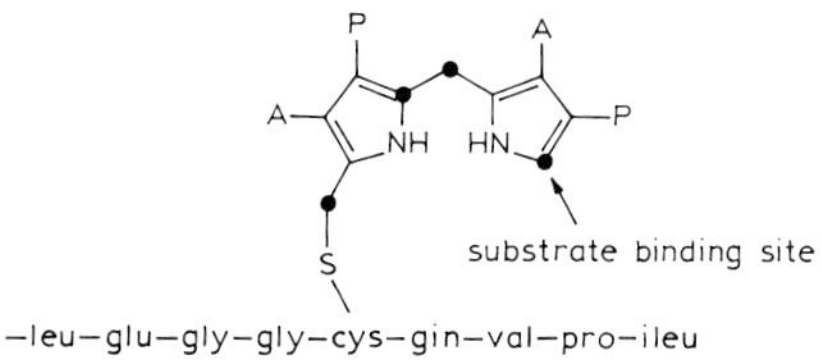

Scheme 17. Structure of the dipyrromethane cofactor showing labelling from 5-amino[5-^{13}C]laevulinic acid and attachment to cysteine 242

one from the methylene group sandwiched between the pyrrole rings and one from the methylene group covalently attached to the enzyme as shown in Scheme 17.

The ^{13}C labelled deaminase was analysed by Scott's laboratory at Texas A & M University and a ^{13}C difference spectrum was determined. The spectrum [241] revealed 4 broad signals: at $\delta=117$ and 128 p.p.m. assigned to the two aromatic carbon atoms and a signal at $\delta=25$ p.p.m. from the bridging methylene position. A very broad fourth signal at $\delta=27$ p.p.m. arising from the methylene carbon atom directly linked to the enzyme was also observed which moved to $\delta=29$ p.p.m. and sharpened when the enzyme was denatured in alkali. The chemical shift of the fourth resonance is around that expected for a methylene group attached to the sulphur atom of cysteine. These findings confirmed the structure of the cofactor unambiguously and highlighted the power of ^{13}C n.m.r. for observing ^{13}C enriched ligands at the catalytic site of a functioning enzyme.

A similar ^{13}C n.m.r approach, but using a different labelling method, was carried out in Battersby's laboratory. Unlabelled enzyme was first treated with HCl to remove the native cofactor and the inactive apoenzyme was incubated with 11-[^{13}C] porphobilinogen. Despite such harsh treatment, a substantial amount of deaminase activity was restored and the ^{13}C n.m.r. difference spectrum of the reconstituted holoenzyme was determined at pH 14 [242] yielding chemical shifts for the protein-bound ^{13}C cofactor similar to those described above and allowing identical conclusions. Because the enzyme was unfolded by the base treatment, the signals were far sharper than those from the catalytically active enzyme [241].

The location of the cysteine residue responsible for the attachment of the dipyrromethane cofactor was determined by protein chemistry and site directed mutagenesis using the *E. coli* deaminase. Firstly it was confirmed that 5,5'-dithiobis(2-nitrobenzoic acid) only titrated with three instead of the four cysteines [241] indicated by the protein sequence. Thus one cysteine was masked, as expected, by the covalently bound cofactor. Of the four cysteines in the *E. coli* enzyme, two are conserved in the human enzyme [243] which made these the prime targets for investigation. These two cysteines, at positions 99 and 242 in the *E. coli* sequence, are separated, providentially, by a single cleavable aspartyl-proline bond at position 103/104. When chemical cleavage of the deaminase, in which the dipyrromethane had been labelled with [^{14}C] as before, was carried out, the [^{14}C] label which remained peptide-bound was located in the carboxyl terminal fragment containing cysteine 242. This indicates that it is cysteine-242 and not cysteine-99 which binds the cofactor in the *E. coli* enzyme [241] as shown in Scheme 17. A more detailed study [244], in which a peptide carrying the cofactor was isolated and sequenced, confirmed subsequently that cysteine-242 is indeed the cofactor attachment site. Site directed mutagenesis of cysteine 99 and 242 to serine residues added additional evidence for cysteine 242 as the cofactor binding site since the C99S mutant enzyme had full catalytic activity and contained the cofactor whereas the C242S mutant was completely inactive [245].

Comparisons of the derived protein sequences for deaminases from *E. coli* [198], human [243], rat [246], mouse [247], *E. gracilis* [248], and *Bacillus subtilis* [249] indicated that the cofactor binding sites are in a conserved region as indicated below (*).

```
                                  *   *   *       *   *
Escherichia coli    M   N   T   R   L   E   G   G   C   Q   V   P   I   G   S   Y   A   E   L
human               F   L   R   H   L   E   G   G   C   S   V   P   V   A   V   H   T   A   M
rat                 F   L   R   H   L   E   G   G   C   S   V   P   V   A   V   H   T   V   M
mouse               F   L   R   H   L   E   G   G   C   S   V   P   V   A   V   H   T   V   I
Euglena gracilis    M   N   R   R   L   N   G   G   C   Q   V   P   I   S   G   F   A   Q   L
Bacillus subtilis   F   L   N   A   M   E   G   G   C   Q   V   P   I   A   G   Y   S   V   L
```

It is interesting that the sequence LEGG is found at the hinge region of immunoglobulins suggesting that a flexible region is also required in the deaminase, possibly to assist in the manipulation of the enzyme intermediate complexes through the catalytic equipage.

The question of how the cofactor is assembled has been addressed by two approaches. Either the apoenzyme may be isolated from recombinant *hemA*$^-$ [229] or *hemB*$^-$ [250] strains of *E. coli* which overproduce the protein or may be generated by treatment with HCl [251]. In both cases incubation of the apoenzyme with porphobilinogen leads to the recovery of part of the enzymic activity suggesting that the deaminase also catalyses the self assembly of its own cofactor. This reaction is merely the same generic reaction as the normal deamination of porphobilinogen which occurs during the construction of the tetrapyrrole product.

6.14. Mechanism of action of porphobilinogen deaminase

The mechanism by which porphobilinogen deaminase catalyses the remarkable tetrapolymerisation reaction can now be deduced from the wealth of experiments described above. The most important experimental observations for consideration are as follows:

1) Porphobilinogen is the only true substrate for the deaminase enzyme which compulsorily assembles preuroporphyrinogen in a stepwise and unidirectional fashion by adding one substrate at a time.

2) The deaminase will accept the aminomethylbilane NH_2CH_2-AP-AP-AP-AP as a pseudosubstrate and will catalyse its deamination to give preuroporphyrinogen [222]. However, the enzyme will neither accept the aminomethyldipyrromethane NH_2CH_2-AP-AP nor the aminomethyltripyrrane NH_2CH_2-AP-AP-AP as a building unit in the presence of porphobilinogen, although these compounds act as inhibitors of the reaction [209,216]. It is also likely that they may be deaminated to their hydroxymethyl equivalents.

3) The enzyme will deaminate hydroxyporphobilinogen (porphobilinogen with the amino group replaced by a hydroxyl group) as a reasonable substrate and will generate preuroporphyrinogen (the hydroxymethylbilane $HOCH_2$-AP-AP-AP-AP) [252].

4) The enzyme catalyses the addition of water to ES_4 to give the product preuroporphyrinogen [229].

5) The purified complexes ES, ES_2 and ES_3, when incubated in buffer alone, release the pyrrole unit at the α-free end as hydroxyporphobilinogen which is then accepted as a substrate leading to the production of preuroporphyrinogen [229].

6) Incubation of the deaminase with porphobilinogen in the presence of high concentrations of the bases NH_3, NH_2OH or NH_2OCH_3 (B=base) has little effect on the consumption of porphobilinogen whereas the formation of preuroporphyrinogen (as uroporphyrinogen I) is greatly inhibited. The base (B) intercepts and releases preferentially the enzyme-bound monopyrrole (ES) as BCH_2-AP or the enzyme-bound bilane (ES_4) as BCH_2-AP-AP-AP-AP. The intermediates ES_2 and ES_3 are also susceptible [215,236,192] but less so. This suggests that the ES and ES_4 complexes have some similarities.

The wealth of evidence above is consistent with a single catalytic site which is able to catalyze the deamination of porphobilinogen and the dehydration of hydroxyporphobilinogen, as well as the reverse reactions which involve the addition of ammonia or water to the enzyme-intermediate complexes [229]. During a single catalytic cycle of the porphobilinogen deaminase four substrate molecules are sequentially deaminated and coupled, the first substrate reacting with the dipyrromethane cofactor and the succeeding substrates reacting with ES, ES_2 and ES_3 complexes. Thus the deamination and coupling reactions are used exclusively to this point when six pyrrole rings are in fact linked together. It is only after the formation of the ES_4 intermediate-complex that the hydrolytic cleavage to yield the hydroxymethylbilane preuroporphyrinogen occurs.

The topography of the catalytic centre in the holoenzyme may be envisaged as having a minimum of two pyrrole recognition sites, one for the incoming substrate, site S, and another, site C, which can accommodate a pyrrole ring with a free α-position for reaction with the deaminated substrate bound at site S. In the holoenzyme, site C is occupied by one of the rings of the dipyrromethane cofactor, hence the designation site C (Scheme 18).

The enzyme reaction is conveniently divided into three stages

i) Assembly of the dipyrromethane cofactor
Before the tetrapolymerisation of substrate can proceed, the apodeaminase has first to self-assemble the dipyrromethane cofactor. This is probably accomplished (Scheme 18) by the deamination of a porphobilinogen molecule (C1) at site S to give a methylene pyrrolenine which reacts with $-S^-$ of cysteine-242 (cys). The binding of the second porphobilinogen unit (C2) of the dipyrromethane cofactor causes the translocation of the first unit to site C which is followed by the deamination and coupling reactions. The final stage in the cofactor assembly process involves a major conformational change by the enzyme so that the terminal cofactor ring (C2) is no longer able to interact productively with site S. This process probably requires a

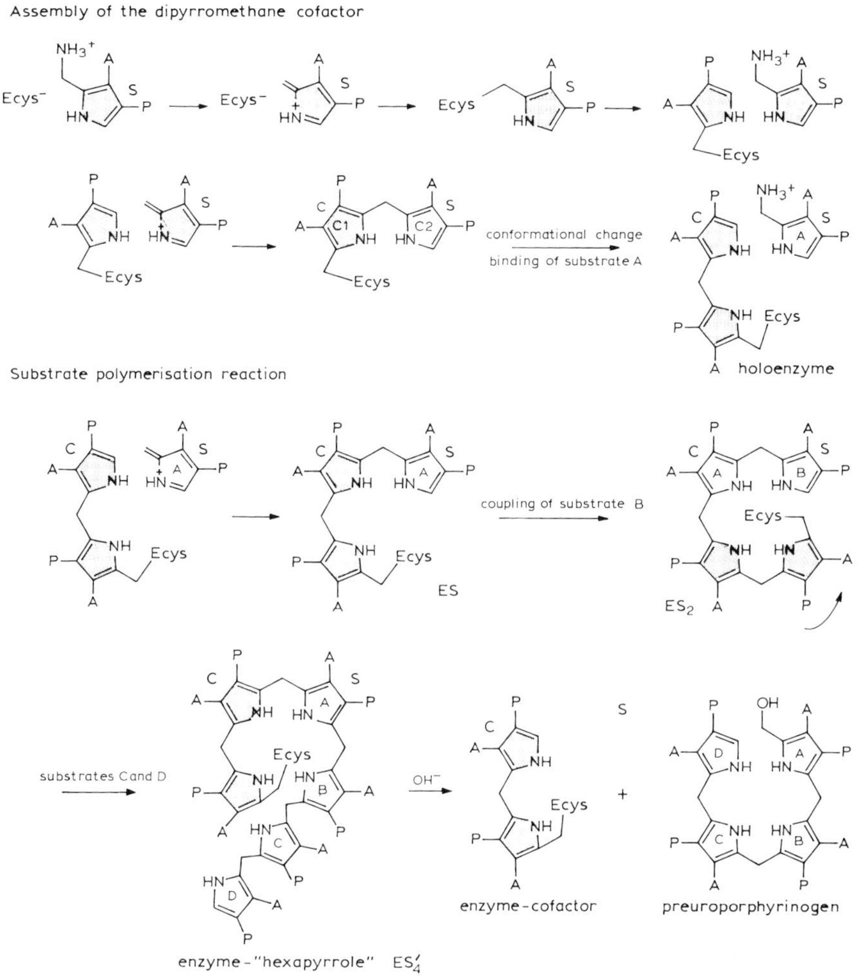

Scheme 18. Mechanism of assembly of the dipyrromethane cofactor and its role in substrate binding and pyrrole chain extension

further molecule of porphobilinogen and results in the formation of ES. Once the deaminase has assumed this conformation the cofactor is permanently linked to the enzyme and is not released by catalytic turn-over [235]. Evidence for a major conformational change during the assembly of the cofactor comes from the fact that the apoenzyme is heat labile whereas the holoenzyme can be heated to 70° with little loss of activity.

ii) The tetrapolymerisation process
The stepwise addition of the four molecules of porphobilinogen required for the as-

sembly of the tetrapyrrole, preuroporphyrinogen, is outlined in Scheme 18. The first substrate (A) binds to the site S and may displace the dipyrromethane cofactor. Deamination to the methylenepyrrolenine followed by reaction with the cofactor and deprotonation results in the intermediate ES. The second substrate molecule (ring B) now binds to site S causing the translocation of the ring A to site C. To achieve this the incoming substrate must have a higher affinity for site S possibly because of the presence of the $\overset{+}{N}H)_3$ group. In the absence of a second molecule of substrate, the pyrrole ring bound in this ES complex remains in the S site and can be released by hydrolysis to yield hydroxyporphobilinogen. In the presence of further substrates (rings C and D), sequential translocation, deamination, condensation and deprotonation events occur leading to ES_2, ES_3 and finally ES_4. Experiments with α-tritiated porphobilinogen have shown that the deprotonation of the α-hydrogen occurs at each coupling stage since ES, ES_2 and ES_3 enzyme intermediate complexes retain only a single tritium label at the terminal free α-position (Jordan and Berry – unpublished results).

iii) Release of preuroporphyrinogen from the enzyme
After the addition of the fourth substrate molecule it is envisaged that steric considerations, caused by the fact that the cofactor is bound both to the enzyme and to the tetrapyrrole, prevent the binding of a 'fifth' substrate. In order for the same active site to catalyse the hydrolytic cleavage between the cofactor C2 ring and the tetrapyrrole (bilane) ring A, the ES_4 intermediate is required to relocate to give ES_4' in which the dipyrromethane cofactor is situated in a position with the A ring of ES_4' occupying site S. It is well established that an enzyme-bound tetrapyrrole can exist in this form since the enzyme is able to deaminate the synthetic aminomethylbilane NH_2CH_2-AP-AP-AP-AP to preuroporphyrinogen [252]. It is likely that ES_4 can only be formed by the tetrapolymerisation reaction proper and that ES_4' is the species involved with the exchange reactions of aminomethylbilanes and hydroxymethylbilanes. Nitrogen bases cause the preferential release of the monopyrrole or the bilane (attached to the appropriate base) from the enzyme intermediate complexes ES and ES_4 [192], respectively, providing evidence for their analogous structures. Although the hydrolytic activity which releases preuroporphyrinogen is normally confined to reaction termination, any of the enzyme-intermediate complexes can be cleaved, however, in these cases it is the α-free terminal ring of the complex which is removed [229]. Thus it is most unlikely that separate active sites are involved with the deamination and hydrolysis reactions, although it is still a possibility.

From the discussion above it may be concluded that the function of the cofactor is 2-fold – firstly to act as a 'primer' on which to assemble the tetrapyrrole chain and secondly to regulate the number of pyrrole rings which can be incorporated to four. This latter property is thought to be due to the fact that the cofactor is permanently linked to the deaminase protein and may thus achieve this latter role by steric means.

6.15. Reaction of porphobilinogen deaminase with 1-aminomethylbilanes

Porphobilinogen deaminase has a high substrate specificity and recognizes pseudosubstrates where the A ring structure resembles porphobilinogen. Thus the enzyme can recognize the aminomethyldipyrromethane NH_2CH_2-AP-AP [209,216], the aminomethyltripyrrane, NH_2CH_2-AP-AP-AP [209] and the aminomethylbilane NH_2CH_2-AP-AP-AP-AP [212] (see Scheme 11). The enzyme appears unable to use the di- and tripyrroles as substrates because they cannot be translocated and therefore they act as inhibitors. The aminomethylbilane NH_2CH_2-AP-AP-AP-AP can act as a pseudosubstrate since it does not depend on translocation to give a meaningful product. Thus it binds, is deaminated and finally released as the equivalent hydroxymethylbilane, preuroporphyrinogen [252]. This accounts for the observation that in the presence of deaminase and synthase, uroporphyrinogen III is formed from NH_2CH_2-AP-AP-AP-AP (see Scheme 13). It should be pointed out however that this aminomethylbilane is not a physiological intermediate, as had been suggested originally [212], and plays no part in the natural biosynthesis of porphyrinogens. The isomeric aminomethylbilane NH_2CH_2-PA-AP-AP-AP [212] cannot act as a substrate because the A ring is not recognized by the enzyme. The inhibition caused by the aminomethylbilane NH_2CH_2-PA-PA-PA-AP with the deaminase/synthase system is potentially the most interesting since this compound may interact with the synthase as an intermediate analogue resembling the structure prior to final ring closure to uroporphyrinogen III (see below and Scheme 21).

6.16. Steric course of the porphobilinogen deaminase and uroporphyrinogen III synthase reactions

One of the most challenging stereochemical problems in bioorganic chemistry is the determination of the absolute configuration of each *meso*-position (5,10,15 and 20) in uroporphyrinogen III in relation to the configuration of the paired hydrogen atoms at position-11 of the precursor porphobilinogen. In order to examine this problem the 11-position of the porphobilinogen needs to be synthesised as a chiral centre by substituting one of the hydrogen atoms with an isotopic label. With the availability of large amounts of deaminase from recombinant strains of *E. coli* [198] the enzymic reaction from chiral porphobilinogen is relatively straightforward, however the chiral analysis of the *meso*-positions in preuroporphyrinogen is a formidable problem.

A major contribution to the problem has been provided by Akhtar and his colleagues by incubating $[11RS-^3H_2;2,11-^{14}C_2]$-porphobilinogen with haem biosynthesis pathway enzymes from avian erythrocytes [254]. These experiments established that exactly 50% of the ^{3}H-label was incorporated into protoporphyrin. Further experiments with [11S-^{3}H;2,11,-^{14}C$_2$]-porphobilinogen showed that a single ^{3}H label was incorporated into position-10 of protoporphyrin IX indicating that three labelled hydrogens had been lost from positions 5, 15 and 20 during the oxidation of protoporphyrinogen IX [254] (see also Chapter 2). One must conclude from this work that porphobilinogen deaminase and uroporphyrinogen III synthase follow mechanisms in which all the four methylene groups of preuroporphyrinogen and uroporphyrinogen III are formed and subsequently manipulated by stereospecific processes [254]. Most importantly these experiments established unambiguously that the hydroxyl group in preuroporphyrinogen must have arisen by an enzymic process at the active site of the deaminase and not by the non-enzymic trapping of a methylenepyrrolenine type intermediate as favoured by some workers. Similar conclusions have been obtained recently by Jackson and his associates using a related approach to that described above in which [11R-^{2}H]-porphobilinogen was transformed enzymically into protoporphyrin IX [255]. As expected, the converse result was obtained, namely, the incorporation of ^{2}H into positions 5, 15 and 20 but not position 10 and the conclusion was the same. Investigation of the chirality at the 1-aminomethyl position of the aminomethylbilane NH$_2$CH$_2$-AP-AP-AP-AP and at the 1-hydroxymethyl position of preuroporphyrinogen has been carried out in Battersby's laboratories [256] from porphobilinogen stereospecifically labelled with deuterium and tritium respectively. Samples of [11R-^{2}H]-porphobilinogen and [11S-^{2}H]-porphobilinogen were transformed into the aminomethylbilane with deaminase in the presence of ammonia. The stereochemistry of the aminomethyl group was determined by reaction with a chiral imidate and the resulting amidine was analysed by n.m.r. The result from this work indicated overall retention of configuration had occurred during the reaction [256]. A similar conclusion was deduced when preuroporphyrinogen was enzymically synthesised from [11R-^{3}H]-porphobilinogen and [11S-^{3}H]-porphobilinogen. Analysis of the chirality of the 1-hydroxymethyl group was performed by ozonolysis to give glycolic acid which was analysed by enzymic oxidation with glycolic acid oxidase [257]. The findings from these experiments are in complete agreement with those of Akhtar and his colleagues [254], namely, that the hydroxyl group of preuroporphyrinogen is

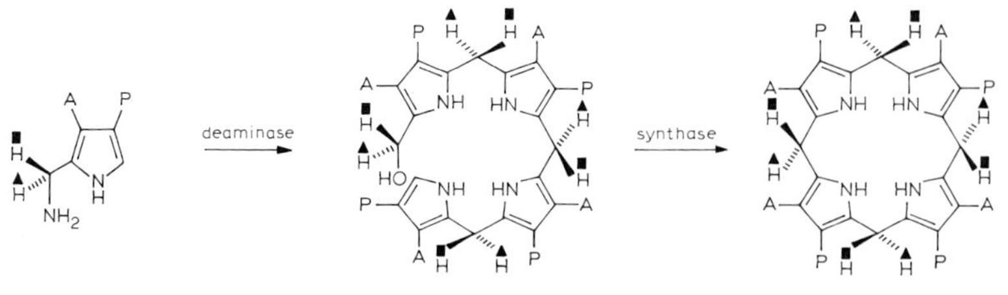

Scheme 19. Stereochemical studies on the formation of uroporphyrinogen III from stereospecifically labelled porphobilinogen

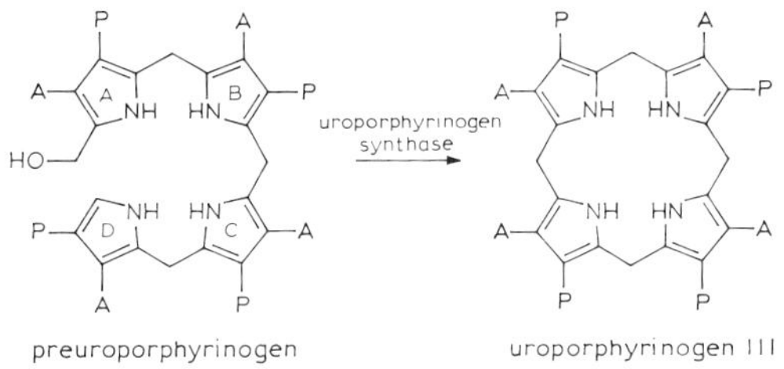

Scheme 20. Reaction of uroporphyrinogen III synthase

added to the tetrapyrrole intermediate, such as a methylenepyrrolenine, whilst it is bound within the confines of the deaminase catalytic site.

Although none of these experiments provide a solution to the absolute stereochemical course at the 5, 10, 15 and 20 positions of uroporphyrinogen III they do provide a foundation for the ultimate experiments. After detailed consideration of several aspects of the reactions the possibility that all the reactions catalysed by both deaminase and synthase enzymes proceed with retention is favoured [254] (Scheme 19).

6.17. Use of synthetic analogues to investigate the uroporphyrinogen III synthase reaction

Few experiments have been carried out on the uroporphyrinogen III synthase in isolation because of the difficulties in purifying the enzyme and preparing the unstable substrate. The physiologically important reaction catalysed by uroporphyrinogen III synthase is the transformation of preuroporphyrinogen into uroporphyrinogen III, a reaction which proceeds with exceptionally high accuracy. Several isomeric hydroxymethylbilanes have been synthesised and their effectiveness as substrates compared with the natural hydroxymethylbilane, preuroporphyrinogen, have been assessed [258,259].

	% enzymic D ring inversion
$HOCH_2$-AP-AP-AP-AP	100 (natural substrate)
$HOCH_2$-AP-PA-AP-AP	0
$HOCH_2$-AP-AP-PA-AP	95
$HOCH_2$-AP-AP-AP-PA	45
$HOCH_2$-PA-PA-PA-AP	0

The results of this study highlight the importance of the A and B rings for recognition by the enzyme since changes in these positions seem to be disastrous. The most interesting finding is that the hydroxymethylbilane, in which the D ring is already 'inverted', acts as a reasonable substrate and, astonishingly, gives rise to the uroporphyrinogen I isomer in good yield. These results point to a mechanism in which the inversion and cyclisation reactions seem to be inexorably linked. Thus the synthase binds preuroporphyrinogen in a conformation close to, or towards, that of the product uroporphyrinogen III so that the ring D may already be 'inverted' in the transition state. The fact that the inversion of the ring D is not complete in HOCH$_2$-AP-AP-AP-PA is hardly surprising since this hydroxymethylbilane is an unnatural substrate and may not bind in the precise conformation to ensure complete inversion. The importance of this result is that it occurs even 45% of the time.

6.18. Mechanism of action of uroporphyrinogen III synthase

The most favoured mechanism by which the synthase enzyme catalyses its audacious reaction was proposed by Mathewson and Corwin in 1961 [260] and is one of the few early postulates that has survived subsequent experimental investigations (Scheme 21). A key feature of this mechanism is the reaction of the substituted α-position of ring D (C-16) to form a methylene bridge (bond a) with the ring A. The resulting *spiro*-pyrrolenine, (or *spiro*-intermediate) is attractive because cleavage of the bond between C-15 and ring D (bond b) followed by ring closure would yield uroporphyrinogen III.

Attempts to synthesise the challenging *spiro*-intermediate have met with partial success and currently the two spirolactam isomers, as shown in Scheme 22, have been prepared [261]. One isomer has been shown to act as a good competitive inhibitor (K$_i$ 1 μM) whereas the other had no inhibitory activity at all. The absolute configurations

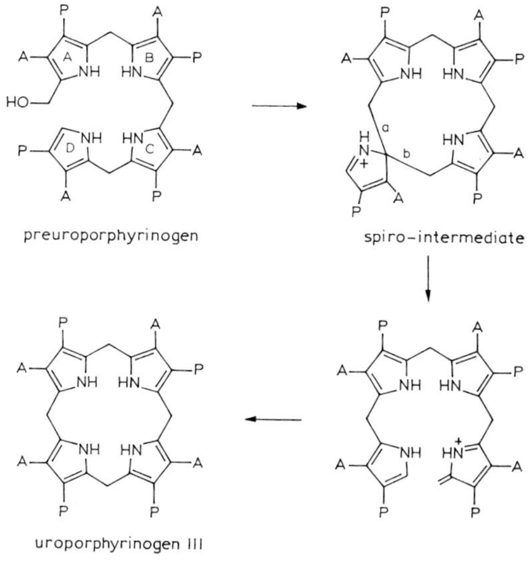

Scheme 21. Mechanism for the biosynthesis of uroporphyrinogen III from preuroporphyrinogen

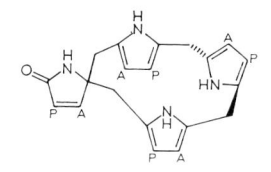

Scheme 22. Spirolactam analogue of the proposed spiro-intermediate in the uroporphyrinogen III synthase reaction

of the two isomers are not known. A review on the deaminase and synthase giving particular emphasis to the work carried out in Battersby's laboratory has been published recently [262]. In this review other possible mechanisms for the rearrangement have been discussed [262].

6.19. Molecular biology and protein structure of porphobilinogen deaminase

Oligonucleotides specifying six porphobilinogen deaminases have been sequenced and thus, for the first time, it is possible to make comparisons between the derived protein primary structures. There is remarkable similarity between the protein sequences suggesting that the deaminases all have similar tertiary structures and operate by a similar mechanism. Sequences are available from *E. coli* [198], human [243], rat [246], mouse [247], *E. gracilis* [248] and *Bacillus subtilis* [249]. The sequencing of the human porphobilinogen deaminase cDNA from erythroid and non-erythroid tissue has revealed that isoforms of the enzyme exist, one specifically for the erythroid system and a general 'housekeeping' protein found in all tissues [263] (see also Chapter 7).

Close examination of the deaminase protein sequences reveals several areas of homology which may have structural or functional significance. Apart from the amino acids incorporating cysteine-242 of the *E. coli* sequence which is involved with binding the dipyrromethane cofactor (see above), there are several other highly conserved areas -G-T-S-S-L-R-R- being the longest. Nine arginine, five lysine and four glutamate residues are conserved in all six amino acid sequences. It is tempting to suggest that several of the basic residues may be involved with the binding and manipulation of the positively charged acetate and propionate side chains of the dipyrromethane cofactor and substrate pyrrole residues. Recently this hypothesis has been put to the test by the generation of lysine mutants in the *E. coli* deaminase. Lysines-55 and 59 were mutated to glutamine [264]. These lysine residues have been shown to interact with pyridoxal 5'-phosphate. Although the K55N mutant was little affected the K59N mutant had a raised K_m but still maintained a substantial amount of enzyme activity. Studies with pyridoxal phosphate had previously identified lysine as a possible active site residue [233,265] although it is not critical for activity [229].

Recently, several experiments carried out in Jordan's laboratory, have revealed that mutation of some arginine residues to histidines especially R131H and R132H can be disasterous, even though they were replaced by another basic group [266a]. These mutations appear in the conserved sequence -G-T-S-S-L-R-R-. Neither R131H or R132H are able to assemble the dipyrromethane cofactor and the enzymes are, as a consequence, completely inactive. Mutants R11H and R155H are unable to add the substrate to the cofactor whereas mutants R149H and R176H are defective in the addition of substrate to ES [266a]. Recent X-ray data explains these findings since arginines 11, 131, 132, 149, 155 and 176 are all located in the catalytic cleft [266b]. Arginine → leucine mutations have also been carried out [266c]. Arginine had been implicated as an important residue at the active site as a result of inactivation experiments with butanedione [267] although the residue/s affected have not been identified.

Interesting observations have been made with the enzyme from *E. coli* with respect to the reaction of cysteine residues. In the absence of porphobilinogen the enzyme is not very reactive to N-ethylmethyl-maleimide (NEM) or other reagents which react with thiols. Reaction of NEM with ES, however, causes substantial inactivation and ES_2 is even more susceptible [229]. Reaction of ES_3 with NEM leads to complete inactivation of the enzyme. These findings suggest that major conformational changes may be occurring as successive substrates bind exposing a cysteine residue normally buried in the native state [229].

6.20. Molecular biology of uroporphyrinogen synthase

The sequence of the *hemD* gene in *E. coli* specifying uroporphyrinogen III synthase was simultaneously determined in two laboratories [204,268,269]. The *hemD* gene is immediately adjacent to the *hemC* gene and under the control of the *hemC* promoter. The *E.coli* synthase has been isolated from an overproducing strain by Jordan and his colleagues and shown to have a M_r of about 29000 [204]. The N terminus of the *E. coli* enzyme has been sequenced, NH_2-S-I-L-V-T-R- and has lost the terminal methionine predicted from the gene sequence. The N-terminus of the human enzyme NH_2-M-L-V-L-L-L- however appears to be intact. The human uroporphyrinogen III synthase cDNA has now been sequenced [270] and comparison of the *E. coli* and human derived protein sequences show very little homology, a surprising fact considering the high degree of similarity with the deaminases

The tremendous advances in molecular biology in recent years has meant that all the genes encoding the enzymes from the early part of tetrapyrrole biosynthesis have been cloned and sequenced. The availability of multi-copy plasmids has allowed the spectacular overexpression of some of the enzymes and permitted, for the first time, the investigation of detailed protein structure. The crystallisation and determination of the X-ray structure of some of the enzymes is already under way in our laboratories and soon it will be possible to have a detailed understanding of how these enzymes catalyse the intricate fabrication of the tetrapyrrole macrocycle.

Acknowledgements

Much of the work carried out by Peter M. Jordan was funded by the Science and Engineering Research Council.

I am grateful to Professor M. Akhtar (Southampton University) for valuable discussions and to Mrs. M. Moran for assistance in the preparation of this manuscript.

References

1. Shemin, D. and Rittenberg, D. (1945) J. Biol. Chem. 159, 567–568.
2. Shemin, D., London, I.M. and Rittenberg, D. (1950) J. Biol. Chem. 183, 757–765.
3. Muir, H.M. and Neuberger, A. (1950) Biochem. J. 47, 97–104.
4. Shemin, D. and Wittenberg, J. (1951) J. Biol. Chem. 192, 315–334.
5. Gibson, K.D., Laver, W.G. and Neuberger, A. (1958) Biochem. J. 70, 71–81.
6. Shemin, D. and Russell, C.S. (1953) J. Am. Chem. Soc. 75, 4873–4875.
7. Shemin, D., Russell, C.S. and Abramsky, T. (1955) J. Biol. Chem. 215, 613–626.
8. Westall, R.G. (1952) Nature 170, 614–616.
9. Kennard, O. (1953) Nature 171, 876–877.
10. Falk, J.E., Dresel, E.I.B. and Rimington, C. (1953) Nature 172, 292–294.
11. Schmid, R. and Shemin, D. (1955) J. Am. Chem. Soc. 77, 506–508.
12. Gibson, K.D., Neuberger, A. and Scott, J.J. (1955) Biochem. J. 61, 618–629.
13. Bogorad, L. (1958) J. Biol. Chem. 233, 501–509.
14. Bogorad, L. (1958) J. Biol. Chem. 233, 510–515.
15. Dolphin, D. (1976) in: The Porphyrins (D. Dolphin, Ed.) Vol. 6, Chapters 1–4, Academic Press.
16. Granick, S. and Beale, S.I. (1978) Adv. Enzymol. 46, 33–203.
17. Battersby, A.R. and Mc Donald, E. (1975) in: Porphyrins and Metalloporphyrins (K.M. Smith, Ed.) Elsevier, Amsterdam, pp. 61–122.
18. Akhtar, M. and Jordan, P.M. (1979) in: Comprehensive Organic Chemistry (D.H.R. Barton and W.D. Ollis, Eds.), vol. 5, pp. 1121–1166, Pergamon Press, Oxford, U.K.
19. Leeper, F.J. (1985) Nat. Prod. Rep. 2, 19–47.
20. Leeper, F.J. (1989) Nat. Prod. Rep. 6, 171–190.
21. Dailey, H. A., ed. (1990) Biosynthesis of Haems and Chlorophylls, McGraw Hill, New York.
22. Kikuchi, G., Kumar, A., Talmage, P. and Shemin, D. (1958) J. Biol. Chem. 233, 1214–1219.
23. Warnick, G.R. and Burnham, B.F. (1971) J. Biol. Chem. 246, 6880–6885.
24. Nandi, D.L. and Shemin, D. (1977) J. Biol. Chem. 252, 2278–2280.
25. Jordan, P. M. and Laghai-Newton, A. (1986) Methods Enzymol. 123, 435–443.
26. Ohashi, A. and Kikuchi, G. (1979) J. Biochem. 85, 239–247.

27 Srivastava, G., Borthwick, I.A., Brooker, J.D., May, B.K. and Elliott, W.H. (1982) Biochem. Biophys. Res. Comm. 109, 305–312.
28 Borthwick, I.A., Srivastava, G., Brooker, J.D., May, B.K. and Elliott, W.H. (1983) Eur. J. Biochem. 129, 615–620.
29 Dzelzkalns, V., Foley, T. and Beale, S.I. (1982) Arch. Biochem. Biophys. 216, 196–203.
30 Volland, C. and Felix, F. (1984) Eur. J. Biochem. 142, 551–557.
31 Urban-Grimal, D., Volland, C., Garnier, T., Dehoux, P. and Labbe-Bois, R. (1986) Eur. J. Biochem. 156, 511–519.
32 Borthwick, I.A., Srivastava, G., Day, A.R., Pirola, B.A., Snoswell, M.A., May, B.K. and Elliot, W.H. (1985) Eur. J. Biochem. 150, 481–484.
33 Bawden, M.J., Borthwick, I.A., Healy, H.M. Morris, C.P., May, B.K. and Elliott, W.H. (1987) Nucleic Acids Res. 15, 8563.
34 Lascelles, J. (1968) Biochem. Soc. Symp. 28, 49–59.
35 Tuboi, S., Kim, H.J. and Kikuchi, G. (1970) Arch. Biochem. Biophys. 138, 147–154.
36 Suwanto, A. and Kaplan, S. (1989) J. Bacteriol. 171, 5850–5859.
37 Tait, G.H. (1972) Biochem. J. 128, 1159–1169.
38 Neuberger, A., Sandy, J.D. and Tait, G.H. (1973) Biochem. J. 136, 477–490.
39 Tschudy, D.P., Marver, H.S. and Collins, A. (1965) Biophys. Res. Comm. 21, 480–487.
40 Strand, L. J., Felsher, B.F., Redeker, A.G. and Marver, H.S. (1970) Proc. Nat. Acad. Sci. 67, 1315–1320.
41 Granick, S. (1966) J. Biol. Chem. 241, 1359–1375.
42 Hayashi, N., Yoda, B. and Kikuchi, G. (1968) J. Biochem. 63, 446–452.
43 Ebert, P.S., Hess, R.A., Frykholm, B.C. and Tschudy, D.P. (1979) Biochem. Biophys. Res. Comm. 88, 1382–1390.
44 Beaumont, C., Deybach, J. C., Grandchamp, B., Da Silva, V., de Verneuil, H. and Nordmann, Y. (1984) Exp. Cell Res. 154, 474–484.
45 Srivastava, G., Brooker, J.D., May, B.K. and Elliot, W.H. (1980) Biochem. J. 188, 781–788.
46 Yamamoto, M., Hayashi, N. and Kikuchi, G. (1982) Biochem. Biophys. Res. Comm. 105, 985–990.
47 Gayathri, A.K., Satyanarayana Rao, M.R. and Padmanaban, G. (1973) Arch. Biochem. Biophys. 155, 299–306.
48 Sassa, S. and Granick, S. (1970) Proc. Nat. Acad. Sci. USA 67, 517–522.
49 Granick, S., Sinclair, P., Sassa, S. and Grieninger, G. (1975) J. Biol. Chem. 250, 9215–9225.
50 Sinclair, P.R. and Granick, S. (1975) Ann. N.Y. Acad. Sci. 244, 509–520.
51 Hayashi, N., Terasawa, M., Yamauchi, K. and Kikuchi, G. (1980) J. Biochem. 88, 1537–1543.
52 Hayashi, N., Watanabe, N. and Kikuchi, G. (1983) Biochem. Biophys. Res. Comm. 115, 700–706.
53 Srivastava, G., Borthwick, I.A., Brookes, J.D., Wallace, J.C., May, B.K. and Elliott, W.H. (1983) Biochem. Biophys. Res. Commun. 117, 344–349.
54 Ohashi, A. and Kikuchi, G. (1972) Arch. Biochem. Biophys. 153, 34–46.
55 Scholnick, P.L., Hammaker, L.E. and Marver, H.S. (1972) J. Biol. Chem. 247, 4132–4137.
56 Baron, J. and Tephly, J.R. (1970) Arch. Biochem. Biophys. 139, 410–420.
57 Sassa, S., Bradlow, H.L. and Kappas, A. (1979) J. Biol. Chem. 254, 1001–10020.
58 Kappas, A., Sassa, S. and Anderson, K.E. (1983) in: The Metabolic Basis of Inherited Disease (J.B. Stanbury et al., Eds.), 5th ed., 4, McGraw-Hill, New York.
59 Granick, S. and Urata, G. (1963) J. Biol. Chem. 238, 821–827.
60 De Matteis, F., Gibbs, A.H. and Smith, A.G. (1980) Biochem. J. 189, 645–648.
61 De Matteis, F., Gibbs, A.H. and Holley, A.E. (1987) Ann. N.Y. Acad. Sci. 514, 30–40.
62 Ortiz de Montellano, P.R., Costa, A.K., Grab, A., Sutherland, E.P. and Marks, G.S. (1986) in: Porphyrins and Porphyrias (Nordmann, Y. Ed.) Vol. 134, pp. 109–117, Colloque INSERM, John Libbey Eurotext, London.
63 Borthwick, I.A., Srivastava, G., Hobbs, A.A., Pirola, B.A., Mattschoss, L., Steggles, A.W., May,

B.K. and Elliot, W.H. (1985) in: Cellular Regulation and Malignant Growth (S. Ebashi, Ed.) pp. 144–151, Japan. Sci. Soc. Press, Tokyo/Springer, Berlin.
64 Fraser, P.J. and Curtis, P.J. (1987) Genes and Devel. 1, 855–861.
65 Sassa, S. (1976) J. Exp. Med. 143, 305–315.
66 Andrew, T.L., Riley, P.G. and Dailey, H.A. (1989) in: Biosynthesis of Haem and Chlorophylls (H.A. Dailey, H.A., Ed.) pp. 163–200, McGraw-Hill.
67 Dierks, P.M. (1989) in Biosynthesis of Haem and Chlorophylls (H.A. Dailey, Ed.) pp. 201–233, Mc-Graw-Hill.
68 Jordan, P.M. and Shemin, D. (1972) in: The Enzymes (3rd edition) (P.D. Boyer, Ed.) 5, pp. 323–356, Academic Press.
69 Matthew, M. and Neuberger, A. (1963) Biochem. J. 87, 601–612.
70 Fanica-Gaignier, M. and Clement-Metral, J. (1973) Eur. J. Biochem. 40, 19–24.
71 Snell, E.E. and di Mari, S. (1970) in: The Enzymes (3rd edition) (P.B. Boyer, Ed.) 2, pp. 335–370.
72 Akhtar, M. and Jordan, P.M. (1968) J. Chem. Soc. Chem. Commun. 1691–1692.
73 Jordan, P.M. and Akhtar, M. (1970) Biochem. J. 116, 277–286.
74 Zaman, Z., Jordan, P.M. and Akhtar, M. (1973) Biochem. J. 135, 257–263.
75 Laghai, A. and Jordan, P.M. (1976) Biochem. Soc. Trans. 4, 52–53.
76 Abboud, M.M., Jordan, P.M. and Akhtar, M. (1974) J. Chem. Soc. Chem. Commun. 643–644.
77 Laghai, A. and Jordan, P. (1977) Biochem. Soc. Trans. 5, 299–301.
78 Nandi, D.L. (1978) J. Biol. Chem. 253, 8872–8877.
79 Emery, V. and Akhtar, M. (1987) in: Enzyme Mechanisms (M.I. Page and A. Williams, Eds.) pp.345–389, The Royal Society of Chemistry.
80 Dunathan, H. (1971) Adv. Enzymol. (F. Nord, Ed.) 35, pp. 79–96, Wiley, New York.
81 Barnard, G.F. and Akhtar, M. (1975) J. Chem. Soc. Chem. Commun. 494–496.
82 Hill, R.K., Sawada, S. and Arsin, S.M. (1979) Bioorg. Chem. 8, 175–190.
83 Neuberger, A. and Turner, J.M. (1963) Biochem Biophys. Acta. 67, 342–343.
84 Maguire, D.J., Day, A.R., Borthwick, I.A., Srivastava, G., Wigley, P.L., May, B.K. and Elliott, W.H. (1986) Nucleic Acids Res. 14, 1379–1391.
85 Schoenhaut, D.S. and Curtis, P.J. (1986) Gene 48, 55–63.
86 McClung, C.R., Somerville, J.E., Guerinot, M.L. and Chelm, B.K. (1987) Gene 54, 133–139.
87 Leong, S.A., Williams, P.H. and Ditta, G.S. (1985) Nucleic Acids Res. 13, 5965–5676.
88 Yamamoto, M., Yew, N.S., Federspeil, M., Dodgson, J.B., Hayashi, N. and Engel, J.D. (1985) Proc. Nat. Acad. Sci. U.S.A. 82, 3702–3706.
89 Riddle, R.D., Yamamoto, M. and Engel, J.D. (1989) Proc. Nat. Acad. Sci., 86, 792–796.
90 Beale, S. I. and Castelfranco, P. A. (1974) Plant Physiol. 53, 291–296, 297–303.
91 Beale, S. I., Gough, S. P. and Granick, S. (1975) Proc. Nat. Acad. Sci. USA 72, 2717–2723.
92 Kannangara. C.G. and Gough, S.P. (1977) Carlsberg Res. Commun. 42, 441–457.
93 Oh-hama, T., Seto, H., Otake, N. and Miyachi, S. (1982) Biochem. Biophys. Res. Commun. 105, 647–652.
94 Porra, R.J., Klein, D. and Wright, P.E. (1983) Eur. J. Biochem. 130, 509–516.
95 Weinstein, J.D. and Beale, S. I. (1984) Plant Physiol. 74, 146–151.
96 Wang, W. Y., Huang, D. D., Stachon, D., Gough, S. P. and Kannangara, C. G. (1984) Plant Physiol. 74, 569–575.
97 Weinstein, J.D. and Beale, S.I. (1985) Arch. Biochem. Biophys. 237, 454–464.
98 Weinstein, J. D. and Beale, S. I. (1983) J. Biol. Chem. 258, 6799–6807.
99 Mayer, S.M., Beale, S.I. and Weinstein, J.D. (1987) J.Biol. Chem. 262, 12547–12549.
100 Einav, M. and Avissar, Y.J. (1984) Plant Sci. Lett. 35, 51–54.
101 Friedmann, H.C., Thauer, R.K. Gough, S.P. and Kannangara, C.G. (1987) Carlsberg Res. Commun. 52, 363–371.

102 Smith, K.M. and Huster, M.S. (1987) J. Chem. Soc. Chem. Commun. 14–16.
103 Li, J-M., Brathwaite, O., Cosloy, S.D. and Russell, C.S. (1989) J. Bact. 171, 2547–2552.
104 Elliott, T., Avissar, Y.J., Rhie, G-E. and Beale, S.I. (1990) J. Bacteriol. 172, 7071–7084.
105 Wang W-Y., Gough, S.P. and Kannangara, C.G. (1981) Carlsberg Res. Commun. 46, 243–257.
106 Kannangara, C.G., Gough, S.P., Oliver, R.P. and Rasmussen, S.K. (1984) Carlsberg Res. Commun. 49, 417–437.
107 Schon, A., Krupp, G., Gough, S.P., Berry-Lowe, S., Kannangara, C.G. and Soll, D. (1986) Nature 322, 281–284.
108 Weinstein, J.D. and Beale, S.I. (1985) Arch. Biochem. Biophys. 239, 87–93.
109 Huang, D-D. and Wang, W-Y. (1986) J. Biol. Chem. 261, 13451–13455.
110 Schneegurt, M.A. and Beale, S.I. (1988) Plant Physiol. 86, 497–504.
111 Schon, A., Kannangara, C.G., Gough, S. and Soll, D. (1988) Nature, 331, 187–190.
112 O'Neill, G.P. and Soll, D. (1990) Biofactors 2, 227–234.
113 Bruyant, P. and Kannangara, C.G. (1987) Carlsberg Res. Commun. 52, 99–109.
114 Weinstein, J.D., Mayer, S.M. and Beale, S.I. (1987) Plant Physiol. 84, 244–250.
115 Chen, M-W., Jahn, D., Schon, A., O'Neill, G.P. and Soll, D. (1990) J. Biol. Chem. 265, 4054–4057.
116 Avissar, Y.J. and Beale, S.I. (1988) Plant Physiol. 88, 879–886.
117 Chen, M-W., Jahn, D., O'Neill, G.P. and Soll, D. (1990) J. Biol. Chem. 265, 4058–4063.
118 Kannangara. G.C. and Gough, S.P. (1978) Carlsberg Res. Commun. 43, 185–194.
119 Houen, G., Gough, S. P. and Kannangara, C. G. (1983) Carlsberg Res. Commun. 48, 567–572.
120 Kannangara, C.G. and Schouboe, A. (1985) Carlsberg Res. Commun 50, 179–191.
121 Gough, S.P., Kannangara, C.G. and Bock, K. (1989) Carlsberg Res. Commun. 54, 99–108.
122 Hoober,J.K., Kahn, A., Ash, D.E., Gough, S.P. and Kannangara, C.G. (1988) Carlsberg Res. Commun. 53, 11–25.
123 Jordan, P.M., Cheung, J., Sharma, R.P. and Warren, M.J. (1991) Tetrahedron Lett. (submitted).
124 Kannangara, C.G., Gough, S.P. and Girnth, C. (1981) Proc. Fifth International Photosynthesis Congress (Akoyunoglou, G. ed.) pp. 117–127, Balaban, Philadelphia.
125 Mau, Y-H,L. and Wang, W-Y. (1988) Plant Physiol. 86, 793–797.
126 Grimm, B. (1990) Proc. Natl. Acad. Sci. U.S.A. 87, 4169–4173.
127 Breu, V. and Dornemann, D. (1988) Biochem. Biophys. Acta. 967, 135–140.
128 Castelfranco, P. and Beale, S. I. (1983) Ann. Rev. Plant Physiol. 34, 241–278.
129 Kannangara, C.G., Gough, S.P., Bruyant, P., Hoober, J.K., Kahn, A. and von Wettstein, D (1988) T. I. B. S. 13, 139–143.
130 Anderson, P.M. and Desnick, R.J. (1979) J. Biol Chem. 254, 6924–6930.
131 Gibbs, P.N.B., Chaudhry, A.G. and Jordan, P.M. (1985) Biochem. J. 230, 25–34.
132 Bevan, D.R., Bodlaender, P. and Shemin, D. (1980) J. Biol. Chem. 255, 2030–2035.
133 Jordan, P.M. and Seehra, J.S. (1986) Methods in Enzymol. 123, 427–434.
134 Nandi, D.L., Baker-Cohen, K.F. and Shemin, D. (1968) J. Biol. Chem. 243, 1224–1230.
135 Clara de Barreiro, O.L. (1967) Biochem. Biophys. Acta. 139, 479–486.
136 Shetty, A.S. and Miller, G.W. (1969) Biochem. J. 114, 331–337.
137 Shibata, H. and Ochiai, H. (1977) Plant Cell Physiol. 18, 421–429.
138 Liedgens, W., Lutz, C. and Schneider, H.A.W. (1983) Eur. J. Biochem. 135, 75–79.
139 Shemin, D. (1972) in: The Enzymes (3rd edition) (P.D. Boyer, Ed.) 7, pp. 323–337, Academic Press.
140 Cheh, A.M. and Neilands, J.B. (1976) Struct. Bonding 29, 123–170.
141 Tsukamoto, I., Yoshinaga, T. and Sano, S. (1979) Biochim. Biophy. Acta. 570, 167–178.
142 van Heyningen, S. and Shemin, S. (1971) Biochemistry 10, 4676–4682.
143 Gurne, D., Chen, J. and Shemin, D. (1977) Proc. Nat. Acad. Sci. U.S.A. 74, 1383–1387.
144 Nandi, D.L. and Shemin, D. (1968) J. Biol. Chem. 243, 1231–1235.
145 Jordan, P.M. and Spencer, P. (1991) Biochem. J. (submitted).

146 Wu, W., Shemin, D., Richards, K.E. and Williams, R.C. (1974) Proc. Nat. Acad. Sci. U.S.A. 69, 2585–2588.
147 Pilz, I., Schwarz, E., Vuga, M. and Beyersmann, D. (1990) Biol. Chem. Hoppe-Seyler 369, 1099–1103.
148 Barnard, G.F., Itoh, R., Hohberger, L.H. and Shemin, D. (1977) J. Biol. Chem. 252, 8965–8974.
149 Seehra, J.S., Gore, M.G., Chaudhry, A.G. and Jordan, P.M. (1981) Eur. J. Biochem. 114, 263–269.
150 Gibbs, P.N.B. and Jordan, P.M. (1981) Biochem. Soc. Trans. 9, 232–233.
151 Gibbs, P.N.B. (1984) Ph.D. thesis. University of Southampton, U.K.
152 Meredith, P.A. and Moore, M.R. (1980) Biochem. Soc. Trans. 6, 760–762.
153 Wetmer, J.G. Bishop, D.F., Cantelmo, C. and Desnick, R.J. (1986) Proc. Nat. Acad. Sci. U.S.A. 83, 7703–7707.
154a Bishop, T.R., Frelin, L.P. and Boyer, S.H. (1986) Nucleic Acids Res. 14, 10115.
154b Myers, A.M., Crivellone, M.D., Koerner, T.J. and Tzagoloff, A. (1987) J. Biol. Chem. 262, 16822–16829.
155 Echelard, Y., Dymetryszyn, J., Drolet, M. and Sasarman, A. (1988) Mol. Gen. Genet. 214, 503–508.
156 Li, J. M., Russell, C.S. and Cosloy, S.D. (1989) Gene 75, 177–184.
157a Berg, J.M. (1986) Science, 232, 485–487.
157b Schaumburg, A., Schneider-Poetsch, A.A.W. and Eckerskorn, C. (unpublished data).
157c Boesse, Q.F., Spano, A.J., Li, J. and Timko, M. (unpublished data).
158 Hasnaim, S.S., Wardell, E.M., Garner, C.D. Schlosser, M. and Beyersmann, D. (1985) Biochem. J. 230, 625–633.
159 Dent, A.J., Beyersmann, D., Block, C. and Hasnaim, S.S. (1990) Biochemistry, 29, 7822–7828.
160 Finelli, V.N., Klauder D.S., Karaffa, M.A. and Petering, H.G. (1975) Biochem. Biophys. Res. Commun. 65, 303–312.
161 Chisholm, J.J. (1971) Sci. Am. 224, (2) 15–23.
162 Bagust, J., Jordan, P.M., Kelley, M.E.M. and Kerkut, G.A. (1985) Neuroscience Letters 21, suppl. S84.
163 Thunell, S., Holmberg, L. and Lundgren, J. (1987) J. Clin. Chem. Clin. Biochem. 25, 5–14.
164 Nordmann, Y. and Deybach, J-C. (1990) in: Biosynthesis of Haems and Chlorophylls (H.A. Dailey, Ed.) McGraw Hill, N.Y.
165 Gibbs, P.N.B., Gore, M.G.G. and Jordan, P.M. (1985) Biochem. J. 225, 573–580.
166 Nandi, D.L. and Shemin, D. (1968) J. Biol. Chem. 243, 1236–1242.
167 Brumm, P.J. and Friedmann, H.C. (1981) Biochem. Biophys. Res. Commun.102, 854–859.
168 Nandi, D.L. (1978) Z. Naturforsch. C. Biosci. 33, 799–800.
169 Gibbs, P.N.B. and Jordan, P.M. (1986) Biochem. J. 236, 447–451.
170 Jordan, P.M. and Seehra, J.S. (1980) J. Chem. Soc. Chem. Commun. 240–242.
171 Jordan, P.M. and Seehra, J.S. (1980) FEBS. Letts 114, 283–286.
172 Jordan, P.M. and Gibbs, P.N.B. (1985) Biochem. J. 227, 1015–1020.
173 Abboud, M.M. and Akhtar, M. (1976) J. Chem. Soc. Chem. Commun. 1007–1008.
174 Chaudhry, A.G. and Jordan, P.M. (1976) Biochem. Soc. Trans. 4, 760–761.
175 Tsukamoto, I., Yoshinaga, T. and Sano, S. (1980) Int. J. Biochem. 12, 751–756.
176 Sommer, R. and Beyersmann, D. (1984) J. Inorg. Biochem. 20, 131–145.
177 Jordan, P.M., Chaudhry, A.G. and Gore, M.G. (1976) Biochem. Soc. Trans. 4, 301–303.
178 Jaffe, E.K. and Hanes, D. (1986) J. Biol. Chem. 261, 9348–9353.
179 Shemin, D. (1976) Phil. Trans. Roy. Soc. Lond. B. 273, 109–115.
180 Seehra, J.S. and Jordan, P.M. (1981) Eur. J. Biochem. 113, 435–446.
181 Jaffe, E.K., Markham, G.D. and Rajagopalau, J.S. (1990) Biochemistry 29, 8345–8350.
182 Bonnett, R. (1973) Ann. N.Y. Acad. Sci. 206, 745–751.
183 Moss, G.P. (1987) Pure Appl. Chem. 59, 779–832.
184 Bogorad, L. and Granick, S. (1953) Proc. Nat. Acad. Sci. U.S.A. 39, 1176–1188.

185 Mauzerall, D. (1960) J. Am. Chem. Soc. 82, 2601–2605.
186 Mauzerall, D. (1960) J. Am. Chem. Soc. 82, 2605–2609.
187 Frydman, R.B., Reil, S. and Frydman, B. (1971) Biochemistry 10, 1154–1160.
188 Sancovich, H.A., Batlle, A.M.C. and Grinstein, M. (1969) Biochem. Biophys. Acta. 191, 130–143.
189 Llambias, E.B.C. and Batlle, A.M.C. (1971) Biochem. Biophys. Acta. 227, 180–191.
190 Frydman, R.B. and Feinstein, G. (1974) Biochem. Biophys. Acta. 350, 358–373.
191 Jordan, P.M. and Shemin, D. (1973) J. Biol. Chem. 248, 1019–1024.
192 Davies, R.C. and Neuberger, A. (1973) Biochem. J. 133, 471–492.
193 Higuchi, M. and Bogorad, L. (1975) Ann. N.Y. Acad. Sci. 244, 401–418.
194 Anderson, P.M. and Desnick, R.J. (1980) J. Biol. Chem. 255, 1993–1999.
195 Shioi, Y., Nagamine, M., Kuraki, M. and Sasa, T. (1980) Biochem. Biophys. Acta. 616, 300–307.
196 Williams, D.C., Morgan, G.S., McDonald, E. and Battersby, A.R. (1981) Biochem. J. 193, 301–310.
197 Williams, D.C. (1984) Biochem. J. 217, 675–678.
198 Thomas, S.D. and Jordan, P.M. (1986) Nucleic Acids Res. 14, 6215–6226.
199 Hart, G.J., Abell, C. and Battersby, A.R. (1986) Biochem. J. 240, 273–276.
200 Jordan, P.M., Thomas, S.D. and Warren, M.J. (1988) Biochem. J. 254, 427–435.
201 Jordan, P.M. (1982) Enzyme 28, 158–169.
202 Tsai, S.F., Bishop, D.F. and Desnick, R.J. (1987) J. Biol. Chem. 262, 1268–1273.
203 Alwan, A.F. and Jordan, P.M. (1988) Biochem. Soc. Trans. 16, 965–966.
204 Jordan, P.M., Mgbeje, I.A.B., Thomas, S.D. and Alwan, A.F. (1988) Biochem. J. 249, 613–616.
205 Kohashi, M., Clement, R.P., Tse, J. and Piper, W.N. (1984) Biochem. J. 220, 755–765.
206 Smythe, E. and Williams, D.C. (1988) Biochem. J. 253, 275–279.
207 Hart, G.J. and Battersby, A.R. (1985) Biochem. J. 232, 151–160.
208 Battle, A.M.C. and Rossetti, M.V. (1977) Int. J. Biochem. 8, 251–267.
209 Frydman, B., Frydman, R.B., Valasinas, A., Levy, E.S. and Feinstein, G. (1976) Phil. Trans. Roy. Soc. Lond. B 273, 137–160.
210 Scott, A.I., Ho, K.S., Kajiwara, M. and Takahashi, T. (1976) J. Am. Chem. Soc. 98, 1589–1591.
211 Battersby, A.R., Fookes, C.J.R., McDonald, E. and Meegan, M. (1978) J. Chem. Soc. Chem. Commun. 185–186.
212 Battersby, A.R., Fookes, C.J.R., Matcham, G.W.J. and McDonald, E. (1978) J. Chem. Soc. Chem. Commun. 1064–1066.
213 Battersby, A.R. and McDonald, E. (1979) Acc. Chem. Res. 12, 14–22.
214 Burton, G., Nordlov, H., Hosozawa, S., Matsumoto, H., Jordan, P.M., Fagerness, P.E., Pryde, L.M. and Scott, A.I. (1979) J. Am. Chem. Soc. 101, 3114–3116.
215 Pluscec, J. and Bogorad, L. (1970) Biochemistry 9, 4736–4743.
216 Battersby, A.R. and McDonald, E. (1976) Phil. Trans. Roy. Soc. Lond. B 273, 161–180.
217 Frydman,R,B., Levy. E.S., Valasinas, A. and Frydman, B. (1978) Biochemistry 17, 110–120.
218 Rossetti, M.V., Juknat de Geralnik, A.A. and Batlle, A.M.C. (1977) Int. J. Biochem. 8, 781–787.
219 Battersby, A.R., McDonald, E., Williams, D.C. and Wurziger, H.K.W. (1977) J. Chem. Soc. Chem. Commun. 113–115.
220 Sburlati, A., Frydman, R.B., Valasinas, A., Rose, S., Priestap, H.A. and Frydman, B. (1983) Biochemistry 22, 4006–4013.
221 Burton, G., Fagerness, P.E. Hosozawa, S., Jordan, P.M. and Scott. A.I. (1979) J.Chem. Soc. Chem. Commun. 202–204.
222 Jordan, P.M., Burton, G., Nordlov, H. Schneider, M., Pryde, L. and Scott, A.I. (1979) J. Chem. Soc. Chem. Commun. 204–205.
223 Jordan, P.M. and Berry, A. (1980) FEBS Lett. 112, 86–88.
224 Battersby, A.R., Fookes, C.J.R., Gustafson-Potter, K.E., Matcham, G.W.J. and McDonald, E. (1979) J. Chem. Soc. Chem. Commun. 1155–1158.
225 Jordan, P.M. and Seehra, J.S. (1979) FEBS Lett. 104, 364–366.

226 Seehra, J.S. and Jordan, P.M. (1980) J. Am. Chem. Soc. 102, 6841–6846.
227 Battersby, A.R., Fookes, C.J.R., Matcham, G.W.J. and McDonald, E. (1979) J. Chem. Soc. Chem. Commun. 539–541.
228 Berry, A., Jordan, P.M. and Seehra, J.S. (1981) FEBS Lett. 129, 220–224.
229 Warren. M.J. and Jordan, P.M. (1988) Biochemistry 27, 9020–9030.
230 Jordan, P.M. and Berry, A. (1981) Biochem. J. 195, 177–181.
231 Battersby, A.R., Fookes, C.J.R. and Pandey, P.S. (1983) Tetrahedron 39, 1919–1926.
232 Hart, G.J., Leeper, F.J. and Battersby, A.R. (1984) Biochem. J. 222, 93–102.
233 Hart, G. H., Abell, C. and Battersby, A.R. (1986) Biochem. J. 240, 273–276.
234 Evans, J.N.S., Burton, G., Fagerness, P.E., Mackenzie, N.E. and Scott, A.I. (1986) Biochemistry 25, 905–912.
235 Jordan, P.M. and Warren, M.J. (1987) FEBS Lett. 225, 87–92.
236 Radmer, R. and Bogorad, L. (1972) Biochemistry 11, 904–910.
237 Warren, M.J. and Jordan, P,M. (1988) Biochem. Soc. Trans. 16, 963–965.
238 Hart, G.H., Miller, A.D., Leeper, F.J. and Battersby, A.R. (1987) J. Chem. Soc. Chem. Commun. 1762–1765.
239 Warren, M. J. and Jordan, P.M. (1988) Biochem. Soc. Trans. 16, 962–963.
240 Umanoff, H., Russell, C.S. and Cosloy, S.D. (1988) 170, 4969–4975.
241 Jordan, P.M., Warren, M.J., Williams, H.J., Stolowich, N.J., Roessner, C.A., Grant, S.K. and Scott, A.I. (1988) FEBS Lett. 235, 189–193.
242 Beifuss, U., Hart, G.J., Miller, A.D. and Battersby, A.R. (1988) Tetrahedron Lett. 29, 2591–2594.
243 Raich, N., Romeo, P.H., Dubart, A., Beaupain, D., Cohen-Sohal, M. and Goossens, M. (1986) Nucleic Acids Res. 14, 5955–5968.
244 Miller, A.D., Hart, G.J., Packman, L.C. and Battersby, A.R. (1988) Biochem. J. 254, 915–918.
245 Scott, A.I., Roessner, C.A., Stolowich, N.J., Karuso, P., Williams, H.J. Grant, S.K., Gonsalez, M.D and Hoshino, T. (1988) Biochemistry 27, 7984–7990.
246 Stubnicer, A.C., Picat, C. and Grandchamp, B. (1988) Nucleic Acids Res. 16, 3102.
247 Beaumont, C., Porcher, C., Picat, C., Nordmann, Y. and Grandchamp, B. (1989) J. Biol. Chem. 264, 14829–14834.
248 Sharif, A.L., Smith, A.g. and Abell, C. (1989) Eur. J. Biochem. 184, 353–359.
249 Petricek, M., Rutberg, L., Schroder, I. and Hederstedt, L. (1990) J. Bacteriol. 172, 2250–2258.
250 Scott, A.I., Clemens, K.R., Stolowich, N.J., Santander, P.J., Gonzalez, M.D. and Roessner, C.A. (1989) FEBS Lett. 242, 319–324.
251 Hart, G.J., Miller, A.D. and Battersby, A.R. (1988) Biochem. J. 252, 909–912.
252 Battersby, A.R., Fookes, C.J.R., Matcham, G.W.J., McDonald, E. and Gustafson-Potter, K.E. (1979) J. Chem. Soc., Chem. Commun. 316–319.
253 Jordan, P.M. (1990) in: Biosynthesis of Haem and Chlorophylls (H.A. Dailey, Ed.) pp. 55–121, Mc-Graw-Hill, N.Y.
254 Jones, C., Jordan, P.M. and Akhtar, M.A. (1984) J. Chem Soc. Perkin Trans. I. 2625–2633.
255 Jackson, A.H., Lertwanawatana, W., Procter, G. and Smith, S.G. (1987) Experentia, 43, 892–894.
256 Neidhart, W., Anderson, P.C., Hart, G.J. and Battersby, A.R. (1985) J. Chem. Soc. Chem. Commun. 924–927.
257 Schauder, J.R., Jendrezejewski, S., Abell, A., Hart, G.J. and Battersby, A.R. (1987) J. Chem. Soc. Chem. Commun. 436–439.
258 Battersby, A.R., Fookes, C.J., Matcham, G.W.J. and Pandey, P.S. (1981) Agnew. Che. Int. Edn. English 20, 293–295.
259 Battersby, A.R., Fookes, C.J. and Pandey, P.S. (1983) Tetrahedron 39, 1919–1926.
260 Mathewson, J.H. and Corwin, A.H. (1963) J. Am. Chem. Soc. 83, 135–137.
261 Stark, W.M., Hart, G.J. and Battersby, A.R. (1986) J. Chem. Soc. Chem. Commun. 465–467.
262 Battersby, A.R. and Leeper, F.J. (1990) Chem. Rev. 90, 1261–1274.

263 Grandchamp, B., de Verneuil, H., Beaumont, C., Chretein, S., Walter, O. and Nordmann, Y. (1987) Eur. J. Biochem. 162, 105–110.
264 Hadener, A., Alefounder, P.R., Hart, G.J., Abell, C. and Battersby, A.R. (1990) Biochem. J.
265 Miller, A.D., Packman, L.C., Hart, G.J., Alefounder, P.R., Abell, C. and Battersby, A.R. (1989) Biochem. J. 262, 119–122.
266a Woodcock, S.J. and Jordan, P.M. (1991) Biochem. J. (in press).
266b Jordan, P.M., Warren, M.J., Wood, S.P. and Blundell, T.L. (in preparation).
266c Louie, G. et al. (unpublished data)
266d Lander, M., Pitt, A.R., Alefounder, P.R., Bardy, D., Abell, C. and Battersby, A.R. (1991) Biochem. J. 275, 447–452.
267 Russell, C.S., Polack, S. and James, J. (1984) Ann. N. Y. Acad. Sci. 435, 202–204.
268 Sasarman, A., Nepveu, A., Echelard, Y., Dymetryszyn, J., Drolet, M. and Goyer, C. (1987) J. Bacteriol. 169, 4257–4262.
269 Jordan, P.M., Mgbeje, I.A.B., Alwan, A.F. and Thomas, S.D. (1987) Nucleic Acids Res. 15, 10583.
270 Tsai, S-F., Bishop, D.F. and Desnick, R.J. (1988) Proc. Nat. Acad. Sci. USA. 85, 7044–7053.

P.M. Jordan (Ed.) *Biosynthesis of Tetrapyrroles*
© 1991 Elsevier Science Publishers B.V.

CHAPTER 2

Mechanism and stereochemistry of the enzymes involved in the conversion of uroporphyrinogen III into haem

MUHAMMAD AKHTAR

Department of Biochemistry, University of Southampton, Medical and Biological Sciences Building, Bassett Crescent, East Southampton SO9 3TU, U.K.

1. Introduction

This chapter deals with the four final stages of the haem biosynthesis pathway involving the transformation of uroporphyrinogen III into haem. The enzymes which catalyse these reactions are called uroporphyrinogen decarboxylase, coproporphyrinogen III oxidase, protoporphyrinogen oxidase and ferrochelatase. Specific emphasis will be placed on the mechanistic and stereochemical aspects of the enzyme reactions and bring up-to-date a previous review on the subject published in 1979 [1]. A comprehensive account of the properties and regulation of these enzymes is available in a recent book [2] and therefore these aspects are covered here only in such details as are necessary for the development of mechanistic arguments. A periodical review of progress in the tetrapyrrole field commissioned by the Royal Society of Chemistry is also available [3].

2. *Uroporphyrinogen Decarboxylase (uroporphyrinogen carboxylyase)*

2.1. Introduction

Early workers in the porphyrin field considered that uroporphyrin III (1a, Fig. 1) was an intermediate in haem biosynthesis. We now know [4] that it is the reduced porphyrin, uroporphyrinogen III, (1, Fig. 1) that is the true enzyme intermediate and that the biosynthesis of haem proceeds through the intermediacy of coproporphyrinogen III (5) and protoporphyrinogen IX and it is only then that the macrocyclic ring is oxidised to a porphyrin. The conversion of uroporphyrinogen III into coproporphyrinogen III by decarboxylation of the acetate side-chains at positions 2, 7, 12, and 18 is catalysed by a single enzyme, uroporphyrinogen decarboxylase (Fig. 1). The en-

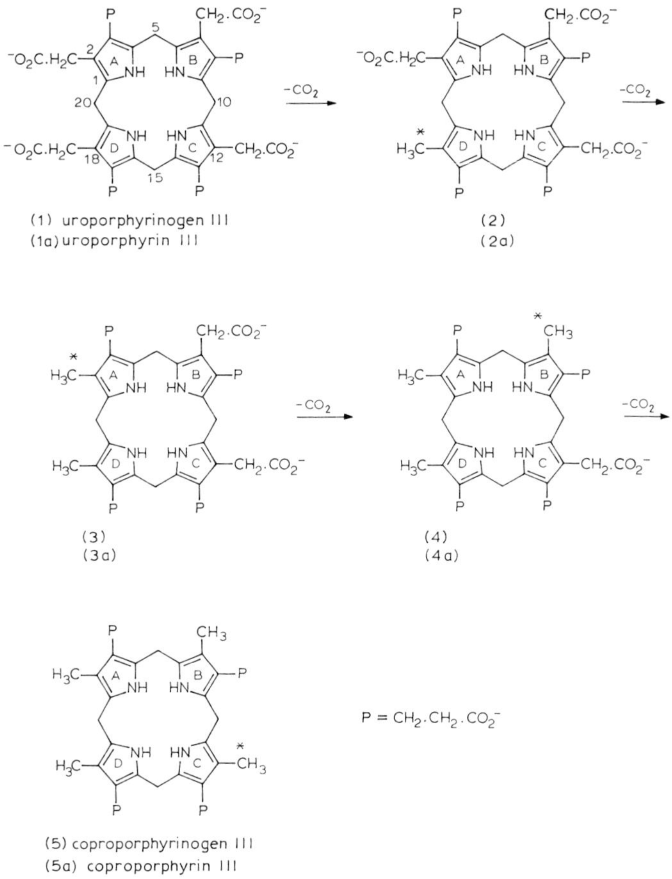

Fig. 1. Postulated sequence for the stepwise decarboxylation of the acetate side-chains of uroporphyrinogen III. In each step the methyl group produced in the decarboxylation reaction is starred, *, throughout structures in the (a) series (e.g. 1a, 2a...) denote the corresponding compounds with the porphyrin nucleus.

zyme also catalyses the decarboxylation of uroporphyrinogen I, but the II and IV isomers are very poor substrates [5–7].

2.2. Intermediates in the decarboxylation

Theoretically there are 14 possible intermediates between uroporphyrinogen III (1) and coproporphyrinogen III (5), containing, seven, six, and five carboxylic acid side-chains [7]. Some of these intermediates have been isolated as their porphyrins from the urine of porphyriac patients [8] or from experimental animals treated with the drug hexachlorobenzene [9,10]. All these intermediates are related to the III series, as shown by their subsequent chemical decarboxylation to coproporphyrin III.

The knowledge of the structures of these intermediates has been used to infer the order in which carboxyl groups from the four acetate side-chains are removed by the decarboxylase. The most extensive study on this aspect has been performed at Cardiff by Jackson, Elder and their colleagues using porphyrins excreted in the faeces of rats which had been poisoned with hexachlorobenzene [10]. The mixture of porphyrins was separated into hepta-, hexa- and pentacarboxylic acid species and the purified porphyrins were subjected to structural determination, predominantly using n.m.r. The heptacarboxylic acid porphyrin (2a) isolated from the faeces was identical to a previously described porphyrin [11] that had been known by several names (phriaporphyrin, porphyrin 208 or pseudouroporphyrin) and this was shown to have the structure 2a.

The isolation of the heptacarboxylic acid porphyrin (2a) suggests that under physiological conditions either it is the D-ring acetate side-chain of uroporphyrinogen III that is first removed by the decarboxylase or that the compound (2) is produced in a minor pathway and accumulates only because it is poorly converted into coproporphyrinogen III. The support for the former proposition 'appeared' to be provided by the observations that the same heptacarboxylic acid porphyrin (2a) was isolated from the urine of patients with different types of porphyrias [12] and also from the incubations in which the conversion of porphobilinogen into haem was studied using an enzyme system from duck erythrocytes [13]. More significantly, Jackson's group synthesised the four possible heptacarboxylic acid porphyrins of the III series and showed that to varying degrees all these derivatives (as their porphyrinogens) were used by the decarboxylase to produce coproporphyrinogen III [10]. Interestingly, however, the urine from normal individuals contained all the isomers of the heptacarboxylic acid porphyrin [12]. In the light of this evidence the accumulation of the heptacarboxylic acid (2a) seems to be the property of an abnormal state. Whether the accumulation of heptacarboxylic acid can be attributed to the operation of an obligatory sequence in which the D-ring acetate side-chain is the first site of action of the decarboxylase remains an interesting possibility.

Of the six hexacarboxylic acid porphyrins which can theoretically be derived from uroporphyrinogen III by decarboxylation of two of its acetate side-chains, the one obtained from the faeces of drug treated rats was shown by n.m.r. and comparison with a specially synthesised compound to have the structure (3a) [10]. The latter structure arises from the decarboxylation of the acetate side-chains in rings D and A of uroporphyrinogen III.

The examination of the pentacarboxylic acid porphyrin fraction isolated from rats treated with hexachlorobenzene revealed it to contain predominantly one species having the structure (4a) corresponding to the decarboxylation of the acetate side-chains from rings D, A, and B of uroporphyrinogen III [10]. With this information in hand Jackson et al. [10] proposed that the conversion of uroporphyrinogen III into coproporphyrinogen III occurs by a preferred sequence whereby the acetate side-chains on rings D, A, B and C are decarboxylated in a clockwise sequence (Fig. 1).

Subsequent careful analysis by Lim et al. [12] confirmed the findings of Jackson and showed that unique isomers of hepta, hexa, and pentacarboxylic acid porphyrins (2a, 3a and 4a) were indeed excreted in the urine of patients suffering from porphyria cutanea tarda. However, the urine of normal individuals contained more than one isomer of the heptacarboxylic acid porphyrin and all the possible isomers of hexa and pentacarboxylic acid porphyrins. Furthermore, the four pentacarboxylic acid porphyrins were excreted in almost equal amounts. The study together with similar data obtained independently by Jackson [14] calls for caution in assigning an intermediary role to a compound merely because it accumulates in a biological fluid. The phenomenon may either be due to the obligatory involvement of the compound in a pathway or conversely that it is a less favoured substrate for subsequent conversion and therefore accumulates.

It could be argued that the most telling argument against the operation of an obligatory sequence for the decarboxylase catalysed conversion is produced by Jackson's own Herculean study in which his group synthesised the fourteen possible intermediates between uroporphyrinogen III and coproporphyrinogen III and showed that all of these were acted upon by the decarboxylase to produce coproporphyrinogen III [7].

Cumulatively, the observations on the isolation of the multiple isomers of each class of carboxylic acid porphyrin from normal urine and the fact that the decarboxylase acts upon a wide range of substrates which include uroporphyrinogen I, four heptacarboxylic acid, six hexacarboxylic acid and four pentacarboxylic acid porphyrinogens of the III series highlight the unusually broad substrate specificity of the enzyme. Indeed it could be argued that from a physiological view point there is no necessity for a strict decarboxylation sequence, since given uroporphyrinogen III as the substrate all different decarboxylation sequences, of which there are 24 (factorial 4), will eventually lead to the formation of the same product, coproporphyrinogen III.

2.3. The interaction between the decarboxylase and the substrates

The types of interactions between the decarboxylase and its substrates which may be involved in molecular recognition must be subtle and can be considered only in general terms. For example, there must be a site on the enzyme which contains a cluster of residues involved in the acetate side-chain decarboxylation. This site is symbolised as S_d (Fig. 2). In close proximity to this may be envisaged a second site for the binding of the neighbouring propionate group (S_p) and a third site for the recognition of the four pyrrolic NH bonds (the site is not shown in any of the illustrations of Fig. 2). The structure 6 in Fig. 2 shows the binding of uroporphyrinogen III to the decarboxylase, with its ring D locked into the catalytic groove. With this mode of binding it is conceivable, indeed likely, that distal interactions also exist between the remaining side-chains of uroporphyrinogen III and six complementary sites on the decar-

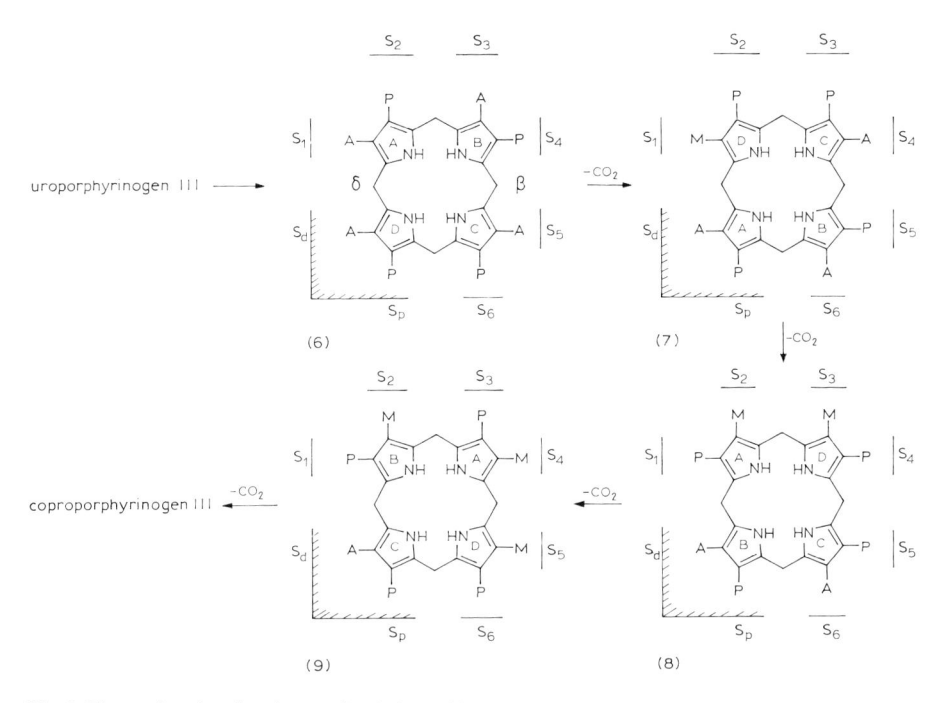

Fig. 2. Illustration showing the proximal (S_d and S_p, hatched) and distal (S_1-S_6) interactions in the reaction catalysed by uroporphyrinogen decarboxylase. S_d is the site containing the cluster of residues for the decarboxylation process and S_p for the binding of the propionate side-chain of the ring engaged in the transformation. The distal sites S_1-S_6 accommodate the remaining six side-chains of the substrates. The conversion (6) → (7) may be envisaged to involve, in addition to the decarboxylation reaction, the rotation of the macrocyclic ring along the C_β-C_δ axis by 180°.

boxylase. These sites in a clockwise arrangement are designated as S_1, S_2, S_3, S_4, S_5 and S_6. With these interactions fully operational and the ring-D acetate side-chain converted into a methyl group, the subsequent rounds of decarboxylation require the dissociation of the product and the binding of the porphyrinogen ring to the enzyme (7) from the face opposite to that used in the complex (6) and amounting to the flipping of the macrocyclic ring around the C_β-C_δ axis by 180°. The flipping allows correct positioning of the acetate and propionate side-chains of the remaining three rings at the S_d and S_p sites. For illustrative convenience we have adhered to the requirements of the Jackson 'clockwise' hypothesis [10] and have shown the next step to involve the decarboxylation of the ring A acetate side-chain, then ring B and finally ring C. In all the four complexes (6–9) identical interactions operate at the S_d, S_p and -NH binding sites, however with respect to distal interactions at the S_1-S_6 sites, the enzyme has a relatively low specificity and the only common feature shared by all the complexes is the presence of a carboxyl group belonging either to an acetate or a propionate side-chain in the S_6 position. In order to accommodate varying combinations of side-chains of different compounds which serve as substrates for the decarboxylase (and there are more than 14 of these) it is important that the overall contribution to binding energy made by these distal interactions is rather small.

2.4. Isolation and structural studies on uroporphyrinogen decarboxylase

The most likely possibility has always been that a single enzyme is involved in catalysing the decarboxylation of all four acetate side-chains of uroporphyrinogen III, however it was not possible to be certain of this until the homogeneous enzyme had been obtained. The isolation of a single enzyme protein capable of catalysing all four decarboxylations from human erythrocytes [15,16], bovine liver [17] and avian erythrocytes [18] has confirmed that the uroporphyrinogen decarboxylase is indeed endowed with a very broad substrate specificity. Similar properties appear to be associated with the yeast enzyme [19]. The enzymes from all the sources have M_r of around 40,000 Da but it is not clear whether these exist as monomers or dimers. That an antibody raised against a homogeneous decarboxylase from human erythrocytes absorbed greater than 96% of the activity from both human liver and erythrocytes further indicated the absence of multiple forms of the enzyme or tissue specific isozymes [20].

Notwithstanding the importance of the above observations, definitive evidence for a single enzyme was provided when the cDNA for the decarboxylase from human spleen was isolated and sequenced [21]. Using the DNA as a probe it was established that the enzyme is encoded by a single copy gene located on the short arm of chromosome 1 [21–23]. This was followed by the cloning and sequencing of the cDNA for the decarboxylase from rat [24]. The comparison of sequences at the DNA and the protein levels showed 85% and 90% similarities, respectively between the decarboxylases from the two species (human and rat). Preliminary sequence information on *Rhodobacter sphaeroides* decarboxylase shows that the N-terminal region has marked similarity with that of the human enzyme [25]. The protein sequence of the decarboxylase, deduced from the nucleotide sequence of the cloned human DNA, shows 10% aromatic amino acids and the presence of 6 cysteine residues, 3 of which are within a cluster of 8 amino acids [21,22]. It is to be noted that in a previous report amino acid compostion of homogeneous decarboxylase from human erythrocytes had shown it to contain 9 half-cystines [15] though only 5 of these were modified by 5,5-dithio*bis*-(2-nitrobenzoic acid) (DTNB). The latter estimation (5 residues) thus more closely approximates the number of cysteine residues (6 residues) revealed by the DNA sequence. That at least one of the cysteines may play an important role in some function of the enzyme is suggested by the fact that a variety of sulphydryl reagents (heavy metals, iodoacetamide, N- ethylmaleimide and DTNB) interfere with the activity of the enzyme or its immunological properties. The inactivation of chicken erythrocyte uroporphyrinogen decarboxylase by diethylpyrocarbonate has been reported thus also indicating the involvement of histidine(s) in catalysis. However, neither the location of the 'essential' histidine nor of the cysteine in the primary sequence has yet been achieved.

2.5. The number of active sites

The evidence from genetic studies and enzymology cited above establishes that a single enzyme is involved in the decarboxylation of all the four acetate side-chains and in the illustration of Fig. 2 we have assumed that the decarboxylation occurs at a single site. The latter conclusion, however, is not universally accepted at present and the existence of 2 or 4 active centres has been advocated [17 and 26, respectively]. Amongst the experiments used in support of the hypothesis the following may be mentioned.

Firstly, the decarboxylase in the presence of certain inhibitors [18], or when subjected to treatments including heat denaturation and modification by N-ethylmaleimide, is inactivated in the overall conversion of uroporphyrinogen III into coproporphyrinogen III yet retains its ability for the decarboxylation of uroporphyrinogen III into the heptacarboxylic acid porphyrinogen. This behaviour is consistent with either the presence of more than one catalytic centre in the enzyme or that the modifications selectively affect the binding of different substrates.

Secondly, de Verneuil and colleagues have performed extensive kinetic studies on the decarboxylation using two types of protocols [26]. In the one they determined the kinetic parameter for the reaction with a variety of substrates in the I as well as the III series. In the second a mixture of two substrates was used to evaluate the effect of one of the components on the rate of decarboxylation of the other. The study showed that hepta and pentacarboxylic acid porphyrinogens III (2 and 4) competitively inhibited the decarboxylation of [$^{14}C_8$] uroporphyrinogen III. In addition, pentacarboxylic acid porphyrinogen III was a competitive inhibitor for the decarboxylation of [$^{14}C_8$] heptacarboxylic porphyrinogen III. Furthermore, a reciprocal inhibition of the decarboxylation of series III porphyrinogens by series I porphyrinogens with the same number of carboxylic groups was found. Although in this study an exhaustive pair-wise comparison was not made, the limited data were interpreted to suggest that porphyrinogens of both isomeric series with the same number of carboxylic groups are decarboxylated at the same active centre.

Finally, it was shown that whereas heptacarboxylic acid porphyrinogen III (2) inhibited the decarboxylation of uroporphyrinogen III, the latter had no effect on the decarboxylation rate of the former (heptacarboxylic acid porphyrinogen III). Cumulatively these observations led to the proposal of 4 separate active centres, each specific for porphyrinogen containing a certain number of carboxylic groups (i.e. one centre each for the octa-, hepta-, hexa- and penta-species) [26]. It should however be noted that the studies with mixed compounds, though performed with care, by necessity involved the use of isotopic techniques to obtain data. Such an approach is susceptible to subtle complications. This is exemplified by reference to experiments in which the decarboxylation rate was studied using [$^{14}C_8$] uroporphyrinogen III. In the absence of any other component, the rate of decarboxylation of the latter is securely measured from the knowledge of its specific radioactivity and the fact that the various products of the reaction (i.e. hepta-, hexa-, penta- and tetra-species) will also have the same specific activity as the original precursor. However, when the decarboxylation of [$^{14}C_8$] uroporphyrinogen III is studied in the presence of the heptacarboxylic acid porphyrinogen III, then the specific radioactivities of the four decarboxylation products will be greatly diminished because the labelled species produced from the first decarboxylation is 'trapped' in the large pool of unlabelled heptacarboxylic acid porphyrinogen. From the crucial Table 3 in the paper [36] it is not clear what allowance was made for the dilution of radioactivity. This criticism apart, realism requires it to be appreciated that the type of questions posed by De Verneuil et al [26] are difficult and can only be answered by using substrates in which a unique carboxyl group is labelled with ^{14}C so that by monitoring the release of $^{14}CO_2$ the kinetics of each of the decarboxylation steps is studied independently. Such a study has not yet been performed and therefore the conclusion that uroporphyrinogen decarboxylase is endowed with multiple active centres should be viewed with caution.

2.6. Mechanistic and stereochemical studies

Light on the mechanism through which the decarboxylation reaction occurs was shed by a stereochemical approach developed in the author's laboratory (Fig. 3) [27,28]. Succinic acid, stereospecifically labelled with deuterium, as well as tritium, at C-2 (10) was used to produce in situ a chiral acetic acid side-chain of porphobilinogen (12) and then of uroporphyrinogen III (13). The absolute stereochemistry at C-2 of succinate used was *R* and when such a species was incorporated via 5-aminolaevulinate into porphobilinogen the C-6 of the latter also had the *R* configuration (12). Since no chemical manipulations are involved at the acetate side-chain of porphoblinogen during its further conversion into uroporphyrinogen III all the four acetate side-chains of the latter (13) retain the original stereochemistry. The mechanism through which these chiral acetate side-chains are handled by the decarboxylase will be reflected in the isotope composition and stereochemistry of the methyl groups of coproporphyrinogen III.

In our studies, the analysis of the methyl groups was performed following the further conversation of coproporphyrinogen III into haem (14). The overall transformation of the succinate (10) into haem was performed in 'one pot' using a cell-free preparation from anaemic chicken erythrocytes. The biosynthetic haem was degraded to obtain its rings A+B as ethylmethylmaleimide (15) and rings C+D as haematinic acid (16). The two imides were separately oxidised to obtain the carbon atoms of interest as acetate. Since in these experiments a reference ^{14}C label was also used, the comparison of the $^3H:^{14}C$ ratio of acetic acid (17) with that of the parent succinate (10) showed that during the multi-step enzymic and chemical manipula-

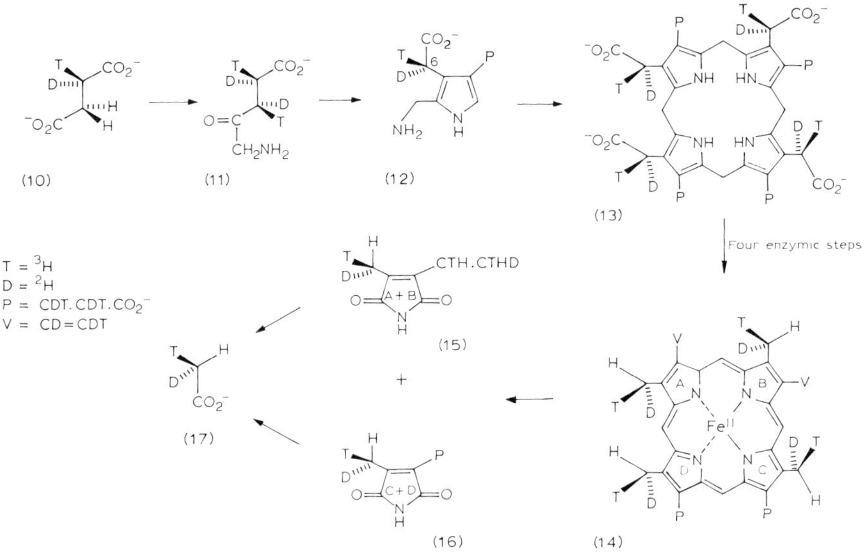

Fig. 3. The stereochemistry of the uroporphyrinogen decarboxylase reaction. The scheme shows the pathway for the conversion of R-[2-^{2}H$_1$,2-^{3}H$_1$] succinate (10) into uroporphyrinogen III (13) and then to the decarboxylated product isolated as haem (14). The latter on degradation, via (15) and (16) gave acetic acid of S-configuration (17).

tions no loss of ^{3}H occurred thus establishing that the decarboxylation reactions occur by a mechanism in which the two methylene hydrogen atoms of the acetate side-chain remain undisturbed and that the methyl group of the acetic acid must contain all the three isotopes of hydrogen (i.e. protium, deuterium and tritium) [27,28; also see 29].

The analysis of the chirality of the methyl group of acetic acid was performed using the elegant approach developed by the schools of Cornforth and Arigoni [30,31]. It was found that both the acetic acid samples (derived from ethylmethylmaleimide and haematinic acid) had the S-configuration (17). A comparison of the structure of S-acetate (17) with that of the appropriate atoms of uroporphyrinogen III allows the conclusion to be drawn that the decarboxylation reaction must have occurred with the retention of configuration. It was shown subsequently that the retention of stereochemistry is also involved in the decarboxylation of the acetate side-chains in bacteriochlorophyll a biosynthesis [32] as well as in the analogous conversion of the C-12 acetate side-chain during corrin biosynthesis [33].

2.7. The mechanism

In the light of the above results a mechanism for the uroporphyrinogen decarboxylase catalysed reaction has been proposed [1,27,28,34]. The surprising fact that no coenzyme or metal requirement has been demonstrated for the enzyme prompted the development of a realistic alternative. It was suggested therefore that the porphyrinogen nucleus of the substrate can be used to generate a conjugated electron sink if a tautomeric form of the pyrrole is produced through protonation of one of its α-positions (18 → 19). The scissile bond is now located in a similar electronic environment (i.e. -NH=C-) which in many enzymic reactions, particularly those operating through the involvement of pyridoxal 5'-phosphate, is known to promote decarboxylation. The overall reaction then occurs through the sequence shown in Fig. 4. In the light of the retention of configuration it was proposed that the enzymic catalytic group involved in either removal of the hydrogen from the O-H bond of the carboxyl group or in the binding of a carboxylate anion may also participate in the protonation of the intermediate (20). The mechanism proposed using a single base to carry out two chemical steps stems from the reasonable, but unproven, assumption that the catalytic sites of enzymes have evolved with due consideration for economy. Other classes of reaction which have been interpreted to use one catalytic group, where two or more could have been employed, include 1,3-allylic and 1,3-aza-allylic rearrangements and aldose-ketose isomerization [for a review see 35].

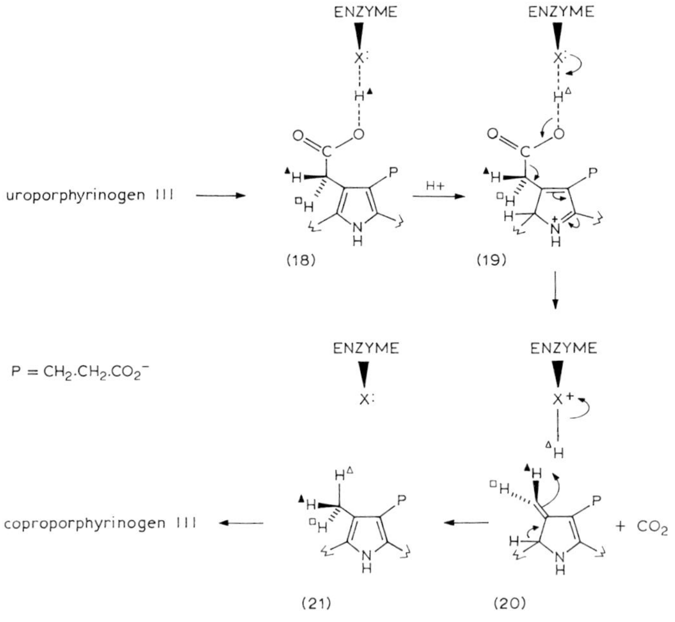

Fig. 4. The postulated mechanism for the uroporphyrinogen decarboxylase reaction. The key features of the postulate are the protonation reaction (18) → (19), and the suggestion that the group involved in the decarboxylation process (19) → (20) also participates in the final protonation (20) → (21).

We have already detailed above the arguments and counterarguments for the proposition whether in the conversion of uroporphyrinogen III into coproporphyrinogen III the decarboxylation of all the four acetate side-chains are catalysed at a single active site or whether several distinct active centres are involved. The stereochemical course of the reaction suggests that irrespective of which of the two alternatives operate the stereochemical mechanism for the decarboxylation of all the four side-chains is the same.

3. Coproporphyrinogen III oxidase

3.1. Introduction

The conversion of coproporphyrinogen III into protoporphyrinogen IX (Fig. 5) involves the oxidative decarboxylation of the two propionate side-chains in rings A and B (positions 3 and 8 in structure 5, Fig. 5) into the vinyl groups. The elaboration of both the propionate groups is catalysed by a single enzyme, coproporphyrinogen III oxidase [36]. The enzyme will not accept coproporphyrinogen I as the substrate, thus explaining why this compound accumulates in inherited diseases, such as congenital erythropoietic porphyria, in which the enzyme uroporphyrinogen III synthase is very

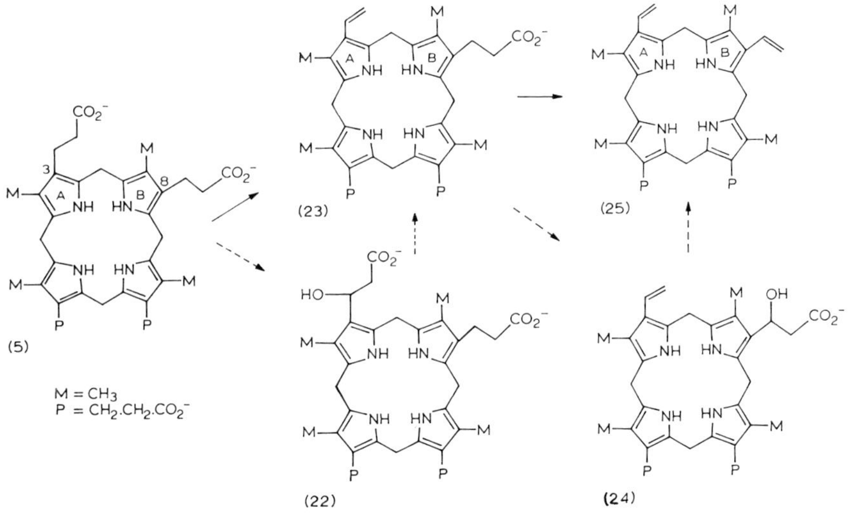

Fig. 5. The sequence for the conversion of coproporphyrinogen III (5) into protoporphyrinogen IX (25). The heavy arrows show the established order in which the two propionate side-chains are oxidatively decarboxylated by coproporphyrinogen oxidase. The broken arrows signify the sequence if the transformation occurred through the involvement of hydroxylated intermediates.

low. Coproporphyrinogen IV, a compound that does not occur in nature is, however, transformed [37].

3.2 Isolation of, and structural studies on, coproporphyrinogen III oxidase

The early sixties were very productive years in Granick's laboratory at the Rockefeller Institute (New York) when work on many new and challenging facets of tetrapyrrole biosynthesis was initiated. Thus Sano and Granick [36] made the first serious attempt to isolate coproporphyrinogen III oxidase from bovine liver and obtained a 21-fold purification of the enzyme. The theme was taken up a few years later by Batlle et al. [38] who, using rat liver, obtained an 18-fold purification of the enzyme and confirmed many of the basic characteristics of coproporphyrinogen III oxidase highlighted by the earlier work [36]. Together these studies provided critical information on the substrate specificity and catalytic properties of the enzyme including the obligatory requirement for O_2. Nearly twenty years later Sano, now working in Japan with Yoshinaga, reported the purification, to homogeniety, of the coproporphyrinogen III oxidase from bovine liver [39,40]. The enzyme required O_2 for catalysis, but no reducing agent. It was found to have a M_r of 71,600 and did not appear to contain any chromophoric prosthetic group or metal ions. Interestingly, all these characteristics are reminiscent of the properties ascribed to the rat liver enzyme by Batlle et al. in 1965 [38]. However, at that time these findings were not given the recognition they deserved because the authors, in keeping with the philosophy of the time, made only modest claims about the state of the purification of the enzyme. The results from two laboratories showing the absence of any prosthetic group or metal ions in coproporphyrinogen III oxidase from two different sources is relevant to the description of the precise chemistry of the reaction mechanism of the enzyme and this facet is debated later in this account.

Yeast has always been a popular source for the isolation of enzymes and the first report of the presence of coproporphyrinogen III oxidase in this organism was by Miyake and Sugimura [41]. Further progress in the field was not made until 1974 when Poulson and Polglase achieved a 150-fold purification of the enzyme and, by gel filtration, estimated its M_r to be 75,000 Da [42]. The partially purified coproporphyrinogen oxidase was shown to function equally well with O_2 or anaerobically with NADP provided that ATP and L-methionine were also present. More recently, the yeast enzyme has been purified to homogeniety and the gene for the enzyme, *hem13*, has also been isolated and its DNA sequence shown to encode for a protein of 328 amino acids with a deduced M_r of 37,673 Da [43]. Since it is now known that the enzyme is a dimer its calculated M_r is in excellent agreement with the value of 75,000 Da reported for the active enzyme by Poulson and Polglase [42]. In contrast, however, the anaerobic activity so convincingly demonstrated by the latter workers could not be confirmed with the homogenous enzyme [43]. Comprehensive reviews on eukaryotic [44] and yeast [19] coproporphyrinogen III oxidases which deal with several historic developments in the field are available and these aspects are not further considered in this chapter.

The obligatory role of O_2 discussed above for the eukaryotic coproporphrinogen III oxidase must be replaced by other oxidants for the corresponding enzyme involved in tetrapyrrole biosynthesis in anaerobic bacteria or facultative anaerobes. Since the original report by Tait [45] that extracts of photosynthetically grown *R. sphaeroides* can oxidise coproporphyrinogen III anaerobically in the presence of NADP, ATP and L-methionine, no substantial progress in the field has been made [for a review see 44]. Although the basic observation of Tait [45] was confirmed by Seehra et al. [46] and Poulson and Polglase [42], other more recent attempts to demonstrate the anaerobic activity have been unsuccessful [43].

3.3. The order in which the propionate side-chains are decarboxylated

There are two main considerations regarding the conversion of coproporphyrinogen III into protoporphyrinogen IX. Firstly, the sequence of the two decarboxylations and secondly, the mechanistic route followed during the formation of vinyl groups including the possible involvement of intermediates in the process. In normal metabolism, the amount of free porphyrinogens and porphyrins found in tissues is extremely small, although free porphyrins are produced in certain genetic diseases which affect the haem biosynthetic pathway or in specialised tissues such as the Harderian gland. Some of these porphyrins, in addition to their interesting structures, have provided valuable information about the route and possible mechanism for the enzymic conversion [47]. One of the most important of these porphyrins is harderoporphyrinogen (23) which resembles coproporphyrinogen except that the 3-position bears a vinyl group. The obvious attractiveness of harderoporphyrin, in its reduced form harderoporphyrinogen, as an intermediate in haem biosynthesis was realised when it was shown to yield protoporphyrinogen IX (25) ten times faster than the isomeric *iso*-harderoporphyrinogen (in this isomer the 8-position rather than the 3-position bears the vinyl group) [47,48]. On the basis of this finding, a preferred biosynthetic sequence (5 → 23 → 25, Fig. 5) in which the decarboxylation of the A ring occurs prior to that of the B ring has been proposed [10].

3.4. Stereochemical and mechanistic studies on coproporphyrinogen III oxidase

Information regarding the mechanism through which coproporphyinogen III is oxidatively decarboxylated into protoporphyrinogen IX has been obtained by studying the conversion of either deuterium or tritium-labelled propionate side-chains into vinyl groups:

$$\overset{\beta}{-CH_2}-\overset{\alpha}{CH_2}-COOH \rightarrow -CH=CH_2 + CO_2$$

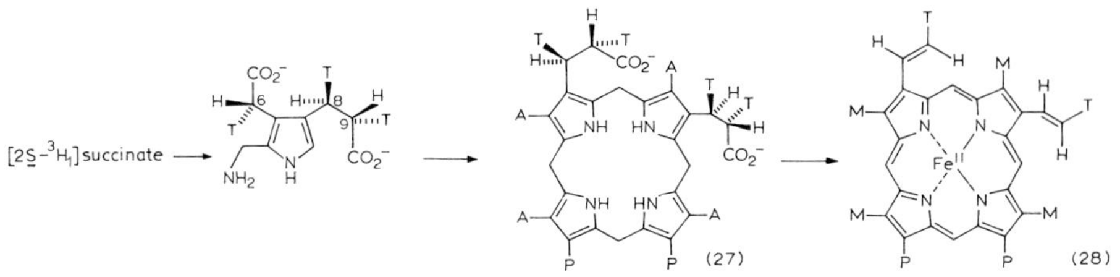

Fig. 6. The stereochemistry of coproporphyrinogen III oxidase reaction. [6S,8S,9S-³H₃] Porphobilinogen (26), enzymically prepared from [2S-³H] succinate, was converted into haem (28). The latter was found to contain ³H only at the terminal methylene position thus showing that in the overall conversion (27) → (28), H_{Si} from the β-position (27) was removed. T = ³H, M = CH₂T, P = CHT.CHT.CO₂⁻

During this conversion the biochemical events involving the α- and β-positions of the propionate side-chain will be reflected in the deuterium or tritium content of the vinyl group. In order to study this facet, porphobilinogen was either chemically synthesised containing deuterium in the side-chain [49] or enzymically labelled with ³H (26) [50–52] and converted into haem with a haemolysed avian erythrocyte preparation (Fig. 6). The biosynthetic material was degraded to haematinic acid and ethylmethylmaleimide for the determination of their isotopic content. These experiments showed that the only hydrogen atom removed from the propionate side-chain was

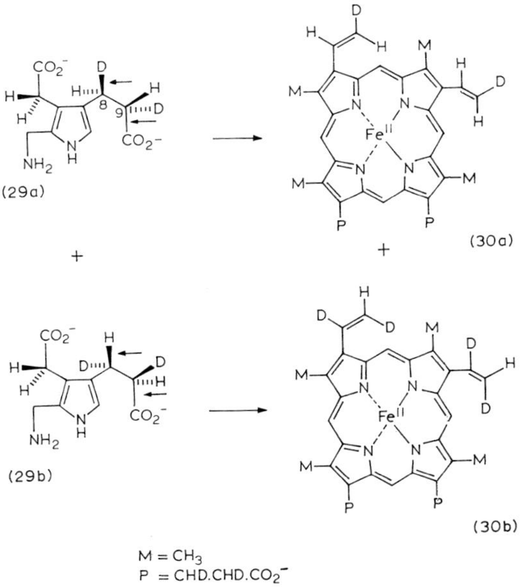

M = CH₃
P = CHD.CHD.CO₂⁻

Fig. 7. The antiperiplanar elimination in the formation of the two vinyl groups. Samples of porphobilinogen containing deuterated species (29a) and (29b) gave haem that was found to contain deuterium atoms in the orientations shown in (30a) and (30b). The results suggest that in the overall conversion the bonds cleaved are those indicated by arrows in (29a) and (29b).

from the β-position [49,50,51,53], occupying the H_{Si} position (conversion 27 → 28) [34,50,51,53]. The stereochemical conclusion was subsequently confirmed using porphobilinogen in which ^3H in the H_{Si} position was introduced by another approach [54]. That the conversion of the propionate into the vinyl groups occurs by an overall antiperiplanar elimination was highlighted when a diastereomeric mixture of 8,9-bis-deutrated porphobilinogen (29a and 29b) was converted into haem and it was shown that the vinyl group contained ^2H and ^1H in orientations shown in the structures (30a) and (30b) (see Fig. 7) [55]. This outcome is best rationalised by assuming that in each case the two bonds cleavaged are those marked by arrows. In other words, the overall process occurs through an antiperiplanar elimination.

3.5 The mechanism of the aerobic coproporphyrinogen III oxidase catalysed reaction

The loss of only one hydrogen atom during the formation of the vinyl group eliminates the participation of acrylic or β-oxo acid intermediates in the process [34,49,50,53,56], although the results do not allow a distinction between the following equally plausible mechanisms (Fig. 8).

Mechanism 1 assumes that reaction proceeds via an oxygen-dependent hydroxylation reaction in which the first step, hydroxylation, provides the energetic requirement for the decarboxylation-elimination process which gives rise to the vinyl group. The support for this mechanism stems from the observation that synthetic 3-(β-hydroxypropionate)-8-propionate deuteroporphyrinogen IX (22) was efficiently con-

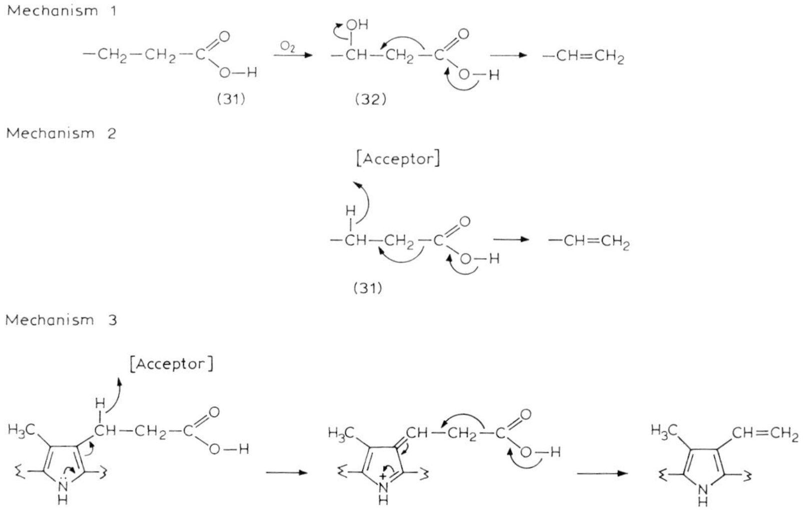

Fig. 8. Possible mechanisms for vinyl group formation. The proposal assumes that the acceptor in aerobic systems is O_2.

verted into protoporphyrinogen IX with an avian erythrocyte preparation as well as by homogeneous coproporphyrinogen III oxidase purified from bovine liver [39,40] The important observation made with the purified enzyme was that while the conversion of the propionate side-chain into the vinyl group was dependent on O_2, the corresponding conversion from the β-hydroxypropionate derivative occurred without the participation of O_2. This behaviour will be broadly consistent with the role of the β-hydroxypropionate derivatives (22 and 24, Fig. 5) as intermediates in the formation of the two vinyl groups. Furthermore, in preliminary experiments Jackson et al. have noted the formation of a β-hydroxy compound of type (22) during the incubation of coproporphyrinogen III with cell-free preparation from chicken erythrocytes [57]. Although the amount of the hydroxy compound isolated was too small to establish unambiguously its role as an intermediate in coproporphyrinogen oxidase catalysed reactions, the results nonetheless may be used in support of Mechanism 1. Another interesting feature was the finding that coproporphyrinogen oxidase, with two of its tyrosine residues modified by treatment with tetranitromethane, although inactive in the overall reaction, could catalyse the formation of the vinyl group from a substrate containing a β-hydroxy propionate side-chain [39,40]. A possible interpretation of this result is that the overall reaction indeed occurs by the sequence in Mechanism 1 and that the modified enzyme had lost selectively the activity for the conversion (31) → (32). An alternative explanation is that the β-hydroxy derivative is decarboxylated artificially by using only a part of the catalytic machinery of the enzyme (see last paragraph of next section). The dilemma posed by Mechanism 1, however, is that it involves a hydroxylation step and as yet the purified enzyme neither appears to contain a chromophoric prosthetic group or a metal ion, nor does it require a reducing agent for catalysis.

$$\begin{array}{c} | \\ \text{H-C-H} \\ | \end{array} \xrightarrow[2e + 2H^+]{O_2} \begin{array}{c} | \\ \text{H-C-OH} + H_2O \\ | \end{array}$$

Three types of enzymic systems for biological hydroxylationare currently are known. These require i) haem b as found in the cytochromes P-450), ii) non-haem iron such as used in the hydroxylation of collagen and iii) a pterin cofactor as required for the hydroxylation of phenylalanine. All these systems use closely related, yet distinct, mechanisms for the activation of oxygen. Therefore, there is no a priori reason why a fourth varient of this should not be used by coproporphyrinogen III oxidase for promoting hydroxylation. The main problem, however, is that in order to obtain a stoichiometrically balanced equation, an oxygen-dependent hydroxylation process **must include** a hydride equivalent. At present such a requirement has not been found for the reaction catalysed by homogeneous coproporphyrinogen III oxidase isolated from rat liver, bovine liver or yeast.

Mechanism 2, originally proposed by Granick and Sano [36], suggests the removal of an hydride ion from the β-position with simultaneous decarboxylation resulting in the formation of the double-bond.

In Mechanism 3 [46,50,53], a variant of Mechanism 2, the transformation occurs by a two-step sequence. In this the crucial C-H bond cleavage in the initial desaturation reaction (33 → 34) is facilitated by electron release from the vinylogous nitrogen with decarboxylation in the second stage (34 → product) being driven by the electron sink provided by the positively charged ring nitrogen -C=$\overset{+}{\text{N}}$H-. It is to be stressed that both these reactions are mechanistically patterned on existing enzymological precedents.

3.6. Concerning the stereochemistry and mechanism of the anaerobic coproporphyrinogen III oxidase

We have already conveyed the general impression that our current knowledge of the process through which coproporphyrinogen III is converted into protoporphyrinogen IX under anaerobic conditions is even scantier than that of the corresponding aerobic reaction. Nothwithstanding this, the combined stereochemical, radiolabelling approach described for the anaerobic reaction has been extended to two types of anaerobic systems. One of these is the cell-free system from *R. sphaeroides*, originally described by Tait [45], in which the anaerobic coproporphyrinogen oxidase activity is expressed using NADP, NADH, ATP and methionine. Using such a system it was shown that, like the aerobic reaction, the anaerobic conversion of the propionate into vinyl side- chains also occurs by the loss of a hydrogen, H_{Si}, from the β-position [46]. Second, and more important, was the evaluation of the stereochemical facet in vivo [58]. For this purpose, two photosynthetic organisms that synthesise bacteriochlorophyll were used, *R. sphaeroides* that biosynthesises bacteriochlorophyll a_{phytyl} (36) and *Rhodospirillum rubrum* that produces bacteriochlorophyll $a_{\text{geranylgeranyl}}$ (37). These organisms [59] were first incubated with stereospecifically tritiated 5-aminolaevulinate (Fig. 9). The chlorophylls were then analysed to trace the path of the α- and β-hydrogen atoms of the 8-propionate side-chains when the latter, via a vinyl group-containing species (partial structure 35), is converted into the 8-ethyl group 35 → 36 or 37). With chlorophylls from both the organisms, only one hydrogen atom, H_{Si}, from the β-position was lost during the conversion under discussion [58].

In summary, therefore, the stereo and regio-specificity of hydrogen elimination during the conversion of the propionate into the vinyl side-chain is the same for all examples studied to date.

In a thoughtful review, Labbe-Bois and Labbe [19] have drawn attention to the fact that anaerobic conditions used for bacterial growth are not rigorous enough to avoid traces of O_2 entering the culture. The level might be sufficient for enzymes which have high affinity for O_2 to function. Nothwithstanding this caution, if the orthodox view is accepted that bacteriochlorophyll *a* biosynthesis is the manifestation of an anaerobic phase, then the isotopic results showing the loss of a single hy-

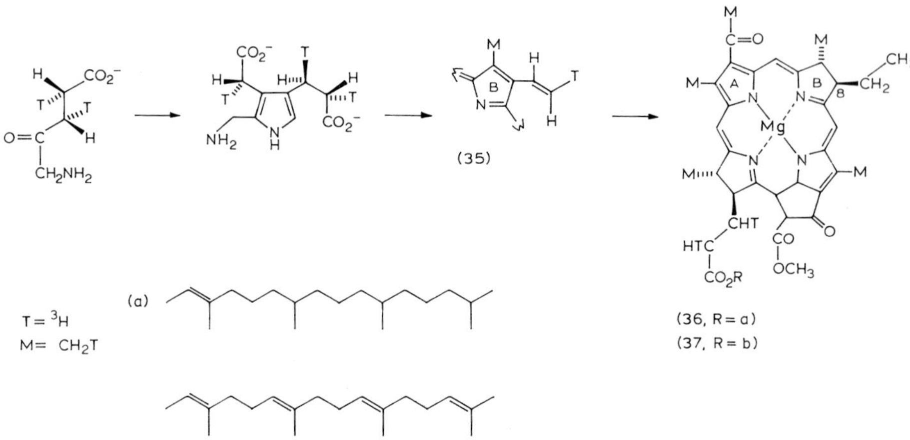

Fig. 9. The incorporation of 5-amino[2S,3S-^{3}H$_2$]laevulinate into bacteriochlorophyll a_{phytyl}(36) and bacteriochlorophyll $a_{geranylgeranyl}$(37). The flow diagram shows the pathway emphasising the fact that in the formation of the vinylic intermediate (35), from the corresponding C-8 propionate side-chain, the only hydrogen atom removed is from the β-position occupying the H$_{Si}$-orientation.

drogen from the β-position are only compatible with Mechanisms 2 and 3 (Fig. 8). Above, we have shown preference for Mechanism 3 and amongst reasons already given in its support [46] the following may be added. The mechanism can be applied to aerobic as well as anaerobic coproporphyrinogen oxidase reactions by merely substituting another electron acceptor for O$_2$. This comment does not imply that the same protein is involved in the two cases; it simply emphasises that once a catalytic mechanism has emerged from natural selection it undergoes subtle rather than dramatic change during subsequent evolution. In addition, Mechanism 3 may be extended to rationalise the conversion of the β-hydroxypropionate derivative (22 or 24) into the vinyl group described by Yoshinaga and Sano [40]. Rather than ascribing it a physiological role, the β-hydroxypropionate derivative may be regarded as a substrate analogue which, by using only a part of the catalytic machinery of the enzyme normally involved in Mechanism 3, undergoes elimination of the hydroxyl group producing the intermediate of the type 34. The latter is then decarboxylated to form the vinyl group.

4. Protoporphyrinogen IX oxidase

4.1. Introduction

The formation of the porphyrin from the porphyrinogen nucleus can occur chemically and thus the possibility existed that in tetrapyrrole biosynthesis the conversion of protoporphyrinogen IX into protoporphyrin IX (Fig. 10) may be a non-enzymic

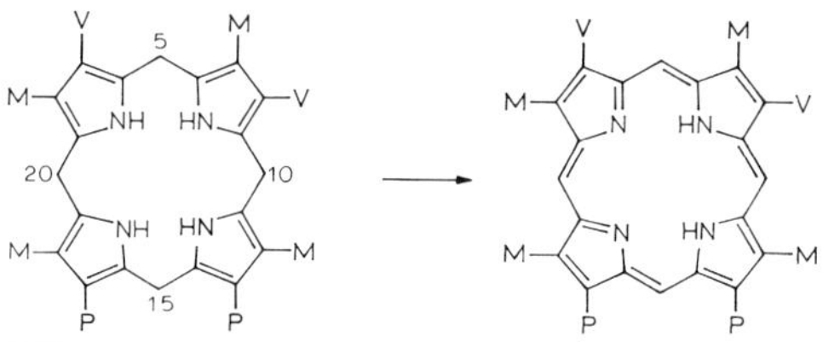

Fig. 10. The conversion of protoporphyrinogen IX (27) into protoporphyrin IX (38)

$A = CH_2.CO_2^-$
$P = CH_2.CH_2.CO_2^-$
$V = CH=CH_2$

phenomenon. However, the pioneering experiments of Sano and Granick [36], performed nearly a quarter of a century ago, pointed to the involvement of an enzyme for the process (27) → (38) when it was found that, in the presence of freeze-thawed rat liver mitochondria, the rate of protoporphyrinogen IX oxidation was significantly increased over the non-enzymic reaction. Contemporaneously, similar activity was described in bovine liver mitochondria and given its current name, protoporphyrinogen IX oxidase [60,61]. The early experiments also showed the requirement of O_2 for the conversion.

4.2. Isolation of, and structural studies on, protoporphyrinogen IX oxidase

After several unsuccessful attempts, protoporphyrinogen IX oxidase has recently been purified to apparent homogeneity from bovine liver [62], mouse liver [63,64] and yeast [unpublished work quoted in ref.19]. Siepker et al. [62] were the first to draw attention to the presence of tightly-bound flavin adenine dinucleotide in the bovine liver enzyme. This claim was subsequently supported for the mouse liver enzyme by Proulx and Dailey [quoted in ref.44], who characterised the flavin in the latter enzyme as FMN (flavine mononucleotide). The bovine and mouse liver enzymes are monomers with M_r around 65,000 Da and are somewhat larger than the yeast enzyme (M_r 56,000–59,000 Da). Although the prosthetic group from the yeast enzyme has not been identified, for all the three enzymes the obligatory electron acceptor was found to be O_2, as had been suggested twenty-eight years previously by Sano and Granick [36]. This requirement for O_2 raises an interesting question regarding haem and chlorophyll biosynthesis which occurs in anaerobic bacteria or facultative anaerobes. The enzyme from these organisms should be able to use electron acceptors other than O_2. The pioneering studies of Jacobs and Jacobs conducted with *E. coli* and *R. sphaeroides* have shown that in these organisms the oxidase activity is present in a membrane-bound fraction and may be solubilised with detergents [65,66,67]. The oxidation of protoporphyrinogen IX by the extracts seems to be intimately coupled to the respiratory chain of the organism and compounds which serve as terminal electron acceptors permit porphyrinogen oxidation. For example, for the *E. coli* enzyme, nitrate and fumarate were successfully used as terminal oxidants [65,66]. Using 2,6-dichlorophenolindophenol as an electron acceptor, considerable progress has been made in the purification of protoporphyrinogen oxidase from *Desulfovibro gigas* [68]. The protein was reported to have an M_r of 148,000 Da and to be composed of three non-identical sub-units. The information currently available therefore suggests that, as far as size is concerned, the prokaryotic enzyme is different from the three eukaryotic enzymes mentioned above.

4.3. Stereochemical and mechanistic studies of protoporphyrinogen IX oxidase

Concurrent with the enzymological studies described above, chemical approaches have been used to prove that during haem biosynthesis the conversion of protoporphyrinogen IX into protoporphyrin IX occurs through an enzyme catalysed process. A noteworthy experiment in this context is the one in which coproporphyrinogen III randomly tritiated at all the 4 *meso*-positions was incubated with a chicken erythrocyte preparation [69]. Protoporphyrin IX produced through the participation of two enzymes (coproporphyrinogen III oxidase and protoporphyrinogen IX oxidase) was found to contain exactly 50% of the tritium present in the precursor [69]. These results which were confirmed later by an independent study at Cambridge highlighted that the four *meso*-hydrogen atoms (C-5, C-10, C-15 and C-20 in Fig. 10) involved in the oxidation must have been removed in a stereospecific fashion and hence manipulated by an enzyme [70]. In a model non-enzymic conversion of the *meso*-tritiated coproporphyrinogen III into coproporphyrin III no appreciable loss of tritium was found, thus indicating that a high isotopic discrimination occurs in the non-enzymic porphyrinogen oxidation [69].

The findings that the conversion of [5,10,15,20-$^3H_8$] coproporphyrinogen III into protoporphyrin IX is attended by retention of 50% of the ^3H were further validated in our laboratory using another approach in which the *meso*-positions of the precursor were labelled from [11RS-$^3H_2$]-porphobilinogen [71,72; also see 70]. The latter approach paved the way for undertaking the more challenging task of evaluating the stereochemical course of the oxidase catalysed reactions. The crucial discovery enabling this goal to be achieved was the availability of a method for the preparation of porphobilinogen containing ^3H in the 11S-position ((39) in Fig. 11) [52]. We had already found that when [2RS-$^3H_2$] glycine, via 5-aminolaevulinate synthase and 5-aminolaevulinate dehydratase, is converted into porphobilinogen the latter contains ^3H in the 11S- position (39) [52 and references cited therein]. When the labelled porphobilinogen is converted into the porphyrinogen nucleus (by way of porphobilinogen deaminase and uroporphyrinogen III synthase) the isotope should be located stereospecifically at the four *meso*-positions of the macrocycle (for example 40).

At present, methods are not available for experimentally assigning the stereochemical orientation to such *meso*-hydrogen atoms and therefore we attempt to

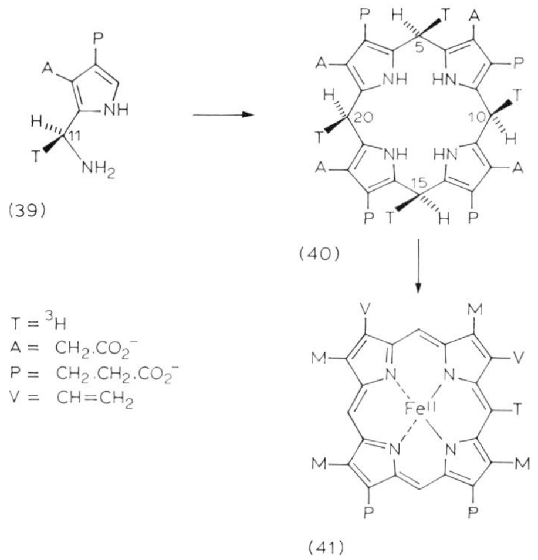

Fig. 11. The incorporation of [11S-^{3}H$_1$] porphobilinogen (39) into haem (41). It is assumed that the overall conversion (39) → (40) occurs through the retention of stereochemistry at the chiral centres. The same stereochemistry will be preserved in the further conversion of (40) into protoporphyrinogen IX.

predict their orientation using enzymological precedents together with chemical intuition [72]. This exercise requires the consideration of the reaction mechanisms of porphobilinogen deaminase and uroporphyrinogen III synthase. The first event in the deaminase catalysed transformation of porphobilinogen into the hydroxymethylbilane, preuroporphyrinogen 47, (Fig. 12) involves the displacement of ammonia and the covalent attachment of the pyrrole moiety to the deaminase (42 → 43). Thereafter, sequential addition of three further molecules of porphobilinogen via (44) and (45) furnishes the enzyme-bound tetrapyrrole (46). All the stages in Fig. 12 involve reactions which superficially appear to be simple displacements. However in view of the known chemistry of porphobilinogen – that its pyrrolic nitrogen participates in the expulsion of substituents at C-11 – we have suggested that these apparent displacements are the indirect ramification of an elimination-addition sequence, a generalised version of which is presented in Fig. 13 [72]. The closest precedent for such a transformation from reactions in enzymology is a class of pyridoxal phosphate-dependent β-replacement reactions which occur by an elimination-addition process and are known to be attended by an overall retention of stereochemistry [73,74]. If the same stereochemical preference was shown by the example under consideration, then all the methylene bridges of the enzyme-bound tetrapyrrole (shown by stars in 46) will be formed by retention mechanisms and hence the steric orientation of ^{3}H at these positions will be identical when the species is generated from [11S-^{3}H] PBG. The release of the linear tetrapyrrole from the porphobilinogen deaminase covalent complex is once again a process identical to the ones involved in earlier transformations of Fig. 12 and hence the argument developed above dictates that the hydroxymethyl group of (47) is also produced by an elimination-addition process maintain-

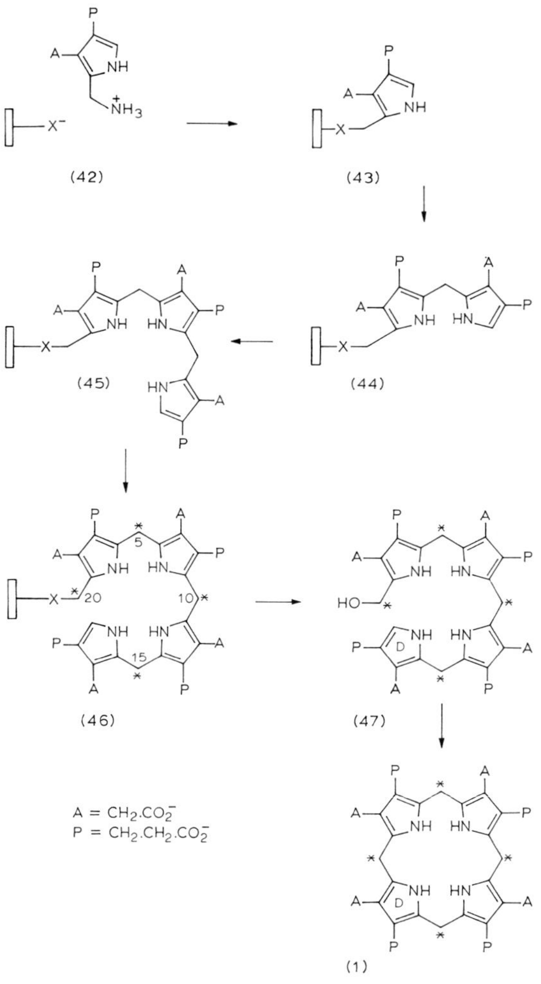

Fig. 12. The postulated mechanism for the conversion of porphobilinogen into uroporphyrinogen III. The overall conversion catalysed by two enzymes (porphobilinogen deaminase and uroporphyrinogen III synthase) is discussed in Chapter 1 which also describes the nature of the dipyrromethane cofactor which forms the binding site for the substrate in porphobilinogen deaminase. The emphasis in the above illustration is to highlight that C-5, C-10 and C-15 of the polypyrrole are formed by the same process, hence these *meso*- positions will have the same stereochemistry up to the bilane (47), when the conversion is performed using porphobilinogen that is rendered chiral by isotopic labelling at C-11. The argument is developed that the further conversion of (47) into uroporphyrinogen III may be attended by the retention of stereochemistry at C-15 and C-20. It should be noted that the numbering system used here for the bilanes (46) and (47) is not standard and is chosen for convenience so that the methylene carbon atoms of bilanes can be directly related to the *meso*-positions of uroporphyrinogen III (5).

ing the same stereochemistry throughout [70]. The subsequent work at Cambridge has established experimentally that indeed the retention of stereochemistry occurs in

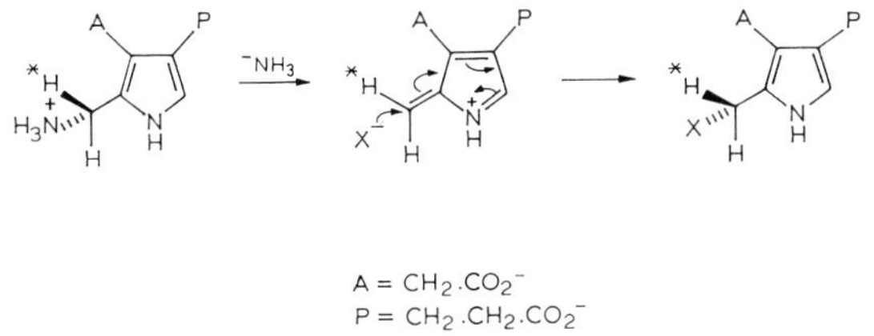

A = CH$_2$.CO$_2^-$
P = CH$_2$.CH$_2$.CO$_2^-$

Fig. 13. The postulated mechanism and stereochemistry for reactions involving displacements at C-11 of porphobilinogen.

the overall process in which the aminomethyl group of porphobilinogen is transformed into the hydroxymethyl group of the bilane, preuroporphyrinogen (47) [75].

In the further conversion of the bilane (47) into uroporphyrinogen III (5) by the synthase, the positions 5 and 10 of the former remain undisturbed and hence the positions will retain their parent stereochemistry in the product. Although we cannot predict, with absolute confidence, the stereochemical outcome of the reactions which occur at C-15 and C-20 during the rearrangement of ring-D, these events, however, may be reduced to a set of elimination-addition sequences, once again requiring the retention of stereochemistry at these centres. The stereochemical outcome of rather complex chemical reactions catalysed by two different enzymes may thus be simplified to one of retention at all the carbon centres. Notwithstanding this broad analysis, it should be emphasised that the key feature of the argument, and one of paramount importance in the future considerations, is that C-5 and C-10 of the growing polypyrrole chain are built into the tetrapyrrole essentially by identical reactions and since there is no reason why these positions should be disturbed during ensuing release of the intermediate and its rearrangement into uroporphyrinogen III, the absolute stereochemistry at C-5 and C-10 must be identical. On the other hand, during the formation of type III macrocycle positions 15 and 20 have been involved in several bond-forming and bond-breaking events and we have made only, what could be described as, a simplistic attempt to guess their stereochemistry.

With this background in hand, [11S-^{3}H] porphobilinogen, admixed with a ^{14}C reference, was incubated with haemolysed red cells from anaemic chicken to produce, in situ, first uroporphyrinogen III which subsequently gave haem performing the crucial transformation of protoporphyrinogen IX into protoporphyrin IX (Fig. 11) [71,72]. The ^{3}H:^{14}C ratio of the precursor porphobilinogen was 7.94. The same ratio must be preserved in uroporphyrinogen III, coproporphyrinogen III and protoporphyrinogen IX. In the oxidase catalysed reaction the fate of the four ^{3}H atoms at the *meso*-positions was to be revealed by the ^{3}H:^{14}C ratio of the biosynthetic haem which was found to be 2.1. This value showed that an average retention of one ^{3}H atom occurred during the porphyrinogen to porphyrin conversion. That the ^{3}H activity remaining in haem was not randomly distributed between various positions but was exclusively resident at C-10 (41) was established by a systematic degradation. A critical reader may like to know that an equivalent series of experiments were performed with [11RS-^{3}H$_2$] porphobilinogen that should introduce ^{3}H at all the eight *meso*-hydrogen atoms, and the biosynthetic haem was found to retain exactly four ^{3}H atoms which were uniformly distributed, as expected, between the four *meso*-positions [72]. The control experiment provided the assurance that out of the 4 ^{3}H atoms incorporated from [11S-^{3}H] porphobilinogen into the porphyrinogen nucleus, the retention of only one ^{3}H at C-10 of haem was indeed the consequence of steric and mechanistic factors operating in the porphyrinogen → porphyrin conversion (Fig. 11).

4.4. The mechanism of the protoporphyrinogen IX oxidase catalysed reaction

Until our involvement in the field, it had been assumed that all the four *meso*-hydrogen atoms are removed from the same face of the macrocyclic ring during the protoporphyrinogen IX oxidase catalysed reaction, however, no mechanism was available to illustrate how this may be achieved [69]. The stereochemical findings described above, embodied in Fig. 11, when taken in conjunction with the assertion that protoporphyrinogen IX biosynthesised from [11S-^{3}H] porphobilinogen (39) has the ^{3}H at C-5 and C-10 co-facial, make it mandatory that these positions (C-5 and C-10) are treated differently in the overall reaction catalysed by protoporphyrinogen IX oxidase. We have suggested, therefore, that one of these positions is the site for the

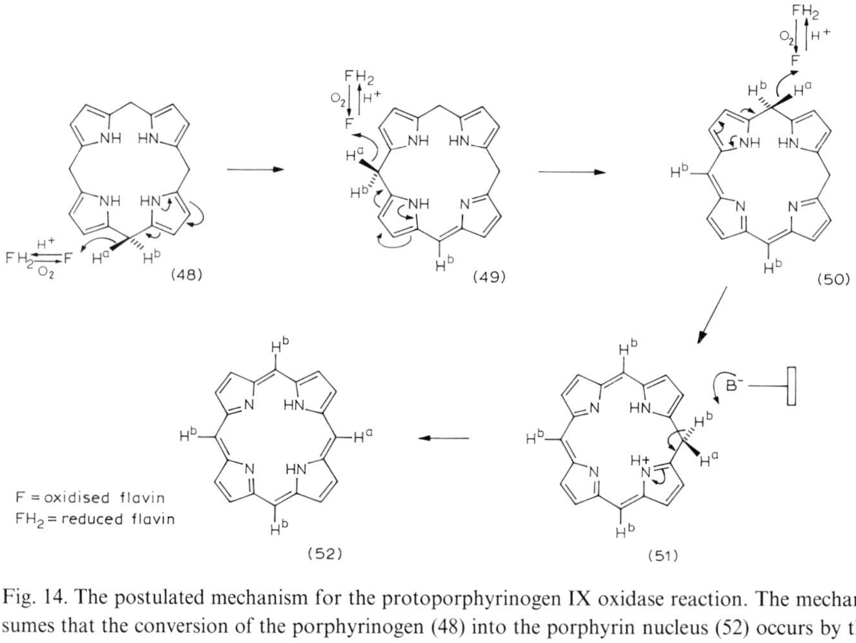

Fig. 14. The postulated mechanism for the protoporphyrinogen IX oxidase reaction. The mechanism assumes that the conversion of the porphyrinogen (48) into the porphyrin nucleus (52) occurs by three desaturation and one prototropic-rearrangement steps. All three desaturations are envisaged to occur at a single active site, though the feature is not obvious from the above illustration. H^a and H^b signify hydrogen atoms above and below the plane of the ring. Emphasis is placed on the fact that the *meso*-hydrogen atoms removed in the three desaturation reactions are co-facial and oriented differently from the one lost in the final rearrangement step.

oxidation reaction (loss of 'hydride') and the other for the tautomerisation process (loss of H$^+$) and that the two processes occur using hydrogen atoms from opposite faces of the macrocycle (Fig. 14). The overall transformation is then viewed to involve three step-wise desaturation reactions, each occurring through the loss of a hydride from the *meso*-position and a proton from the pyrrole NH to give, after three rounds, the intermediate of the type (51). Tautomerisation to the porphyrin then occurs through the stereospecific loss of the fourth *meso*-hydrogen as a proton (51 → 52). Since, with the information currently available, it is not possible to define the precise order in which the various positions of protoporphyrinogen IX are handled by the oxidase, we have illustrated (Fig. 14) our proposed mechanism using an unsubstituted porphyrinogen nucleus. The illustration however does reveal our bias that the three *meso*-hydrogen atoms removed in the desaturation process are located at C-5, C-15 and C-20 and are co-facial. It may be inferred that the three hydrogens which are labelled with ^{3}H from [11*S*-^{3}H] porphobilinogen and are lost in the oxidation phases of the transformation occupy these positions. Since, we have also argued that the chiral porphobilinogen will place the ^{3}H at C-5 and C-10 co-facially, means that the labelling pattern of the porphyrinogen nucleus biosynthesised from (39) is as shown in (40). The retention of ^{3}H at C-10 dictates that the hydrogen atom lost from the position in the tautomerism reaction is the protium shown to be located below the plane of the ring.

Attention has already been drawn to the enzymological studies which have revealed that the two well characterised protoporphyrinogen oxidases, from mouse and bovine liver, contain flavin as the prosthetic group [44,62]. The three desaturation reactions of Fig. 14 are mechanistically analogous to the insertion of unsaturated linkages between C-N and C-C bonds catalysed by D-amino acid oxidases and acyl-CoA dehydrogenases, respectively. Both of these classes of enzyme contain flavin as the prosthetic group which acts as the acceptor of a 'hydride' equivalent. The reduced flavin thus produced is reoxidised by O$_2$ to regenerate the active enzyme. In the illustration of Fig. 14 a similar scenario is envisaged for the desaturation reactions catalysed by eukaryotic protoporphyrinogen oxidases. At present it is not known whether a flavin is involved in the corresponding enzymes from anaerobic bacteria but, assuming that it is, then its reoxidation must be performed by an alternative redox agent, presumably that involved in the respiratory chain of the organism (see above).

The mechanism of the oxidase catalysed reaction shown in Fig. 14 is the consequence of our efforts to rationalise the unexpected stereochemical results. We do, however, acknowledge that the same mechanism could have been deduced from first principles by that 'mythical man from Mars' as being the simplest explanation for achieving the enzymic conversion of the porphyrinogen into the porphyrin nucleus.

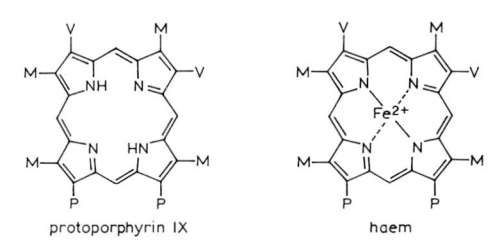

Fig. 15. The ferrochelatase reaction.

5. Ferrochelatase

5.1. Occurrence and isolation of ferrochelatases

The final step in the haem biosynthesis pathway, the insertion of ferrous iron into protoporphyrin IX (Fig. 15) is catalysed by the enzyme ferrochelatase (EC 4.9.9.1.1.). The enzyme is also referred to as haem synthase and, more correctly, as protohaem ferro-lyase. Ferrochelatase has been reported in avian and mammalian systems including duck erythrocytes [76], rat liver [77], pig liver [78] and yeast [79]. The enzyme has also been demonstrated in bacterial cells including R. spheroides [80], Spirillum itersonii [81] and Cyanidium caldarium [82].

The eukaryote enzyme appears to be an intrinsic inner membrane protein in the mitochondria [83,84]. The most convincing evidence for this has come from chemical modification studies which have shown that the enzyme spans the inner mitochondrial membrane [84]. The purification of ferrochelatases thus requires the use of detergents and has been achieved with great difficulty in some cases. Nevertheless the homogeneous enzyme has been isolated from a number of sources including rat liver mitochondria [85,86], bovine liver [87,88], bovine kidney [113], chicken erythrocytes [89], mouse liver [90] and human liver [91].

Some investigators have found that lipids activate the enzymes from pig mitochondria [92], rat mitochondria [93,86] and chicken erythrocytes [94,89]. Such phenomena are often observed with membrane-bound enzymes and presumably reflect the fact that, when surrounded by lipids, an environment similar to the natural situation in the membrane is created. Alternatively, the lipids could be increasing the solubility, or facilitating the accessibility, of the substrate for the active site. In contrast, the bovine enzyme does not appear to be stimulated by lipids.

The purified enzyme has also been isolated from the photosynthetic bacterium R. sphaeroides [80,95,96] and from S. itersonii [97] where it also appears to be an intrinsic membrane protein.

5.2. Properties of ferrochelatases

Eukaryote ferrochelatases all appear to be monomeric proteins with broadly similar

Mrs of 40–44,000 Da. This similarity also extends to their immunogenicity, the enzymes from chicken, murine and bovine origin all showing cross reactivity. Studies on the mouse enzyme have provided evidence for the existence of a pro-enzyme with a Mr of 43,000 Da which, after import into the mitochondria, is processed to the mature protein of Mr 40,000 Da [98]. The enzyme from *R. spheroides* is very different from the mammalian ferrochelatases with a Mr of 115,000 Da.

The Km for protoporphyrin ranges from 0.35 μm for the human liver enzyme [99] to 54 μm for the enzyme from bovine liver [87]. The differences in values of Km may reflect the method of preparation of the substrate, its purity and the assay method employed (for a review see Dailey, 1990 [45]).

5.3. Substrate specificity

Ferrochelatase exhibits a broad substrate specificity and will catalyse the insertion of iron into a range of porphyrins in addition to protoporphyrin IX [100,86,87,89]. Detailed investigations have been carried out with the enzymes from mouse [101] and bovine liver [102]. The affinity of the enzyme for substrate decreases in the order protoporphyrin, hematoporphyrin, mesoporphyrin and deuteroporphyrin (Fig. 16). Thus the enzyme can accommodate a range of small, uncharged substituents such as methyl (M), hydroxyethyl (HEt) and hydrogen (H) groups in place of the vinyl (V) groups at the 3- and 8-positions on the a and b rings. Uroporphyrins I and III, with charged substituents on the a and b rings, are not accepted as substrates by the enzyme. The finding that protoporphyrin XIII, which has the methyl and vinyl groups

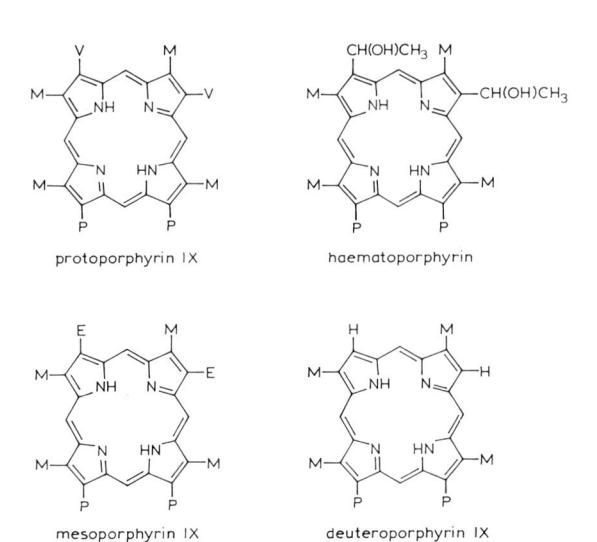

Fig. 16. Structures of hematoporphyrin, mesopoporphyrin and deuteroporphyrin compared to protoporphyrin IX.

switched at the 2- and 3-positions on ring a [100], is a good substrate for the ovine enzyme suggests that the rings a and b do not have specific recognition sites on the enzyme but that general, non-specific hydrophobic interactions are involved.

In contrast to the loose specificity of the enzyme to substituents on the a and b rings, the substituents on the c and d rings are vital for substrate recognition [100]. The propionic acid side chains at the 13- and 17-positions are obligatory and protoporphyrin derived from the type I isomer of coproporphyrin, with propionic acid side chains at the 13- and 18-positions, is not a substrate for the enzyme. The ability of various naturally occurring and synthetic porphyrins to act as substrates is summarised below although species differences may mean that this is only a general guide.

β-substituent position

'old' numbering system	1	2	3	4	5	6	7	8	substrate
'new' numbering system	2	3	7	8	12	13	17	18	activity
protoporphyrin IX	M	V	M	V	M	P	P	M	++++++
protoporphyrin XIII	V	M	M	V	M	P	P	M	+++++
hematoporphyrin	M	V	M	HEt	M	P	P	M	++++
mesoporphyrin IX	M	E	M	E	M	P	P	M	+++
deuteroporphyrin IX	M	H	M	H	M	P	P	M	++
protoporphyrin I	M	V	M	V	M	P	M	P	−
mesoporphyrin I	M	E	M	E	M	P	M	P	−
coproporphyrin I	M	P	M	P	M	P	M	P	−
coproporphyrin III	M	P	M	P	M	P	P	M	−

5.4. Metal requirements

The metal requirements for the enzyme are quite rigid with ferrous iron being the main substrate for the rat enzyme [83]. It is thought that the enzyme system for generating ferrous iron is coupled to mitochondrial electron transport [103,104] but the details of this system need to be elucidated. The divalent cations cobalt and zinc also act as reasonable substrates. Ferric ion is not a substrate [96]. Other divalent cations such as manganese, cadmium, lead and mercury act as inhibitors [105]. The most potent metal inhibitor of the ferrochelatase is mercury which is thought to interact with reactive sulphydryl groups. This aspect is discussed more fully below.

5.5. Inhibition of ferrochelatases by N-alkylporphyrins

The most spectacular inhibition of ferrochelatase is exibited by N-methyl porphyrins related to protoporphyrin IX (Fig. 17) which have K_i values in the low nM range. These compounds were first characterised by Smith and De Matteis [106] as the 'green pigments' generated in the livers of rats administered 3,5-diethoxycarbonyl-

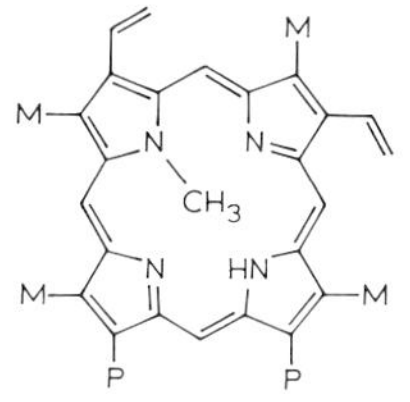

N-methylprotoporphyrin IX

Fig. 17. Structure of N-methylprotoporphyrin IX.

1,4-dihydro- 2,4,6-trimethyl pyridine (DDC). The pigments arise from the haem group of cytochrome P450 which is alkylated by the substrate in a 'suicide' reaction [107]. The four N-methyl porphyrin isomers have been synthesized [108] and all have approximately similar inhibitory activity with ferrochelatase [109]. In general, as the N-alkyl substituent is increased in size the ability of the inhibitor to inhibit the enzyme decreases. The position of N-alkylation also becomes more important when the N-sustituent is larger. Thus the c or d ring N-propylprotoporphyrins are very poor inhibitors compared with the a or b ring isomers [110]. Investigations with numerous N-alkyl porphyrins related to synthetic substrates have also been investigated [110].

5.6. Kinetics, active site and mechanism of ferrochelatases

Ferrochelatase exhibits ordered bi-bi kinetics with the following reaction profile [88].

Fe^{2+} $2H^+$ protoporphyrin IX   haeme
↓   ↑   ↓   ↑

Initially Fe^{2+} binds to the enzyme and displaces two H^+ ions. Protoporphyrin IX then binds, the iron is inserted and haem is released. The reaction requires no energy source such as ATP.

Ferrochelatases are all inactivated by sulphydryl reagents such as N-ethyl maleimide and related compounds. Ferrous ion protects the bovine enzyme against inactivation by N-ethylmaleimide and enzyme inactivated by this reagent is unable to bind Fe^{2+} [111]. These findings suggest that the ferrous ion may bind directly to, or in the vicinity of, an active site sulphydryl residue [111]. The possibility of a bidentate ferrous iron binding site with two sulphydryl groups, of the type $-S-Fe^{2+}-S-$, has been proposed on the basis of reversible inhibition by arsenite [111]. Protoporphyrin IX, in contrast to ferrous iron, affords no protection to the reaction with inhibitor.

Investigations with the bovine ferrochelatase [84] have provided evidence for the location of the active site with respect to the membrane. These investigators observed that dimaleimidylstilbene disulphonate inactivated the solubilised enzyme but had little effect on the enzyme in intact mitochondria. It was therefore suggested that the active site of the ferrochelatase is located on the matrix side of the inner membrane.

Attempts to identify other potential catalytic or binding residues have shown that arginine may also play an important role. Modification of the bovine enzyme with phenylglyoxal leads to a reduction of the affinity of protoporphyrin IX for the enzyme [112]. The inhibition by this reagent is prevented by protoporphyrin IX. It is possible that the positively charged side chains of arginine assist in the alignment of the paired propionic acid substituents on the porphyrin c and d rings.

It has been suggested that the remarkable sensitivity of ferrochelatases to inhibition by N-alkyl porphyrins may provide an insight into the mechanism of iron insertion into the porphyrin ring. This is based on the fact that the addition of the N-alkyl substituent causes a major change in conformation of the macrocyclic ring. Thus the macrocycle is distorted from the essentially planar conformation of a normal porphyrin so that the N-substituted ring assumes a conformation some 40° out of the plane of the macrocycle. It is possible that such a conformation actually mimics the conformation adopted by the natural substrate in the porphyrin-ferrous iron transition state complex at the active site of the enzyme [44].

The isolation, cloning and sequencing of the cDNA for mouse [114] and yeast [115] ferrochelatases has provided valuable insight into the primary structure of the enzymes. It is clear from a comparison of the derived protein sequences that the enzymes are closely related in evolutionary and structural terms. The availability of amino acid sequence information will greatly stimulate further investigations to establish the mechanism of action of the enzyme.

Acknowledgement

This Chapter could not have been completed without the encouragement, support and persistence of Professor P.M. Jordan. He made available many of the references and with him I have had the benefit of extensive discussions which enabled choices to be made between conflicting alternatives

References

1. Akhtar, M. and Jordan, P.M. (1979) in: D.H.R. Barton and W.D. Ollis (Eds.), Comprehensive Organic Chemistry, Vol.5, Pergamon Press, Oxford, pp.1121–1166.
2. Dailey, H.A. (Ed) (1990) Biosynthesis of Heme and Chlorophylls, McGraw-Hill, New York.
3. Leeper, F.J. (1985) Natural Product Reports 2, 19–47 and 561–580; (1987) 5, 441–469.
4. Neve, R.A., Labbe, R.F. and Aldrich, R.A. (1956) J. Am. Chem. Soc.78, 691–692.
5. Mauzerall, D. and Granick, S. (1958) J. Biol. Chem. 232, 1141–1162.
6. Smith, A.G. and Francis, J.E. (1979) Biochem. J. 183, 455–458.
7. Jackson, A., Sancovich, H. and Ferramola, A. (1980) Bioorganic Chem. 9, 71–120.

8. San Martin De Viale, L.C. and Grinstein, M. (1968) Biochim. Biophys. Acta 158, 79–91.
9. San Martin De Viale, L.C., Viale, A.A., Nacht, S. and Grinstein, M. (1970) Clin. Chim. Acta 28, 13–23.
10. Jackson, A.H., Sancovich, H.A., Ferramola, A.M., Evans, N., Games, D.E., Matlin, S.A., Elder, G.H. and Smith, S. (1976) Phil. Trans. Roy. Soc. Lond. B273, 191–206.
11. Batlle, A.M. Del C. and Grinstein, M. (1964) Biochim. Biophys. Acta 82, 13–20.
12. Lim, C., Rideout, J. and Wright, D. (1983) J. Chromatog. 282, 629–641.
13. Battersby, A.R., Hunt, E., Ihara, M., McDonald, E., Paine III, J.B., Satoh, F. and Saunders, J. (1974) J. Chem. Soc. Chem. Commun. 994–995.
14. Smith, S., Rao, K. and Jackson, A.H. (1980) Int. J. Biochem. 12, 1081–1084.
15. de Verneuil, H., Sassa, S., Kappas, A. (1983) J. Biol. Chem. 258, 2454–2460.
16. Elder, G.H., Tovey, J.A. and Sheppard, D.M. (1983) Biochem. J. 215, 45–55.
17. Straka, J. and Kushner, J. (1983) Biochemistry 22, 4664–4672.
18. Kawanishi, S., Seki, Y. and Sano, S. (1983) J. Biol. Chem. 258, 4285–4292.
19. Labbe-Bois, R. and Labbe, P. (1990) in: H.A. Dailey (Ed.) Biosynthesis of Heme and Chlorophylls, McGraw-Hill, New York, pp. 235–286.
20. Elder, G.H. and Urquhart, A. (1984) Biochem. Soc. Trans. 12, 663–664.
21. Romeo, P-H., Raich, N., Dubart, A., Beaupain, D., Pryor, M., Kushner, J., Cohen-Solal, M. and Goossens, M. (1986) J. Biol. Chem. 261, 9825–9831.
22. de Verneuil, H., Grandchamp, B., Beaumont, C., Picat, C. and Nordmann, Y. (1986) Science 234, 732–734.
23. de Verneuil, H., Grandchamp, B., Foubert, C., Weil, D., N'Guyer, V.C, Gross, M.S., Sassa, S. and Nordmann, Y. (1984) Human Genet. 661, 202–205.
24. Romana, M., Le Bonlch, P., Romeo, P-H. (1987) Nucleic Acids Res. 15, 5487, 7211.
25. Jordan, P.M. and Jones, R.M. (Unpublished work).
26. de Verneuil, H., Grandchamp, B. and Nordmann, Y. (1980) Biochim. Biophys. Acta 611, 174–186.
27. Barnard, G.F. and Akhtar, M. (1975) J. Chem. Soc. Chem. Commun. 494–496.
28. Barnard, G.F. and Akhtar, M. (1979) J. Chem. Soc. Perkin Trans. I, 2354–2360.
29. Smith, A.G. and Francis, J.E. (1981) Biochem. J. 195, 241–250.
30. Cornforth, J.W., Redmond, J.W., Eggerer, H., Buckel, W. and Gutschow, C. (1969) Nature 221, 1212–1213.
31. Luthy, J., Retey, J. and Arigoni, D. (1969) Nature 221, 1213–1215.
32. Battersby, A.R., Gutmann, A.L., Fookes, C.J.R. (1981) J. Chem. Soc. Chem. Commun. 645–647.
33. Battersby, A.R., Deutscher, K.R., Martinoni, B. (1983) J. Chem. Soc. Chem. Commun. 698–700.
34. Akhtar, M., Abboud, M.M., Barnard, G., Jordan, P. and Zaman, Z. (1976) Phil. Trans. Roy. Soc. Lond. B 273, 117–136.
35. Hanson, K.R. and Rose, I.A. (1975) Accounts of Chem. Res. 8, 1–10.
36. Sano, S. and Granick, S. (1961) J. Biol. Chem. 236, 1173–1180.
37. Battersby, A.R., Mombelli, L. and McDonald, E. (1976) Tetrahedron Letts. 1037–1040.
38. Batlle, A.M. Del C., Benson, A. and Rimington, C. (1965) Biochem. J. 97, 731–740.
39. Yoshinaga, T. and Sano, S. (1980) J. Biol. Chem. 255, 4722–4726.
40. Yoshinaga, T. and Sano, S. (1980) J. Biol. Chem. 255, 4727–4731.
41. Miyake, S. and Sugimura, T. (1968) J. Bacteriol. 96, 1997–2003.
42. Poulson, R. and Polglase, W.J. (1974) J. Biol. Chem. 249, 6367–6371.
43. Camadro, J.M., Chambon, H., Jolles, J. and Labbe, P. (1986) Eur. J. Biochem. 156, 579–587.
44. Dailey, H.A. (1990) in: H.A. Dailey (Ed.) Biosynthesis of Heme and Chlorophylls, McGraw-Hill, New York, pp. 123–161.
45. Tait, G.H. (1972) Biochem. J. 128, 1159–1169.
46. Seehra, J.S., Jordan, P.M. and Akhtar, M. (1983) Biochem. J. 209, 709–718.

47 Kennedy, G.Y., Jackson, A.H., Kenner, G.W. and Suckling, C.J. (1970) FEBS Lett. 6, 9–12; 7, 205–206.
48 Cavaleiro, J.A.S., Kenner, G.W. and Smith, K.M. (1974) J. Chem. Soc. Perkin Trans. I, 1188–1194.
49 Battersby, A.R., Baldas, J., Collins, J., Grayson, D.H., James, K.M. and McDonald, E. (1972) J. Chem. Soc. Chem. Commun. 1265–1266.
50 Zaman, Z., Abboud, M.M. and Akhtar, M. (1972) J. Chem. Soc. Chem. Commun. 1263–1264.
51 Abboud, M.M. and Akhtar, M. (1978) Nouv. J. Chim. 2, 419–425.
52 Akhtar, M. and Jones, C. (1986) Methods Enzymology 123, 375–383.
53 Zaman, Z. and Akhtar, M. (1976) Eur. J. Biochem. 61, 215–223.
54 Battersby, A.R. (1978) Experientia 34, 1–13
55 Battersby, A.R., McDonald, E., Wurziger, H.K.W. and James, K.J. (1975) J. Chem. Soc. Chem. Commun. 493–494.
56 Battersby, A.R. and McDonald, E. (1975) in: K.M. Smith (Ed.) Porphyrins and Metalloporphyrins, Elsevier, Amsterdam, pp. 61–122.
57 Jackson, A.H., Jones, D.M., Philip, G., Lash, T.D. and Batlle, A.M. Del C. (1980) Int. J. Biochem. 12, 681–688.
58 Ajaz, A.A. (1982) Ph.D. Thesis, University of Southampton, Southampton (U.K.).
59 Ajaz, A.A., Corina, D.L. and Akhtar, M. (1985) Eur. J. Biochem. 150, 309–312.
60 Porra, R.J. and Falk, J.E. (1961) Biochem. Biophys. Res. Commun. 5, 179–184.
61 Porra, R.J. and Falk, J.E. (1964) Biochem. J. 90, 69–75.
62 Siepker, L.J., Ford, M., de Kock, R. and Kramer, S. (1987) Biochim. Biophys. Acta 913, 349–358.
63 Dailey, H.A. and Karr, S.W. (1987) Biochemistry 26, 2679–2701.
64 Ferreira, G.C. and Dailey, H.A. (1988) Biochem. J. 250, 597–603.
65 Jacobs, N.J. and Jacobs, J.M. (1975) Biochem. Biophys. Res. Commun. 65, 435–441.
66 Jacobs, N.J. and Jacobs, J.M. (1976) Biochim. Biophys. Acta 449, 1–9.
67 Jacobs, N.J. and Jacobs, J.M. (1981) Arch. Biochim. Biophys. 211, 305–311.
68 Klemm, D.J. and Barton, L.L. (1987) J. Bact. 169, 5209–5215.
69 Jackson, A.H., Games, D.E., Couch, P.W., Jackson, J.R., Belcher, R.V. and Smith, S.G. (1974) Enzyme 17, 81–87.
70 Battersby, A.R., Staunton, J., McDonald, E., Redfern, J.R. and Wightman, R.H. (1976) J. Chem. Soc. Perkin Trans. I, 266–273.
71 Jones, C., Jordan, P.M., Chaudhry, A.G., and Akhtar, M. (1979) J. Chem. Soc. Chem Commun. 96–97.
72 Jones, C., Jordan, P.M. and Akhtar, M. (1984) J. Chem. Soc. Perkin Trans. I, 2625–2633.
73 Akhtar, M., Emery, V.C. and Robinson, J.A. (1983) in: M.I. Page (Ed.) New Comprehensive Biochemistry: The Chemistry of Enzyme Action, Vol. 6, Elsevier, Amsterdam pp. 303–372.
74 Emery, V.C. and Akhtar, M. (1987) in: M.I. Page and A. Williams (Eds.) Enzyme Mechanisms, The Royal Society of Chemistry, London pp. 345–389.
75 Neidhart, W., Anderson, P.C., Hart, G.J. and Battersby, A.R. (1985) J. Chem. Soc. Chem Commun. 924–927.
76 Yoneyama, Y., Ohyama, H., Sugeta, Y. and Yoshikawa, H. (1962) Biochim. Biophys. Acta 62, 261–268.
77 Yoneyama, Y., Tamai, A., Yasuda, T. and Yoshikawa, H. (1965) Biochim. Biophys. Acta 105, 100–105.
78 Porra, R.J. and Jones, O.T.G. (1963) Biochem. J. 87, 181–185.
79 Labbe, R.F. and Hubbard, N. (1960) Biochim. Biophys. Acta 41, 185–191.
80 Jones, M.S. and Jones, O.T.G. (1970) Biochem. J. 119, 453–462.
81 Dailey, H.A., Jr. and Lascelles, J. (1977) J. Bacteriol. 129, 815–820.
82 Brown, S.B., Holroyd, J.A., Vernon, D.I. and Jones, O.T.G. (1984) Biochem. J. 220, 861–863.
83 Jones, M.S. and Jones, O.T.G. (1969) Biochem. J. 113, 507–514.

84 Harbin, B.M. and Dailey, H.A. (1985) Biochemistry, 24, 366–370.
85 Mailer, K., Poulson, R., Dolphin, D. and Hamilton, A.D. (1980) Biochem. Biophys. Res. Comm., 96, 777–784.
86 Taketani, S. and Tokunaga, R. (1981) J. Bio. Chem., 256, 12748–12753.
87 Taketani, S. and Tokunaga, R. (1982) Eur. J. Biochem, 127, 443–447.
88 Dailey, H.A. and Fleming, J.E. (1983) J. Biol. Chem., 258. 11453–11459.
89 Hanson, J.W. and Dailey, H.A. (1984) Biochem. J. 222, 695–700.
90 Dailey, H.A., Fleming, J.E. and Harbin, B.M. (1986) Methods Enzymol., 123, 401–408.
91 Mathews-Roth, M.M., Drouin, G.L. and Duffy, L. (1987) Arch. Dermatol., 123, 429–430.
92 Mazanowska, A.M., Neuberger, A. and Tait, G.H. (1966) Biochem. J. 98, 117–127.
93 Simpson, D.M. and Poulson, R. (1977) Biochim. Biophys. Acta 482, 461–469.
94 Sawada, H., Takeshita, M., Sugita, Y. and Yoneyama, Y. (1969) Biochim. Biophys. Acta 178, 145–155.
95 Dailey, H.A. (1982) J. Biol. Chem., 257, 14714–14718.
96 Dailey, H.A. (1986) Methods Enzymol., 123, 408–415.
97 Dailey, H.A., Jr. (1977) J. Bacteriol., 132, 302–307.
98 Karr, S.R. and Dailey, H.A. (1988) Biochem. J. 254, 799–803.
99 Camadro. J.M., Ibraham, N.G. and Levere, R.D. (1984) J. Biol. Chem., 259, 5678–5682.
100 Honeybourne, C.L., Jackson, J.T. and Jones, O.T. (1979) FEBS Lett., 98, 207–210.
101 Dailey, H.A., Jones, C.S. and Karr, S.W. (1990) Biochim. Biophys. Acta 999, 7–11.
102 Dailey, H.A. and Smith, A. (1984) Biochem. J. 223, 441–445.
103 Barnes, R. and Jones, O.T.G. (1973) Biochim. Biophys. Acta 304, 304–308.
104 Taketani, S., Tanaka-Yoshioka, A., Masaki, R., Tashiro, Y. and Tokunaga, R. (1986) Biochim. Biophys. Acta 883, 277–283.
105 Dailey, H.A. (1987) Ann. N.Y. Acad. Sci., 514, 81–86.
106 Smith, A.G. and De Matteis, F. (1980) Clin. Haematol., 9, 399–425.
107 De Matteis. F., Gibbs, A.H. and Holley, A.E. (1987) Ann. N.Y. Acad. Sci., 514, 30–40.
108 Ortiz de Montellano, P.R., Kunze, K.L., Cole, S.P. and Marks, G.S. (1980) Biochem. Biophys. Res. Comm., 97, 1436–1442.
109 De Matteis, F., Jackson, A.H., Gibbs, A.H., Rao, K.R., Atton, J., Weerasinghe, S. and Hollands, C. (1982) FEBS Lett., 142, 44–48.
110 De Matteis, F., Gibbs, A.H. and Harvey, C. (1985) Biochem. J. 226, 537–544.
111 Dailey, H.A. (1984) J. Biol. Chem., 259, 2711–2715.
112 Dailey, H.A. and Fleming, J.E. (1986) J. Biol. Chem., 261, 7902–7905.
113 Nakanashi, Y., Taketani, S., Sameshuray, Y. and Takunaga, R. (1990) Biochim. Biophys. Acta 1037, 321–327
114. Taketani, S., Nakanashi, Y., Osumi, T. and Takunaga, R. (1990) J. Biol. Chem., 265, 19377–19380.
115 Labbe-Bois, R. (1990) J. Biol. Chem., 265, 7278–7283.

P.M. Jordan (Ed.) *Biosynthesis of Tetrapyrroles*

CHAPTER 3

The biosynthesis of vitamin B_{12}

A.I. SCOTT and P.J. SANTANDER

Department of Chemistry, Texas A & M University, College Station, Texas 77843-3255, U.S.A.

1. Introduction

Other chapters in this volume deal with the iron (heme) and magnesium (chlorophyll) pathways. In this chapter we will discuss the cobalt branch which leads to the cobalamins, a specific group of corrinoids in which the lower axial ligand attached to cobalt is 5,6-dimethylbenzimidazole (DMBI). The coenzyme forms, adenosyl cobalamin (1) and methyl cobalamin (2), in which the 6th upper axial ligands are deoxy-

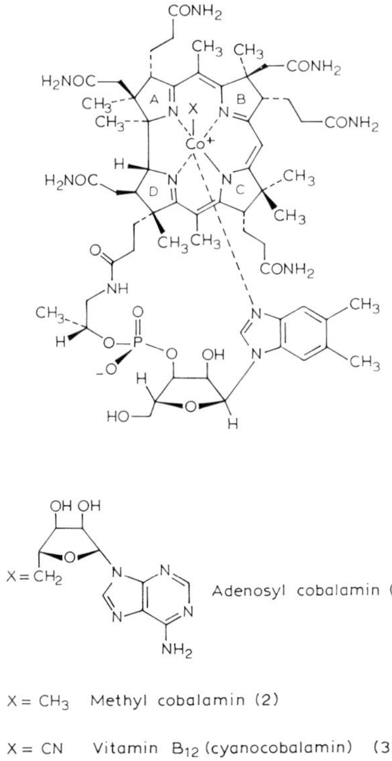

Adenosyl cobalamin (1)

X = CH_3 Methyl cobalamin (2)

X = CN Vitamin B_{12} (cyanocobalamin) (3)

Fig. 1

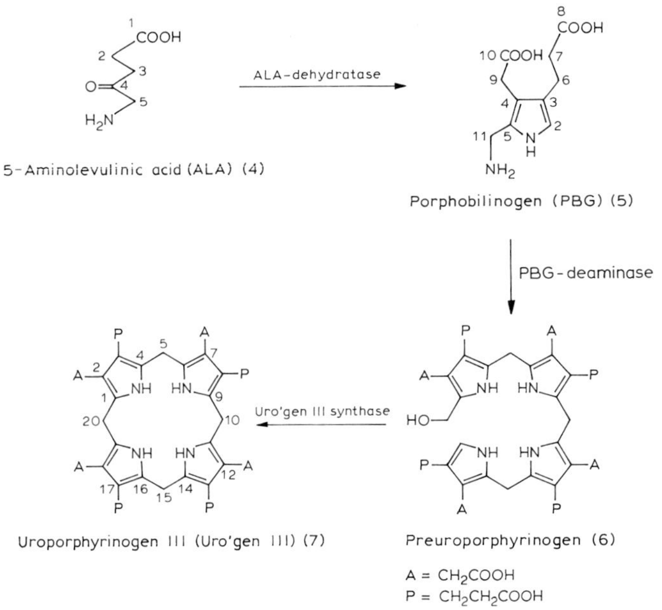

Fig. 2

adenosine and methyl respectively, (Fig. 1), are the biologically active forms of vitamin B_{12} (cyanocobalamin) (3). These coenzymes mediate an important and interesting series of rearrangements and group transfers which depend on initial homolysis of the cobalt carbon bond (Co(III) → Co(II)). Detailed discussion of these topics is outside the scope of this chapter but several excellent reviews of the field are available in the monograph 'B_{12}' published in 1982 [1].

The biosynthesis of B_{12} [2–8] involves approximately twenty enzymes and the phases can be divided into three main sequences as follows:

1. Synthesis of uro'gen III (7) from 5-aminolevulinic acid (ALA;4) and porphobilinogen (PGB;5) via the preuroporphyrinogen, hydroxymethylbilane, (6) (Fig. 2).
2. C-methylation of uro'gen III (7) by a series of distinct methyl transferases utilizing S-adenosyl methionine (SAM), decarboxylation, ring contraction and insertion of cobalt to form cobyrinic acid (8) (the simplest member of the natural corrins) (Fig. 3).
3. Conversion of cobyrinic acid (8) via cobyric acid (9) and cobinamide (10) to cobalamin (11) and then to the coenzyme forms of B_{12} (Fig. 4).

The biosynthetic pathway follows those of heme and chlorophyll as far as uro'gen III at which stage the three main routes diverge in one direction to oxidative/decarboxylative pathways leading to heme and chlorophyll, and in the other to a series of C-methylations leading from uro'gen III to sirohydrochlorin (and on to cobalamin) in which the 'cobalt' route maintains the same oxidation level throughout. A further (iron) branch connects sirohydrochlorin to siroheme. The biosynthesis of uro'gen III

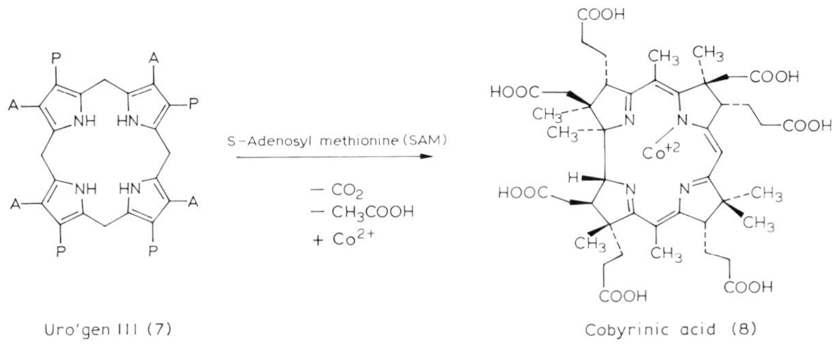

Fig. 3

is discussed in chapter 1 and a knowledge of the mechanisms involved in the formation of this pivotal intermediate is a prerequisite for understanding the ultimate fate of each carbon atom in B_{12} following administration of the labeled building blocks (ALA,PBG) as they are transferred to the later intermediates and finally into the corrinoid template.

From the outset of the work on vitamin B_{12} biosynthesis, ^{13}C NMR spectroscopy has played a dominant role, first in defining the carbon balance between the precursors 5-aminolevulinic acid and methionine and, more recently, in the elucidation of the intermediates which forge the link between the porphyrins and corrins.

The discovery that corrins, like heme and the chlorophylls, are formed in Nature from 5-aminolevulinic acid (ALA,4) and, by implication, porphobilinogen (PBG,5) was made over thirty years ago [9,10]. With the knowledge that the corrin template is constituted from the building blocks of ALA, methionine [11,12] and cobalt ion, the experiments necessary for the observation of regiospecifically enriched carbons in the CMR spectra of corrins derived from *Propionibacterium shermanii* were initiated

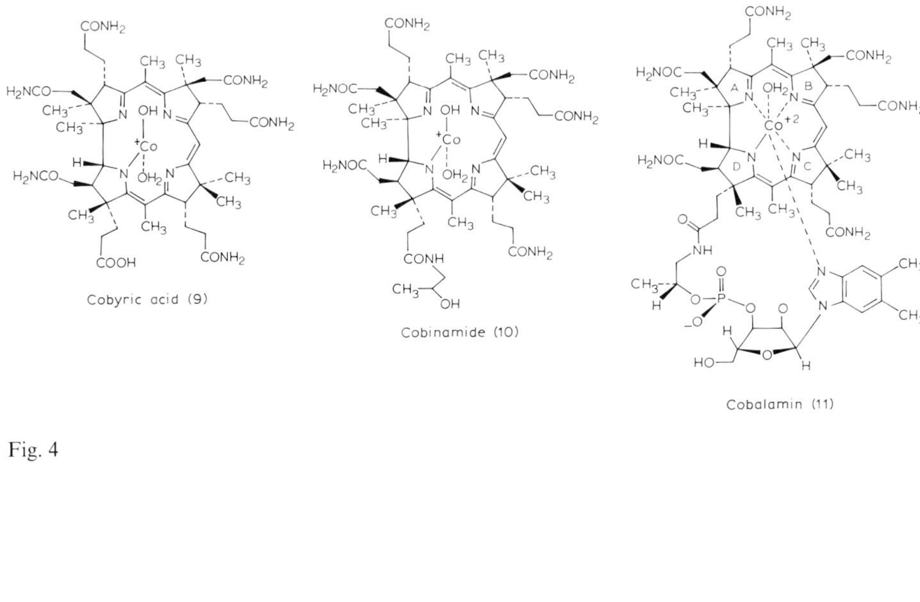

Fig. 4

and laid the basis for determining the carbon balance. These were followed by studies on the sequence leading from PBG via uro'gen III to cobyrinic acid and thence to the cobalamins. Most of our discussion will focus on the biosynthetic steps from uro'gen III to cobyrinic acid.

2. The carbon balance

Administration of [2-^{13}C]-ALA (13●) to *P. shermanii* (ATCC 9614) afforded a sample of cyanocobalamin (16a●) in which eight high-field signals in the CH_2 and CH_3 region were enriched [13,14] in the proton noise-decoupled ^{13}C FT-NMR spectrum. Assignments of the eight ^{13}C resonances were made to the seven CH_2CONH_2 methylenes and one of the *gem*-dimethyl groups of ring C in full accord with earlier ^{14}C studies (see Fig. 5). A sample of B_{12} (16a■) enriched by feeding [5-^{13}C]-ALA (13■) provided the surprising result that, of the eight anticipated enriched carbons, only seven signals appeared in the low field region associated with sp^2 (C=C and C=N) functions. The distribution of label is illustrated in Fig. 5. Such an array is in harmony with current ideas on the mechanism of type III uro'gen formation (vide infra) and this result [13,14] was simultaneously and independently confirmed in Shemin's laboratory [15] and the following year at Cambridge [16]. However, there was no ^{13}C-enhanced signal above 95 ppm downfield from HMDS showing that no enrichment of the C-1 methyl occurred. This indicates that one of the original $^{13}CH_2NH_2$ termini of ALA (and hence of PBG and the corresponding C-20 methylene group of uro'gen III) has been extruded in the formation of the vitamin. The origin of the 'missing' C-1 methyl group turned out to be methionine[13]. Thus the ^{13}C FT NMR spectrum of dicyanocobalamin obtained by feeding [$^{13}CH_3$]-methionine (12*) revealed seven signals highly enriched above natural abundance (C* in 16a Fig. 5) between 15 and 23 ppm.

3. Stereochemistry of methyl group insertion in corrinoid biosynthesis

Before developing further mechanistic proposals for corrin biosynthesis, which appears to be controlled by both steric and electronic consequences of methyl group insertion via S-adenosylmethionine (SAM), resolution of the problem of the stereochemistry of methylation at C-12 in ring C became necessary. Although it had been rigorously demonstrated that one of the methyl groups at C-12 is derived from methionine (12) and the other from C-2 of ALA (13●), the stereospecificity of this process had not been established. The following experiments provided a solution to this problem and demonstrate the advantage of using ^{13}C-enriched chemical shifts for the determination of carbon isotope chirality.

Labeled specimens of dicyanocobinamide (16b*) and neocobinamide (16c*) were obtained, where **one** of the C-12 methyl groups was specifically enriched, by feeding

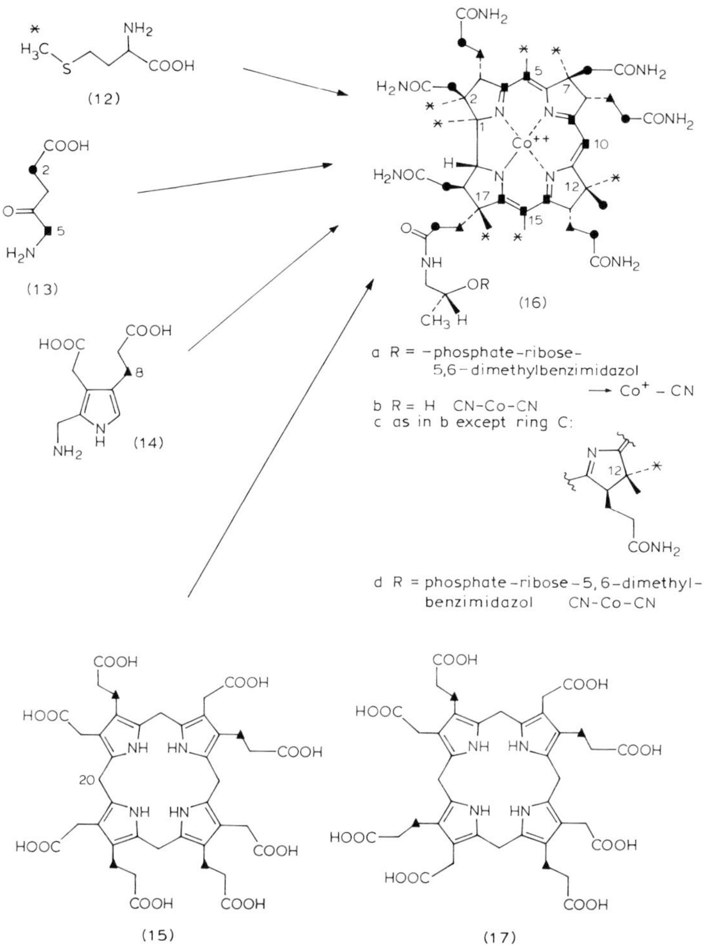

Fig. 5

[^{13}CH$_3$]-methionine (12*) to *P. shermanii* cells followed by hydrolysis (CF$_3$CO$_2$H) and separation. If the methionine-derived methyl at C-12 is α-oriented in the *neo* series, it will bear an anti-periplanar relationship to the propionamide side chain and the concomitant removal of the *gamma* effect should result in a down field shift of the enriched methyl resonance signal. This was clearly shown by the downfield shift of approximately 12 ppm in the ^{13}C FT-NMR spectrum for one of the methyl resonance lines in going from cobinamide to neocobinamide [17].

These results established that the [^{13}CH$_3$]-methionine methyl (*) is inserted into the corrin nucleus at C-12 from the α-face and that the absolute configuration at C-12 is (R) when isotopic chirality is introduced in this way to the *gem* dimethyl grouping. Furthermore, the ^{13}C results rationalize the apparent anomaly (observed previously) that the β-methyl group[13] of the *gem*-dimethyl grouping at C-12, derived from C-2 of ALA (13●), resonates at substantially lower field (31.6 ppm) than the methyl region of cobalamin tentatively assigned by Doddrell and Allerhand [18]. It should be noted that all of the remaining methyl groups at sp^3 carbons appear at higher field (15–23 ppm), i.e. within the 'normal' methyl region proposed by Doddrell and Allerhand, because of *gamma* interactions. An independent, different proof of the absolute stereochemistry of the methylation at C-12 was obtained at Cambridge [19]. At this stage of the investigation (1972–1973) the technique of cell-free corrin biosynthesis with *P. shermanii* was developed. A supernatant fraction (100,000 g) with appropriate additives [20] is capable of transforming ALA, methionine, PBG and, as described below, uro'gen III to cobyrinic acid (8). A similar system from *Clostridium tetanomorphum* which makes cobyrinic acid, but not heme, was developed simultaneously by Müller [21].

4. Concerning the fate of the methyl group protons

Work at Yale and at Cambridge indicated that virtually intact methyl transfer from S-adenosylmethionine takes place and in developing a theory for the mechanism of A → D ring fusion, the fate of the original protons of methionine assumes an important role with respect to the C-1 methyl group. Therefore, a sample of $^{13}C^2H_3$-methionine (90% in ^{13}C; 98% in ^{2}H) was administered to resting whole cells of *P. shermanii* and the resultant purified cyanocobalamin (16a) examined by a technique developed in our laboratory to provide a maximum sensitivity for studying ^{2}H-^{1}H exchange phenomena [22].

Analysis of the carbon-coupled ^{1}H FT spectrum of the enriched sample shows no exchange of the $^{13}CD_3$ groups C-2α, C-12α and C-17β [23], thus confirming an earlier conclusion [19]. These results impose strict requirements on the mechanism of corrin synthesis with respect to retention of the C-1 methyl protons and were confirmed by independent methods at Zurich [24] and Cambridge [25]. Thus, unless an exchanged proton from C-1 CH_3 is returned regio- and presumably stereo-specifically, structures involving an *exo*-methylene at C-1 can be eliminated as part of the biosynthetic assembly of corrin from *seco*-corrin as envisaged in the schemes developed at the end of this discussion.

The stereochemistry of the methyl transfers has been studied with chiral (CHDT) methionine. The isolated corrins were degraded and the fragments oxidized (Kuhn-Roth) to acetic acid which was analyzed enzymatically. At all centers so far examined (C-5, C-7, C-12, C-15) stereochemical inversion has taken place [26].

5. Uro'gen III is a precursor of vitamin B_{12}

With the establishment of proper pH control and feeding conditions, it was possible to achieve reproducible conversion of [8-^{13}C]-PBG (14▲) and of chemically synthe-

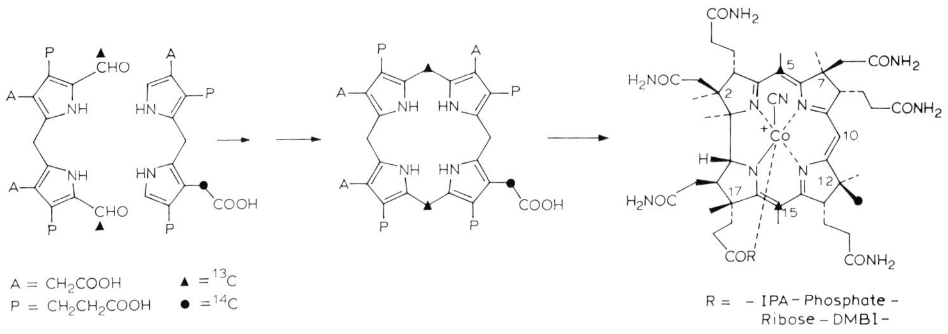

Fig. 6

sized type I-IV uro'gen mixture labeled in the propionate side chains to specimens of dicyanocobalamin (16d▲) whose ^{13}C-NMR enrichment occurred with identical chemical shift (Fig. 1). Since uro'gen I (17) had been shown to be quite ineffective in labeling B_{12}, the conclusion became inescapable – urogen III must be the precursor of vitamin B_{12} [27,28]. This finding restored considerable confidence and led to confirmation by a further set of ^{13}C labels inserted into uro'gen III, this time by total regiospecific synthesis [29] at C-5 and C-15, and the appearance of enrichments at C-5, C-15 (108.4 and 105.2 ppm) in the cyanocobalamin derived from whole-cell feeding of this precursor (Fig. 6), a result later confirmed by others using ^{14}C uro'gen III [30,31].

The clear demonstration of the porphyrinoid-corrinoid connection was a particularly important result in view of several previous failures to demonstrate the intermediacy of uro'gen III and set the stage for the development of cell-free methodology for the isolation and identification of the post-uro'gen III intermediates and for the preliminary assignments of the ^{13}C NMR spectrum [28]. Since most of the recent work on the biosynthesis of corrins depends on the correct NMR assignment for the enriched centers following administration of labeled precursors, it is important to have both proton and ^{13}C chemical shifts available for the corrin series. Complete assignments (^{1}H, ^{13}C) have been made for cobester and for cyanocobalamin which are the main biosynthetic corrins derived from cell-free extracts and whole cell preparations respectively [32–37]. The complete spectrum of coenzyme B_{12} has recently been assigned by 2D-NMR methods [38].

6. Characterization and intermediacy of the isobacteriochlorins of P. shermanii

As soon as uro'gen III was defined as the precursor of the corrin nucleus (Fig. 6), the search for partially methylated intermediates on the way to B_{12} began and it was

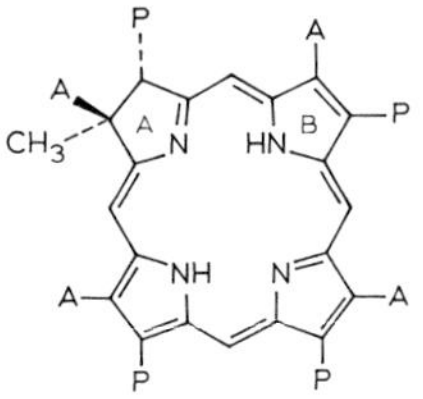

Factor I (18)
A = CH₂COOH
P = CH₂CH₂COOH

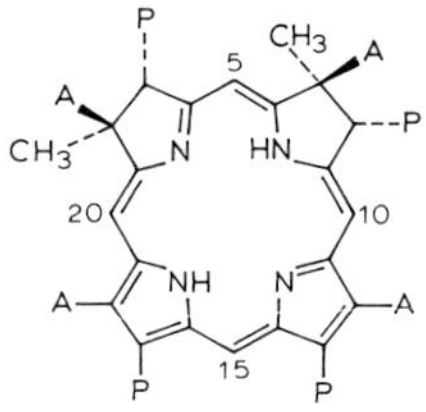

Factor II (19)
(SIROHYDROCHLORIN)

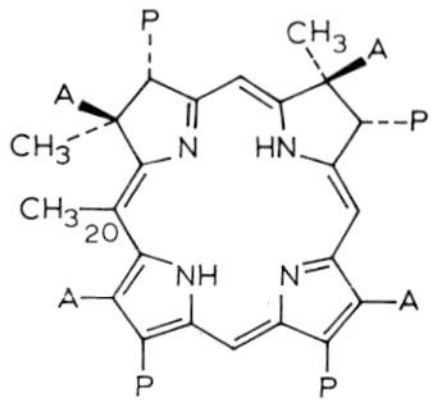

Factor III (20)

Fig. 7

noted [39] as early as 1973 that a proposed structure for sirohydrochlorin, an iron-free prosthetic group of siroheme from *E. coli*, could be modified to accommodate its possible role as a biosynthetic intermediate. At first sight this would have seemed an extraordinary coincidence but as the study of sirohydrochlorin developed in parallel with the isolation of orange-fluorescent substances from cobalt- deficient, ALA-supplemented incubations of *P. shermanii*, complete identity of the *dimethyl* isobacteriochlorin (Factor II) from the six-electron sulfite and nitrite reductases and Factor II from the B_{12}-producing organism was established. Thus, inspection of the UV, CD, mass and PMR spectral data for the methyl ester of the *P. shermanii* metabolite and of sirohydrochlorin from *E. coli* sulfite reductase left no doubt that sirohydrochlorin and Factor II are identical in every respect [40,41]. The molecular constitution was also confirmed by high resolution mass determination of the molecular ion [40,41].

The complete stereostructure (19) for sirohydrochlorin (Fig. 7) was deduced by biosynthetic labeling experiments as described below for the similar case of factor III (20) and the role of sirohydrochlorin as a precursor for cobyrinic acid (8) in the cell-free system established [40,41]. As the search continued, two closely related intermediates, Factor I and Factor III, were isolated (Fig. 7). Factor I (18) is a *mono* methyl chlorin derivative corresponding to C-methylation in ring A of uro'gen III [42,43].

Factor III, while undoubtedly a trimethylisobacteriochlorin to which the structure (21) had been assigned on PMR evidence, was found to be another isomer on the basis of careful NMR studies (Fig. 8) which showed that the 'extra' methyl group is added to sirohydrochlorin (Factor II) at C-20 rather than at C-5 leading to the revised structure (20) for Factor III, i.e. 20-methylsirohydrochlorin, which suffers loss of both C-20 and its attached methyl group as acetic acid during biotransformation to cobyrinic acid [44–48]. The techniques used in the structure determination of Factor III are given in detail since they are typical of the strategies employed to solve the structures of the intermediates of the B_{12} pathway.

Factor III was isolated from aminolevulinic acid and methionine-supplemented cobalt-free incubations of *P. shermanii* (ATCC 9614) and from a B_{12}-deficient mutant of this organism. High resolution FD mass spectrometry of the octamethyl ester (22) established the formula $C_{51}H_{64}N_4O_{16}$ (988.4343) and analysis of the PMR spectrum (300 MHz) revealed only three signals at 6.43, 7.21, and 8.33 ppm in contrast to the four signals in this region in the spectrum of the octamethylester (23) which have been assigned to the four *meso* protons at C-5, C-10/C-20, and C-15 (6.78, 7.36/7.46, and 8.54, respectively). Factor III is therefore 10- or 20-methylsirohydrochlorin. In order to decide between these alternatives a specimen of factor III (400 μg) was prepared from a suspended-cell incubation from *P. shermanii* containing [$^{13}CH_3$]-methionine and [5-^{13}C]-ALA. When this ^{13}C-enriched species (as the octamethyl ester) (24) was examined by microprobe CMR spectroscopy (Fig. 9), it became possible to deduce the complete structure (22) [47,48]. First, the downfield position of the C-15 *meso*-carbon triplet at δ 108.98 (J=72 Hz) confirms that rings A and B are methylated, and since the *meso*-carbon signals at 89.5 and 95.4 each showed ^{13}C-^{13}C coupling to an enriched neighbour (J=70 Hz) these are assigned to C-5 and C-10, respectively, by analogy with the corresponding resonances in sirohydrochlorin derived by biochemical enrichment with [5-^{13}C]-ALA. Thus, the remaining *meso*-carbon resonance at δ 104.8 which consists of a doublet (J=44.5 Hz) must correspond to C-20, the additional fine structure being due to long-range coupling with C-4 and C-16. That the ^{13}C-^{13}C coupling constant of 44.5 Hz for C-20 is due to substitution by a methionine-derived methyl group is confirmed by inspection of the *methyl* region of the CMR spectrum which displays three enriched species consisting of singlets at 20.17 and 19.62 and a doublet at 18.79 (J=44.5 Hz). It can be seen that, due to different efficiencies of incorporation of ^{13}C-SAM and of [5-^{13}C]-ALA, the enrichments in the methyl groups and in the ALA-derived sp^2 carbons are not identical, the satellite intensities reflecting a greater enrichment in C-20 than in its pendant methyl group. Thus Factor III is (20), i.e. 20-methylsirohydrochlorin, rather than the C-5 or C-10 methylated isobacteriochlorin. The absolute stereochemistry of Factor III and its relationship to cobyrinic acid was obtained by converting doubly labelled ($^3H/^{14}C$) 20-methylsirohydrochlorin to cobyrinic acid with loss of the C_2 unit consisting of C-20 (derived from C-5 of ALA) and the attached methyl group originating from methio-

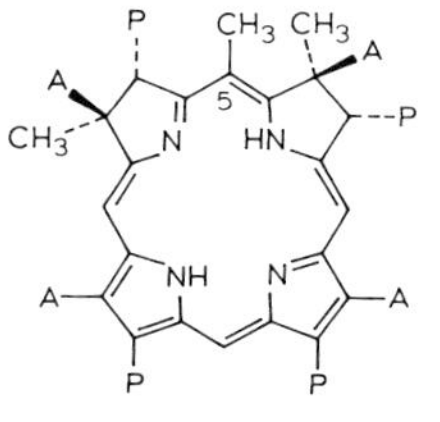

(21) A = CH_2CO_2H
P = $CH_2CH_2CO_2H$

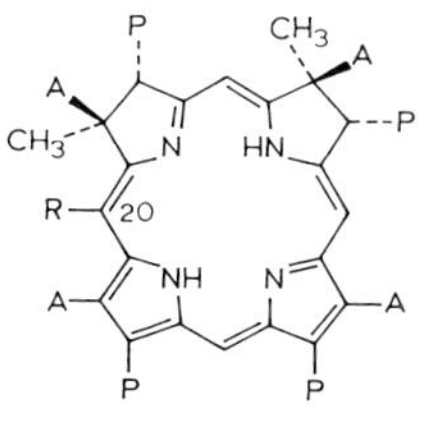

(22) A = $CH_2CO_2CH_3$ R = CH_3
P = $CH_2CH_2CO_2CH_3$
(23)   R = H

Fig. 8

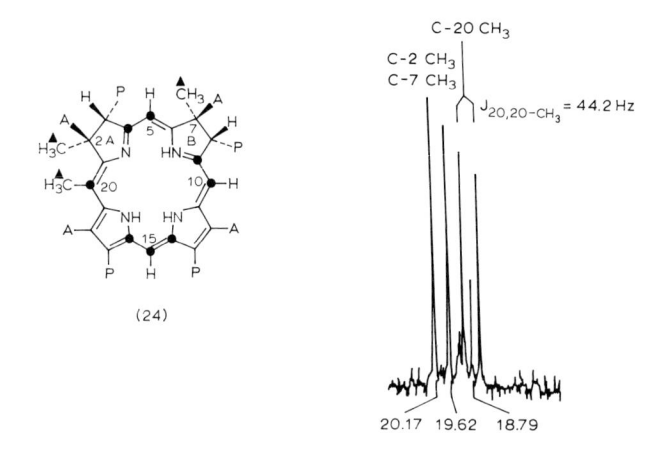

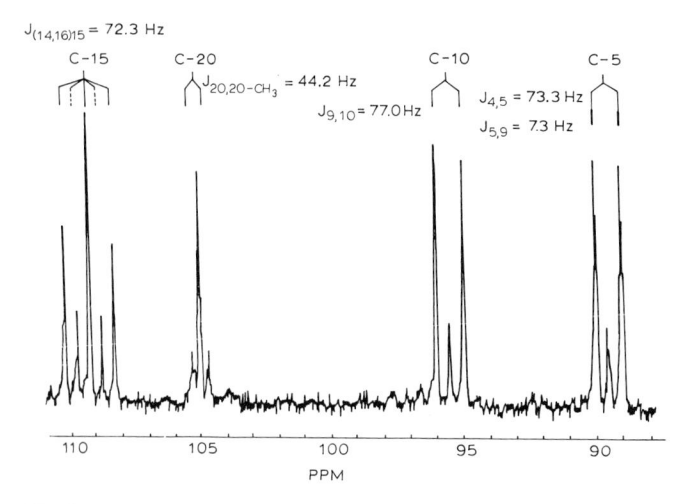

Fig. 9

nine in the form of acetic acid [45,46,48]. Since the chirality at C-2 and C-7 is not changed in this process the absolute stereochemistry at these positions corresponds in Factor III and cobester.

The bioconversion of uro'gen III to cobyrinic acid can now be summarized in Fig. 10 where it is assumed that the oxidation level of the various intermediates is maintained at the same level as that of uro'gen III. The lability of the reduced isobacteriochlorins, the fact that they are normally isolated in the oxidized form (Fig. 7), and the requirement for chemical reduction of Factor I (but not Factors II or III) before incorporation into corrin, lends credence to this idea but rigorous proof for this suggestion must await non-invasive studies using in vivo techniques. From this point in our discussion we will assume that incorporation of both Factors II and III involves the *dihydro* versions whose presence and intermediacy have been detected [49] in vitro

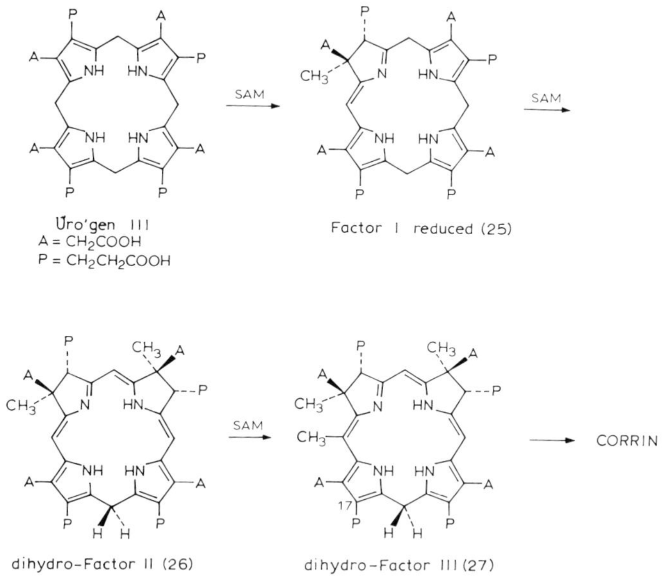

Fig. 10

for the case of Factor II (Fig. 10). In order to distinguish between the oxidation levels of the isolated chlorins and isobacteriochlorins and to avoid possible confusion in using the term 'Factor' (which has other connotations in B_{12} biochemistry, e.g. intrinsic Factor/Factor III) the term precorrin-n has been suggested [50] for the actual structures of the biosynthesized intermediates after uro'gen III where n denotes the number of SAM derived methyl groups. Thus, although the names Factors I-III for the isolated species will probably survive (for historical reasons) the true intermediates have the same oxidation as uro'gen III viz; tetrahydro Factor I (precorrin-1) (25), dihydro Factor II (precorrin-2) (26) and dihydro Factor III (precorrin-3) (27). Subsequent intermediates can be named accordingly. From this point in the chapter we have adopted this nomenclature (even for postulated structures) in order to preserve consistency.

In spite of intensive search spanning the last 15 years, no new intermediates containing four or more methyl groups (up to a possible of eight) have been isolated, but the proven biochemical conversion of Factor III to cobyrinic acid (8) must involve the following events (Fig. 11), not necessarily in the order indicated: (1) The successive addition of five methyls derived from S-adenosyl methionine (SAM) to reduced Factor III (precorrin-3) (30). (2) The contraction of the permethylated (seven or eight methyls) macrocycle to corrin. (3) The extrusion of C-20 and its attached methyl group leading to the isolation of acetic acid [44–48]. (4) Decarboxylation of the acetic acid side chain in ring C (C-12). (5) Insertion of Co^{+++} after adjustment of oxidation level from Co^{++}. In order to justify the continuation of the search for such inter-

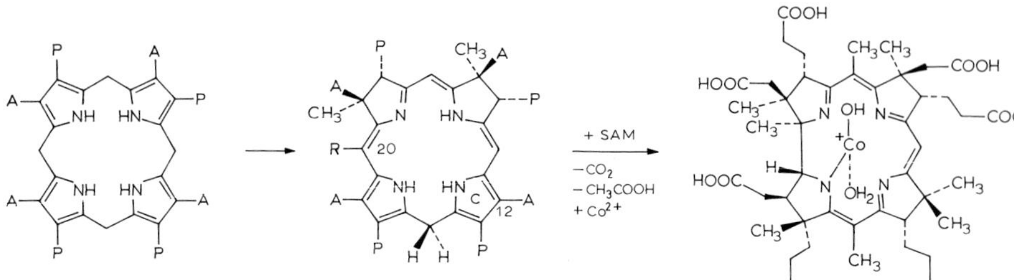

Fig. 11

Uro'gen III (7)
A = CH$_2$COOH
P = CH$_2$CH$_2$COOH

R = H precorrin 2 (26)
R = CH$_3$ precorrin 3 (27)

Cobyrinic acid (8)

mediates whose inherent lability is predictable, ^{13}C pulse labelling methods were applied to the cell-free system which converts uro'gen III (7) and precorrin-2 (26) to cobyrinic acid (8), a technique used previously in biochemistry to detect the flux of radio-labels through the intermediates of a biosynthetic pathway.

7. The methylation sequence: Pulse experiments

It was argued that the full methylation cascade could be differentiated in time, provided that *enzyme-free* intermediates accumulated in sufficient pool sizes to affect the resultant methyl signals in the CMR spectrum of the target molecule, cobyrinic acid (8), when the cell-free system is challenged with a pulse of ^{12}CH$_3$-SAM followed by a second pulse of ^{13}CH$_3$-SAM (or vice versa) at carefully chosen intervals in the total incubation time (11 hours). By this approach it should be possible to 'read' the biochemical history of the methylation sequences as reflected in the dilution (or enhancement in the reverse experiment) of ^{13}CH$_3$ label in the seven methionine-derived methyl groups of cobyrinic acid after conversion to cobester (28), whose ^{13}C-NMR spectrum has been assigned [34,35].

The validity of the method was tested in a preliminary experiment using a two-phase system. ^{13}CH$_3$-methionine (90% ^{13}C; 30 mg) was added to a whole-cell (100 g) suspension of *P. shermanii* in phosphate buffer containing 5-aminolevulinic acid (20 mg), *in the absence of Co^{++}*. Under these conditions the cells produce uro'gen III (7), precorrin-2 (26) and precorrin-3 (22), but not corrin. Cell disruption and incubation of the extract with added Co^{++} and pulses of ^{12}CH$_3$-SAM at varied time intervals was monitored for differential methylation by isolation of cobyrinic acid, conversion to cobester (28), extensive purification, and finally CMR analysis of the enriched samples (normally 50–150 μg). Optimization of these conditions led to the spectrum shown in Fig. 12 which reveals a clear distinction between the peak heights of the seven methyl groups of cobester when compared with a spectrum obtained by adding ^{13}CH$_3$-SAM at the outset. The high relative intensities of the C-2 and C-7 signals bear testimony to the initial formation in whole cells of the precorrin-2 (26) in the absence of Co^{++}. The differential dilution of methyl intensity in cobester not only suggests that free intermediates have accumulated, but that even in this qualitative experiment, the sequence of methylation is revealed as C-2=C-7 > C-17 > C-12 α > C-1 > C-5=C-15 since the timing of the addition of $^{12}CH_3$-SAM dilutes the pool sizes of each [$^{13}CH_3$]-enriched methylated intermediate in the order in which it is formed. Thus C-17 is the site of the fourth, C-12 the fifth, and C-1 the sixth methylation [51], confirmed by a reverse pulse experiment using Factor II as substrate.

Methylase activity has been further resolved temporally by pulsing with [$^{13}CH_3$]-SAM 4 hours after beginning the incubation with precorrin-2. The isolated cobester now shows virtually no enrichment for the C-17 methyl group (Fig. 13), this indicates that an intermediate bearing a C-17 $^{12}CH_3$ group has accumulated to sufficient pool size to dilute the observed enrichment in cobester almost to natural abundance in ^{13}C. By leaving out $^{13}CH_3$-SAM from the incubation almost until the end (45 minutes) a series of experiments [52] has confirmed the late addition of the C-5 and C-15 methyl groups in a dramatic demonstration of the technique whereby all the signals, except those for C-5 and C-15, have been removed, as portrayed in Fig. 14. None of these experiments, of course, defines the structures of the intermediates since the latter are only observed indirectly as a function of the methyl signal intensities in cobester. The sequence C-17 > C-12α > C-1 has been found in *Clostridium tetanomorphum* [50] and further differentiation between C-5 and C-15 insertion suggested the order C-15 > C-5. Thus, while a series of statistically analysed spectra of cobester from cell-free extracts of *C. tetanomorphum* indicated that the last two methylations take place in

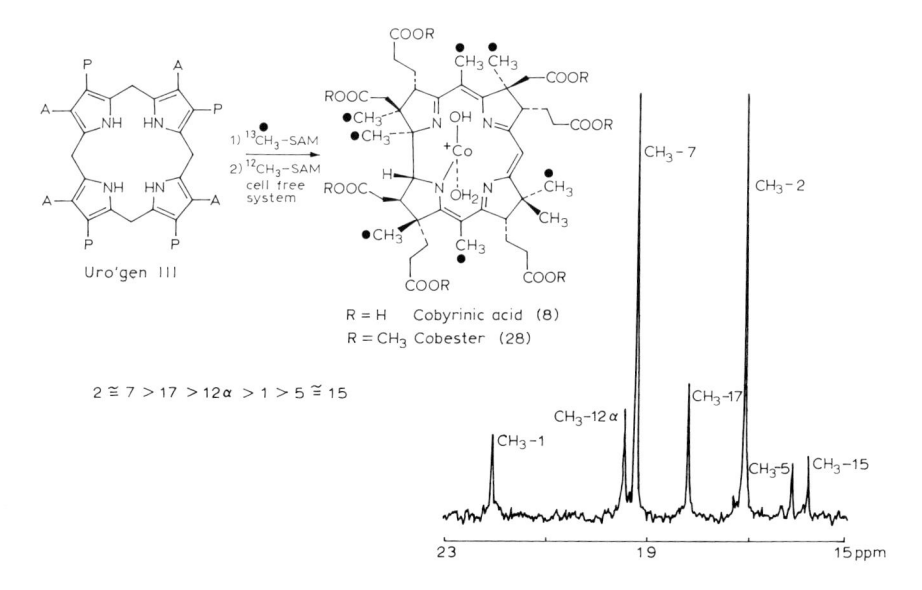

Fig. 12

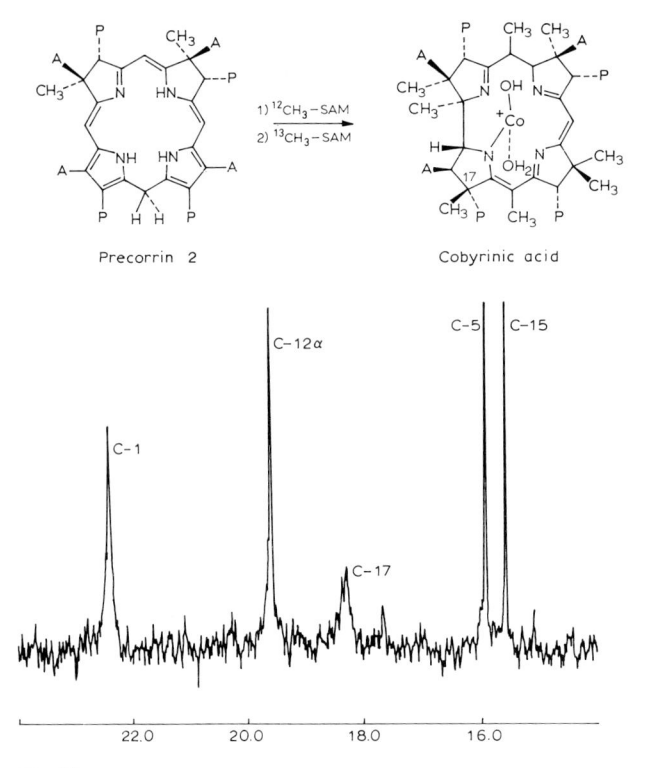

Fig. 13

the order C-15 > C-5, in *P. shermanii* C-methylation occurs at C-5 before C-15 as shown by the following experiments.

A sample of cobyrinic acid labelled equally in all seven methyl groups was isolated from a cell-free homogenate from *P. shermanii* cells (which had previously been fed with $^{13}CH_3$-SAM) followed by esterification to give cobester (Fig. 15) whose NMR spectrum revealed (Fig. 16a) that the intensity of the ^{13}C signal for the methyl group at C-5 is 5% lower than that for the C-15 methyl group (due to small T_1 differences for these centres) in contrast to the edited ^1H spectrum (Fig. 16e) in which protons of the C-5 and C-15 methyls produce equal signal intensities. By studying over 50 temporal variants of the 'two-phase' sequence in which precorrin-2 is accumulated in whole cells containing an excess of SAM in the absence of Co^{++}, the following protocol gave consistent results. The cells were sonically disrupted and Co^{++} added immediately, followed by a pulse of $^{13}CH_3$-SAM (90 atom%) after 4 hrs. After a further 1.5 hr. cobyrinic acid was isolated as cobester. The ^{13}C NMR spectrum of this specimen obtained by the 'normal pulse' method (Fig. 16b) defines the complete methylation sequence, beginning from precorrin-2, as C-20 > C-17 > C-12α > C-1 > C-5 > C-15, with a differentiation of 25% (± 5%) in the relative signal intensities for the C_5 and C_{15} methyl groups. Methylation at C-20 of precorrin-2 to give precorrin-3 is not

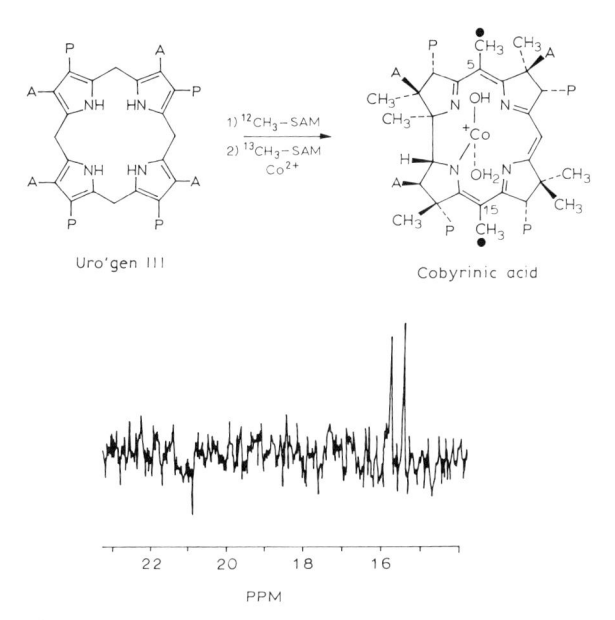

Fig. 14

recorded in the spectrum of cobester since C-20 is lost on the way to cobyrinic acid, together with the attached methyl group, in the form of acetic acid [45,46,48].

A 'reverse pulse' experiment ($^{13}CH_3$-SAM added at the outset, $^{12}CH_3$-SAM after 3 hr.) inverts the relative intensities of the C-5 and C-15 methyls (Fig. 16d) which are now virtually equal, although this technique involves kinetics which preclude perfect inversion of the 'normal pulse' profile (as in Fig. 16b). Independent confirmation of these results can be clearly seen in the edited ^1H spectra of cobester obtained both by the 'normal pulse' technique (Fig. 16f) and the 'reverse pulse' method (Fig. 16g) which reveal the identical order (C-5 > C-15).

It is now apparent that several discrete methyl transferases are involved in the biosynthesis of cobyrinic acid from uro'gen III, since enzyme-free intermediates must accumulate in order to dilute the ^{13}C label. A rationale for these events is given in Fig. 17, which takes the following facts into account: (a) the methionine derived methyl group at C-20 of precorrin-3 does not migrate to C-1 and is expelled together with C-20 from a late intermediate (as yet unknown) in the form of acetic acid, (b) neither 5,15-norcorrinoids nor descobalto-cobyrinic acid are biochemical precursors of cobyrinic acid (c) regiospecific loss of ^{18}O from [1-^{13}C, 1-$^{18}O_2$]-5-aminolaevulinic acid-derived cyanocobalamin from the ring A acetate occurs [53], in accord with the concept of lactone formation, as portrayed in Fig. 17, where precorrin-5 is methylated at C-20 followed by C-20 → C-1 migration and lactonization. If this mechanism is operative, the C-20 → C-1 migration must be stereospecific, since precorrin-3 labelled at

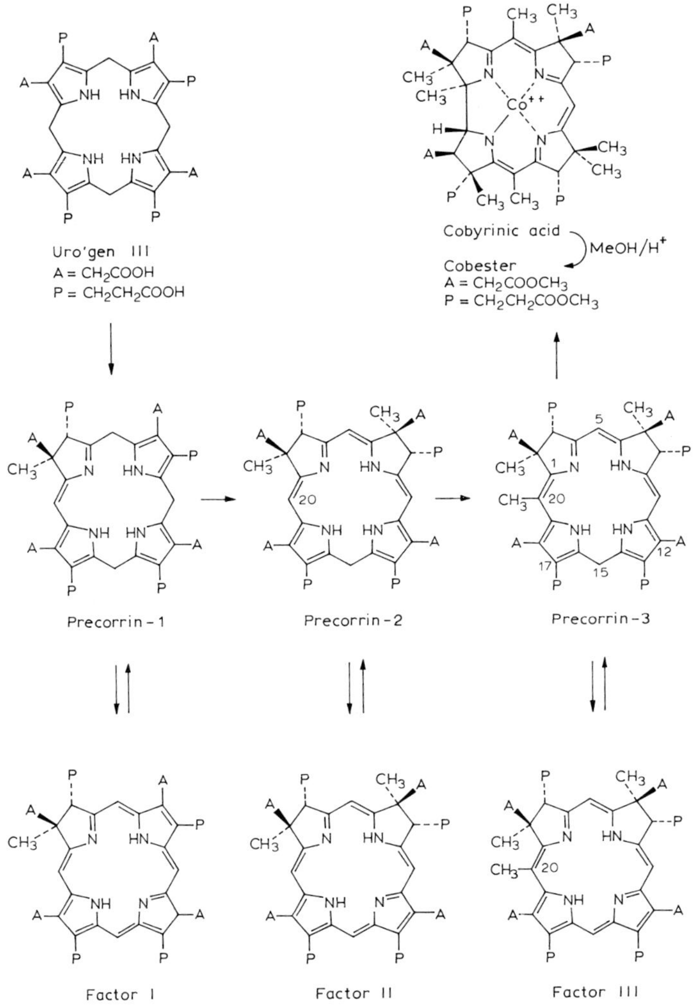

Fig. 15

C-20 with $^{13}CH_3$ is transformed to cobyrinic acid with complete loss of label, a less attractive alternative being direct methylation at C-1.

8. Timing of the decarboxylation step

It is also possible to define the point in the biosynthetic sequence where ring C-decarboxylation occurs by using synthetic [5,15-$^{14}C_2$]-12-decarboxylated uro'gen III (29) as a substrate for the non-specific methylases of *P. shermanii* to prepare the 12-methyl analogs (30) and (31) of factors II and III respectively. Reduction and incuba-

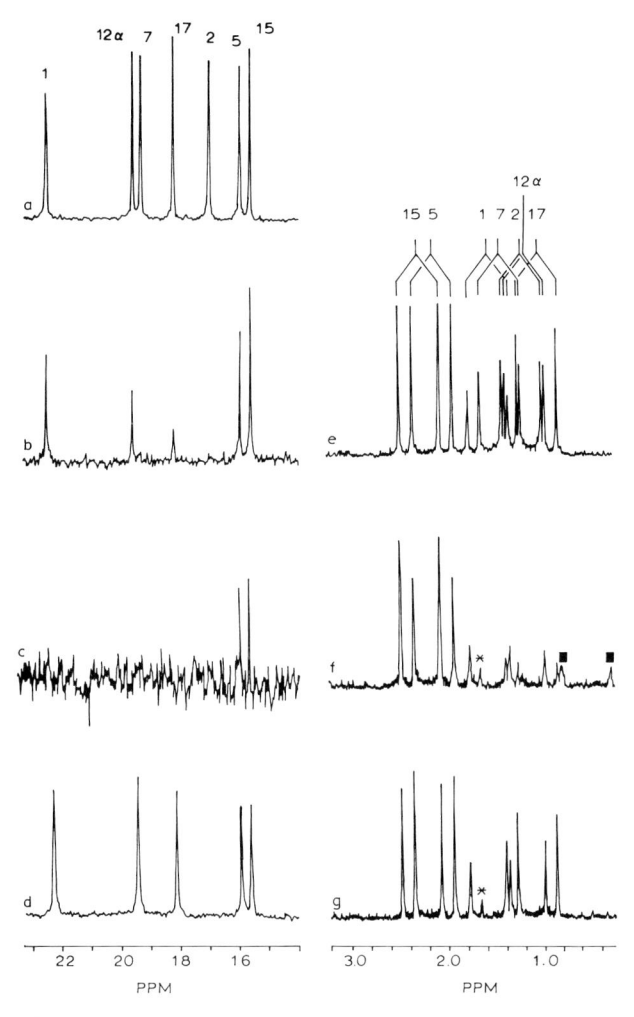

Fig. 16

tion of these possible intermediates with the corrin-synthesizing cell-free system in the presence of SAM afforded (after esterification) samples of cobester whose specific activities were compared in a control experiment with those of cobester derived from [^{14}C]-precorrin-2 and from a mixture of [^{3}H]-precorrin-3 and [^{14}C]-precorrin-2. It was found that the ring C-decarboxylated analogs are not substrates for the enzymes of corrin biosynthesis. Independent work with *P. denitrificans* using purified methylases was carried out [54] and again it was found that the analogs (29) and (30) were not intermediates in corrin biosynthesis. Thus it was concluded that (a) in normal biosynthesis, precorrin-3 is *not* the intermediate which is decarboxylated (b) decarboxylation occurs at some stage after the fourth methylation (at C-17) and by mechanis-

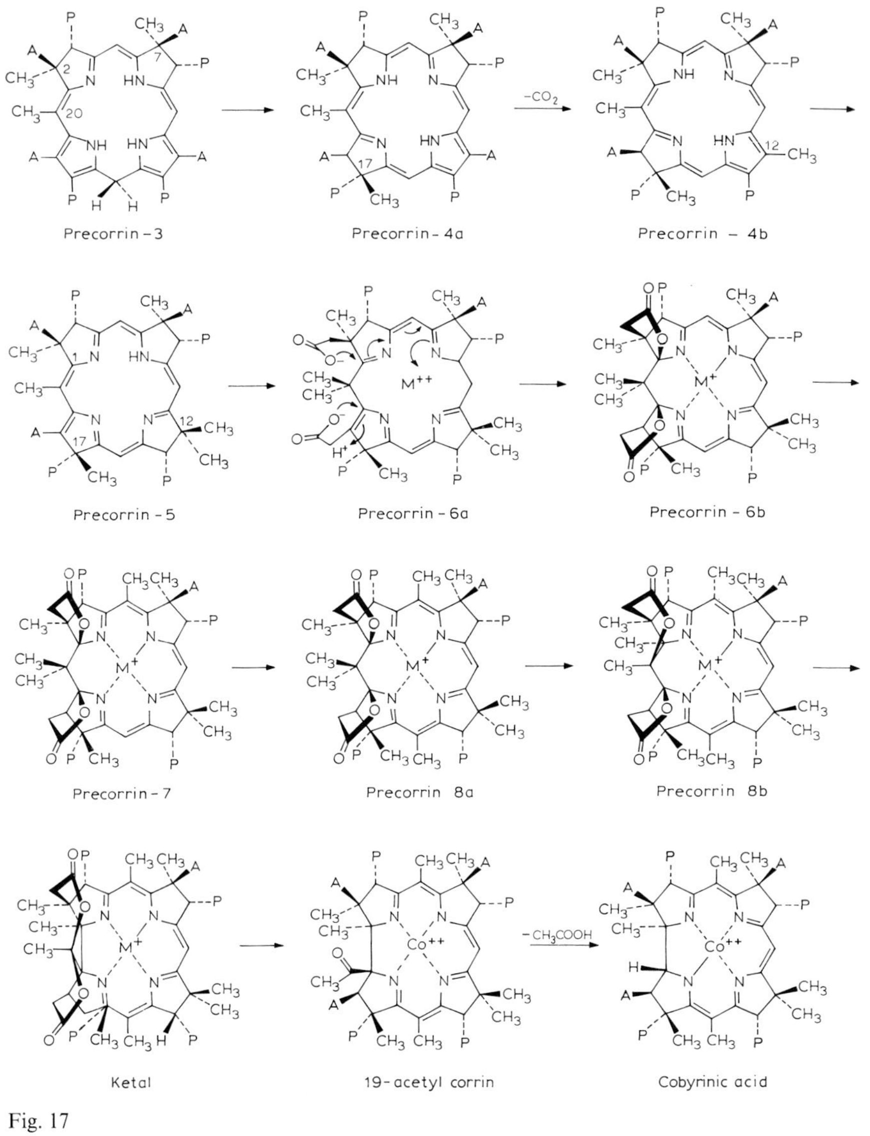

Fig. 17

tic analogy, before the fifth methylation at C-12. Hence, two isolable pyrrocorphin intermediates viz; precorrins 4a, 4b (Fig. 17) should intervene between precorrins-3 and -5 i.e. precorrin-3 is C-methylated at C-17 to give precorrin-4a followed by decarboxylation (→ 4b) and subsequent C-methylations at C-12α, C-1, C-5, C-15, in that order, a sequence which differs in the penultimate stages from that found in *C. tetanomorphum* [50], (C-15 > C-5).

The decarboxylation mechanism most plausibly involves a ring C exomethylene species. The stereospecificity (retention) of this step has been demonstrated by Kuhn-Roth oxidation of chirally-labelled C-12β methyl group of cobester following incorporation of ALA chiral (^{2}H,^{3}H) on the methylene group, which becomes incorporated into the C-12 acetate side chain [55]. Although factors I-III when adjusted to the requisite oxidation level each serve as excellent substrates for conversion to cobyrinic acid in cell-free systems from *P. shermanii* and *C. tetanomorphum*, direct evidence for their sequential interconversion was obtained only recently. In the presence of ^{13}CH$_3$-SAM a specimen of Factor II is converted to Factor III containing a single ^{13}C-enriched methyl at C-20 (δ 19.2 ppm). The demonstration of this conversion removes any doubt that separate pathways could exist for the biosynthesis of vitamin B$_{12}$ from precorrins-2 and 3 respectively [52].

We now return to the construction of a working hypothesis for corrin biosynthesis. We suggest that dihydrofactor III (Precorrin-3) (Fig. 17) is β-methylated at C-17 to precorrin-4a which could exist as a pyrrocorphin or as the isomeric corphin. Decarboxylation at the C-12 acetate side chain (→ precorrin-4b) is followed by methylation to the corphin, precorrin-5. At this stage the insertion of the eventual C-1 methyl (to give precorrin-6a) can be envisaged as an alkylation at the electron-rich C-20 to be followed later by stereospecific C-20 → C-1 migration. The formulations precorrins 6b, 7, 8a, 8b take into account the idea of lactone formation using rings A and D acetate side chains. The lack of enrichment of ^{13}C at the C-1 methyl following conversion of 20-^{13}CH$_3$ factor III to cobyrinic acid shows that no loss of stereospecificity could have occurred at C-20 during such a hypothetical process. The migration C-20 → C-1 could be acid or metal ion catalyzed and the resultant C-20 carbonium ion quenched with external hydroxide or by the internal equivalent from the carboxylate anion of the C-2 acetate in ring A precorrin 8a → 8b as suggested by the results of labelling with ^{18}O discussed above [53]. In any event the resultant dihydrocor-

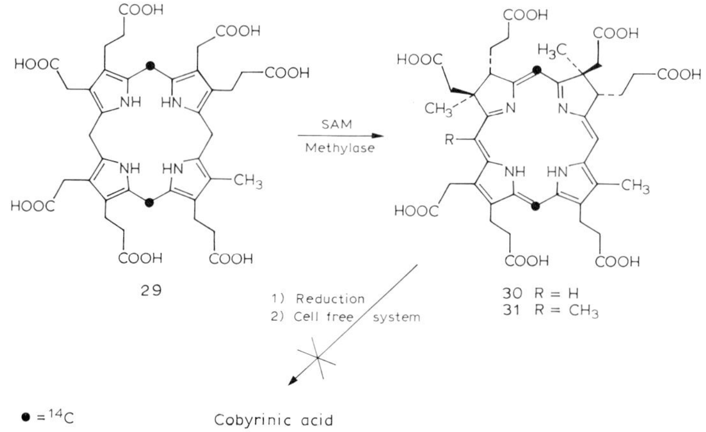

Fig. 18

phinol-bislactone precorrin-8b, is poised to undergo the biochemical counterpart of Eschenmoser ring contraction [56] to the 19-acetyl corrin. Before this happens we have suggested that the final methyl groups are added at C-5 (precorrin-7), then C-15 (precorrin-8a) to take account of the non-incorporation of the 5, 15-norcorrinoids. The resultant precorrin-8b (most probably with cobalt in place) then contracts to 19 acetyl corrin which by loss of acetic acid leads to cobyrinic acid. The notion that cobalt is inserted either very late in the sequence, i.e. after all the methyl groups have been inserted, or at the outset, i.e. into precorrin-2 is tentatively supported by preliminary 'cobalt pulse' experiments [57]. The valency change Co(II) → Co(III) during or after cobalt(II) insertion has so far received no explanation. A seperate pathway for cobalt-free corrin synthesis may exist (see p. 135).

9. The proton balance of vitamin B_{12} biosynthesis

With the knowledge that methyl transferases insert the SAM derived methyl groups intact [22] into the substrate precorrin-3 in the order C-17 > C-12 > C-1 > C-5 > C-15 [51,52] and that acetic acid is lost from C-20 and its original attached methyl group [47], probably at the end of the pathway, experiments have been carried out to trace the origin of the methine protons in vitamin B_{12}. A possibility existed that the proton at C-18 (together with that at C-19) is introduced via reduction of 18, 19 dehydrocobyrinic acid [58,59] although early experiments at Yale using NAD^{3}H and NADP^{3}H showed no incorporation of ^{3}H radioactivity at C-18 and C-19 [60]. On the other hand, adoption of Eschenmoser's in vitro model for the dihydrocorphinol → acetylcorrin ring contraction and deacetylation [56] as a construct for B_{12} biosynthesis would leave the oxidation level of the pathway unperturbed from uro'gen III to cobyrinic acid. In order to examine the stage at which deacetylation occurs kinetic and equilibrium probes for the protonations at C-18 and C-19 as well as at the other β-positions carrying protons, viz; C-3, C-8, and C-13 were developed. Thus the presumed precursors (Fig. 17) such as precorrins 4–7 retain the propensity for equilibrium exchange at all of these positions, as well as at C-10. However, on reaching the putative penultimate intermediate precorrin-8, C-19 is already quaternary and the proton at C-18 no longer labile. The loss of acetic acid from precorrin-8 is biochemically irreversible so that position 19 may be susceptible to a large kinetic isotope effect. Thus by allowing the production of B_{12} to take place in D_2O enriched medium, the early precursors should have reached equilibrium with reference to isotopic replacement at C-3, C-8, C-13, and C-18 and possibly C-10. When cyanocobalamin (16a) was isolated from a fermentation carried out in 50% D_2O in the presence of [4-^{13}C]-ALA the observation of large isotopic shifts on the ^{13}C-enriched positions could be used to assign the position of deuterium incorporation (Fig. 19). Thus, as expected, C-3, C-8 and C-13 showed large α-isotopic shifts. C-17 shows a β-shift (11.24 Hz; Fig. 19) in consonance with considerable enrichment at C-18. In contrast,

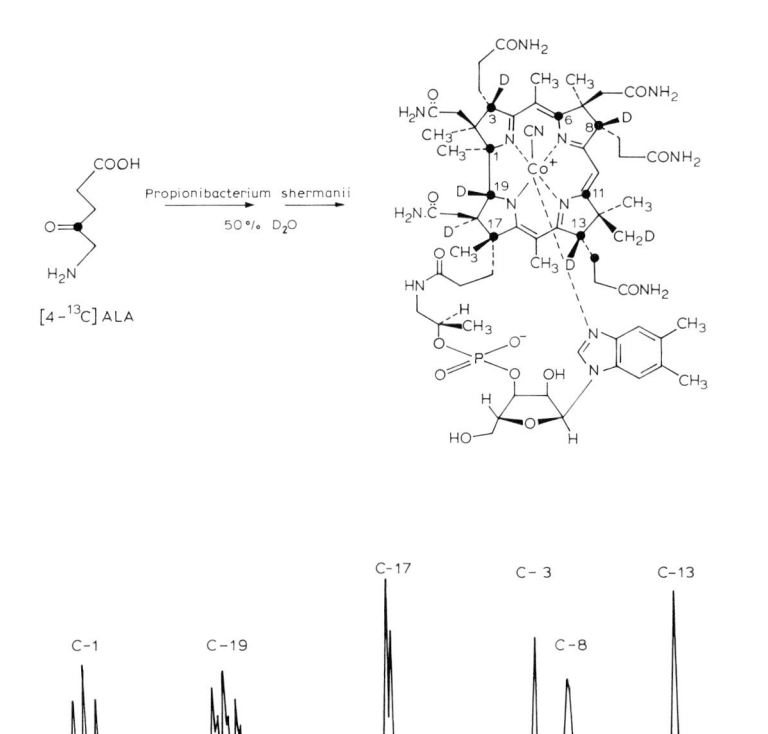

Fig. 19

the deuterium enrichment at C-19 although observable was on approximately 30% of the content at the other centres while C-10 was devoid of enrichment. A second incubation in D_2O, using [3-^{13}C]-ALA as substrate, furnished a specimen of B_{12} also containing enrichments in the nuclei of interest i.e. C-17a and C-18 (together with C-2, C-3a, C-7, C-8a, C-12, and C-13a). Inspection of the signals in Fig. 20 show substantial deuteration at C-2, C-7, and C-12 (note the double effect from C-13 and CH_2D). In Fig. 20 a small β-isotopic shift to higher field is observed at C-18 (39 ppm; $\delta - 10Hz$) together with a normal intensity α-shift of 42Hz at this carbon. No corresponding β-shift could be observed on the propionate side chain (C-17a) but in view of the relatively weak β-shifts on propionate carbons at C-3a, 8a, and 13a, this is most probably a function of the side chain geometry. In any case, the relative ratios of ^2H at C-18 and C-19 (approximately 3:1) compared with the consistently higher deuterium content at C-3, C-8, and C-13 are in accord with the concept that not only is deacetylation a late step but that it takes place irreversibly under kinetic control,

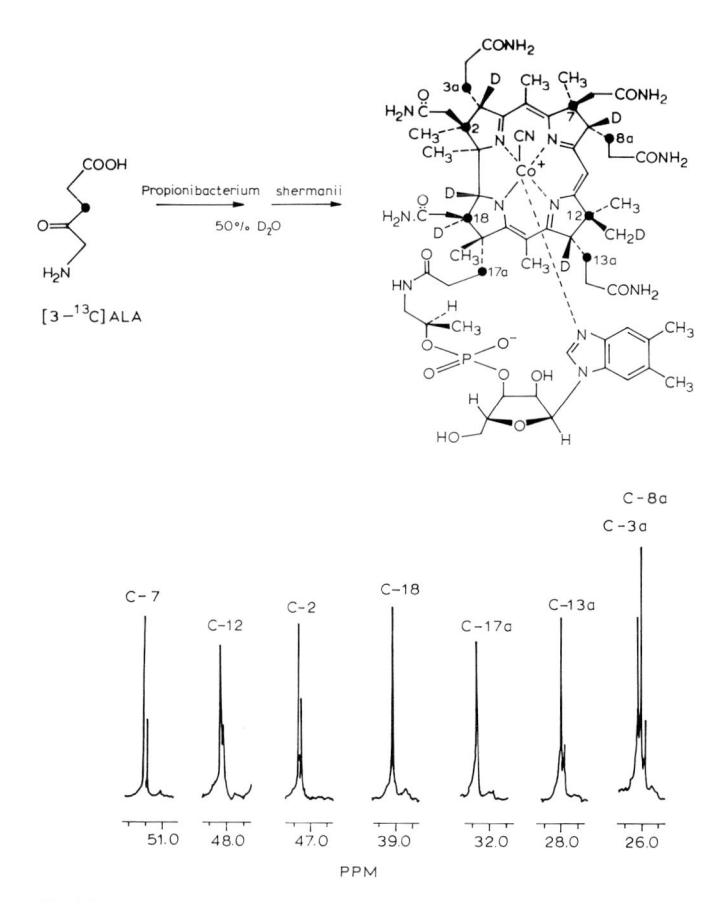

Fig. 20

leading to the observed isotope effect [61]. Since the chemical counterpart [26] of 19-deacylation involves a nickel or cobalt corrin it is tempting to suggest that the insertion of cobalt takes place before this step. However, as mentioned, above, the non-incorporation of cobalt into cobalt free cobyrinic acid in presence of the corrin synthesizing enzymes[57] is surprising and suggests that although corrin biosynthesis can be achieved in the absence of cobalt, metal insertion is not the last step on the way to cobyrinic acid.

In summary, all of the recent information gained from the $^{13}CH_3$-SAM pulse experiments [50–52] is in accord with the pathway shown in Fig. 17 where C-5 and C-15 methylations precede deacylation and where cobalt is inserted either after methylation at C-2 and C-7 or just before the ring contraction step. The experiments performed in D_2O [61] rule out the delivery of hydrogen to C-18 and C-19 from sources other than water and suggest that deacetylation is a rate determining step under kinetic control. It had earlier been shown [62] that in a cell-free system, C-18 and C-19 in cobyrinic acid are deuterated in a similar ratio. In Fig. 17 the portrayal of several intermediates as divalent metal complexes rather than the free pyrrocorphins and corphins takes into account the preparation of metal complexes of isobacteriochlorins [63] and the possibility that a divalent cation e.g. Zn^{++} can mediate in the various methylations and rearrangements leading from the porphinoid to the corrinoid template. Support for this concept comes from the discovery of a remarkable new set of corphinoid metabolites isolated as their zinc complexes as described in the next section.

10. Factors S_1-S_4, isomeric, tetramethylated corphinoids derived from uro'gen I

The knowledge gained from the many incubations designed to probe the sequence of methylation of uro'gen III has led to the isolation of a novel set of isomeric substance which at first sight appeared to be tetramethylated versions of uro'gen III, containing corphin-like chromophores. Thus short incubation of cobalt-deficient cell-free extracts of *P. shermanii* with ALA in the presence of SAM and Zn^{++} followed by esterification (MeOH/H_2SO_4) and extensive preparative TLC afforded four isomeric compounds (Factors S_1-S_4), each containing a UV/vis chromophore reminiscent of the synthetic model zinc corphinate (32) [64,65] (see Fig. 22). The most abundant of these isomers, Factor S_3 (300 µg), was first chosen for structural studies. FD- and FAB-MS revealed a molecular weight m/z 1102 ($C_{52}H_{67}O_{16}N_4ZnCl$). Factor S_3 is an octacarboxylic acid (FD-MS m/z octaethyl ester – octamethyl ester=112 mass units)

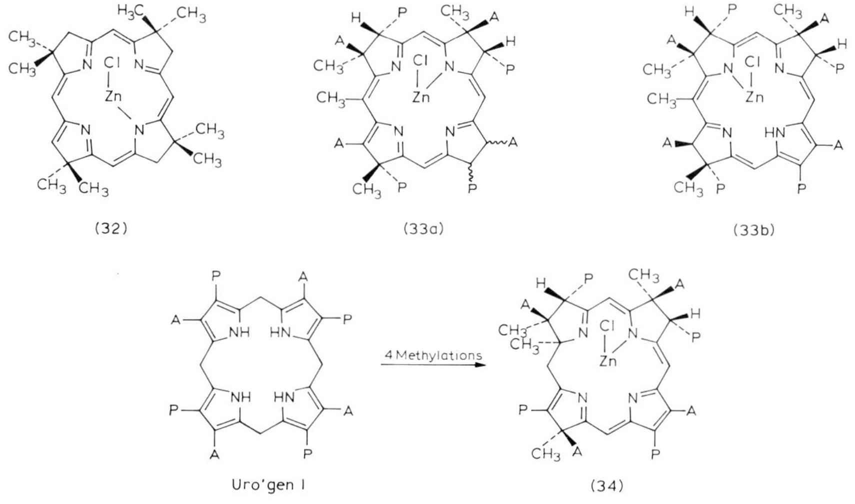

Fig. 22

containing four methionine-derived methyl groups ($^{14}C/^{3}H$ ratios of factors derived from [$^{14}CH_3$]SAM+[2,3-$^{3}H_2$]ALA: F II:F III:F S_3, 2 (1.99):3 (3.02):4 (3.98); FD-MS on a sample derived from CD_3-SAM showed addition of 12 mass units). Based on the above considerations, the most likely structure on biogenetic grounds is the zinc corphinate (33a) i.e. the long-sought precorrin-4a (Fig. 17). However, the absence of strong fluorescence and the ratios of the UV/vis absorption maxima in the 300–560-nm region suggested that Factor S_3 is, in fact, a close relative rather than a full member of the corphin family, a proposal which was confirmed by the following experiments.

Incubation of a cell suspension of *P. shermanii* in the presence of [4-^{13}C]ALA (90% ^{13}C) and [$^{13}CH_3$]L-Met (90% ^{13}C) and isolation of the major pigments afforded, after esterification (MeOH-H_2SO_4) and multiple TLC, the chloro-zinc complex of Factor S_3 (100 μg) enriched (~70%) with ^{13}C (FAB MS m/z 1114=M + 12). The proton decoupled ^{13}C NMR spectrum showed temperature dependence and was recorded in $CDCl_3$ at −38 °C to remove the maximum number of tautomeric species [66]. Under these conditions signals for four methyl groups (δ 16–24 ppm) and for eight enriched carbons derived from C-4 of ALA were observed. Of particular note are the signals at 79 ppm (>C-N) and two resonances in the sp^3 region (45–50 ppm) shown to be methines (>C-H) by off-resonance decoupling. The remaining five ALA-derived resonances (●) are assigned to sp^2 hybridized carbons (130–180 ppm) (Fig. 21a). A ^{13}C INADEQUATE NMR experiment (Fig. 21b) reveals that only one SAM-derived methyl group is coupled (*J*=54 Hz) to the quaternary carbon (∋C-N) at 79 ppm (▲). Further structural assignments were made by preparing two additional versions of the ^{13}C-enriched zinc complex. The NMR spectrum of Factor S_3 octamethyl ester (70 μg) derived from [^{13}C-5]ALA and [$^{13}CH_3$]SAM surprisingly exhibited a ^{13}C-^{13}C coupling pattern expected for a Type I rather than a Type III derivative of porphyrinogen, since the eight [5-^{13}C]ALA derived resonances (■) were observed in the sp^2/sp^3 region as coupled pairs. No contiguous trio, the hallmark of Type III porphyrinogens [67] and of isobacterochlorins, was present. The molecule therefore belongs to the symmetrical Type I series. This result, taken together with the absence of any H_3C-C coupling and a [5-^{13}C]ALA-derived resonance at 35 ppm (CH_2) coupled (*J*=37 Hz) to > C=N (147 ppm), leads to structure 34 for Factor S_3, (see Fig. 22) i.e. a tetramethylated Type I-derived isomer of a zinc corphinatochloride. The stereochemistry shown in (34) is based on analogy with the structures of Factors I-III and remains to be proved. Rigorous confirmation that three of the four extra SAM-derived methyl groups are attached to quaternary carbons bearing acetate side chains was forthcoming when a third isotopomer of Factor S_3 (70 μg) was isolated from an incubation with [3-^{13}C]ALA and $^{13}CH_3$-SAM. The ^{13}C INADEQUATE spectrum of the octamethyl ester of this specimen (Fig. 21c) reveals three one-bond ^{13}C-^{13}C couplings (*J*=35 Hz) corresponding to methylation at the acetate termini of three of the β-positions of the original pyrrolic rings of uro'gen I (38). The fourth methyl group (at C-1 or C-11) and the five remaining enrichments (●) from [3-^{13}C]ALA are 'silent'

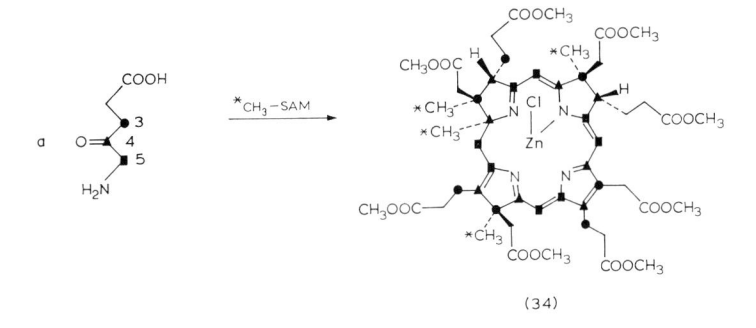

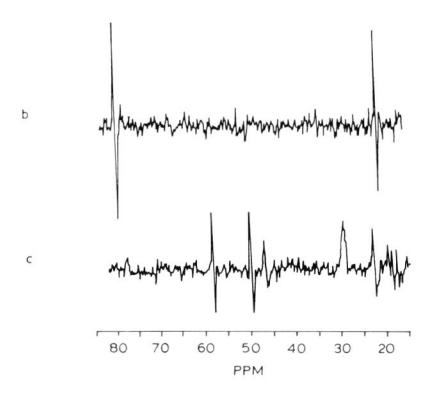

Fig. 21

in the INADEQUATE spectrum as required by structure 34 or an isomer methylated at C-11. Incorporation experiments with ^{3}H and ^{14}C labelled Factor S$_3$ as a potential precursor of vitamin B$_{12}$ were negative, in accord with the proposed structure.

Although not on the pathway to vitamin B$_{12}$, Factor S$_3$ represents the first example of biological methylation of uroporphyrinogen I [68] and is highly suggestive of the mode of incorporation of methyl groups into the more familiar type III nucleus, including both β- (C-2, C-7, C-17) and α-alkylation (C-1 or C-11). The fact that its ^{13}C-NMR spectra display temperature-dependent tautomeric flux places Factor S$_3$ in the same category as Eschenmoser's chloro-Zn-corphinate (32) whose ^{13}C spectrum is completely resolved only at −35 °C [66], and this phenomenon forms an important part of the structure proof for 34 [67]. At present the biological roles of uroporphyrinogen I and Factors S_1-S_4 are unknown, but the methylations responsible for the production of Factor S_3 in cell-free extract and in whole cells of *P. shermanii* must surely be related to the B_{12} pathway in that an isomer of the putative precorrin-4a has been produced as a zinc complex. The prediction can now be made that precorrin-4a, the elusive tetramethyl intermediate of vitamin B_{12} biosynthesis, will have the structure of a 2, 7, 17, 20-tetramethyl zinc corphinate (33a) or more likely the isomeric pyrrocorphinate complex (33b).

From the same incubation, the isomers Factors S_1, S_2, and S_4 were isolated as the octamethyl esters. Examination of the ^{13}C-INADEQUATE spectrum of Factor S_1 octamethylester derived from [3-^{13}C]-ALA and [^{13}CH$_3$]SAM showed (Fig. 23b) that in contrast to the pattern found for Factor S_3, all four methyl groups were coupled to ^{13}C-ALA derived quaternary carbons. The template is again type I as shown in Figure 23. Since the chromophore is that of a zinc corphinate, the methylase has in this case delivered all four methyl groups to acetate termini. The constitution of Factor S_1 is therefore (35) [70]. These experiments leave open the question of relative and absolute configurations since like Factor S_3, Factors S_1 cannot be correlated biochemi-

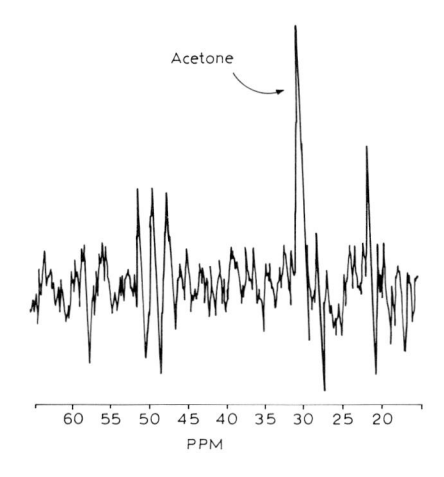

Fig. 23

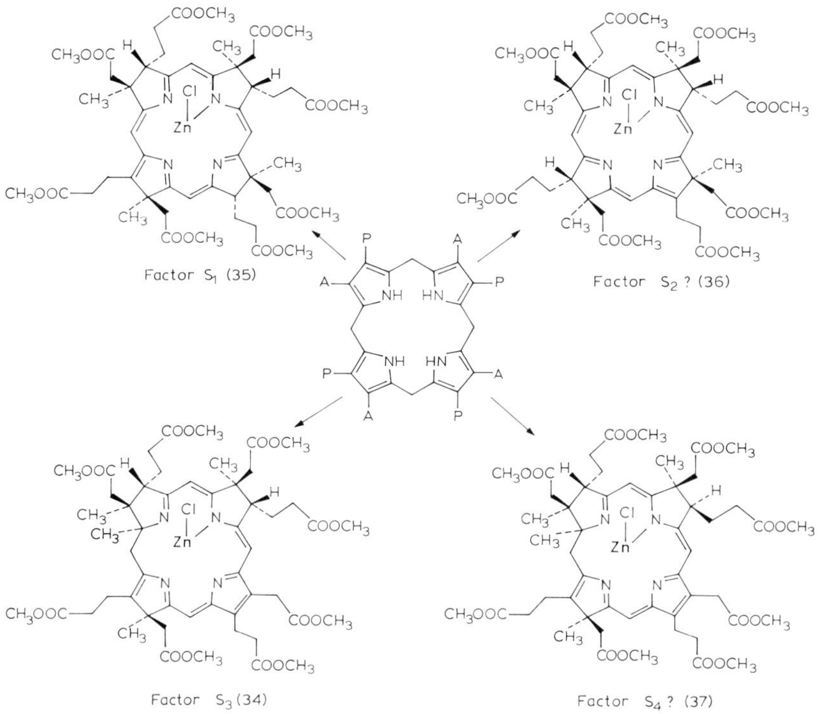

Fig. 24

cally with vitamin B_{12}. Factors S_2 and S_4 are also uro'gen-I derived and at present are regarded as close relatives (epimers ?) of Factors S_1 and S_3. The discovery of this new family of corphinoids based on a type I template (Fig. 24) raises the intriguing possibility that Nature can make corrins from Uro'gen I.

The suggestion that corphins may be intermediates in B_{12} biosynthesis was first made in connection with the idea that C_{20} of uro'gen III could become the C_1 methyl group of B_{12} [71]. Although this proved to be biochemically invalid [20], the corphins were synthesized and their subsequent chemistry has fostered a true renaissance, due in part to the newly discovered biochemical nature of this class of molecule e.g. Factor F430 [72], to their involvement (via the isomeric pyrrocorphins) in biomimetic peripheral C-methylation [73], and to the dihydrocorphinol → 19-acetyl corrin transformation [56] and subsequent deacylation. We note that the concept of a corphinoid intermediate in B_{12} biosynthesis (as a result of successive methylation of uro'gen III) was first invoked almost 15 years ago [74]. Of course at that time it could not have been predicted that corphins of the type I family would be found in Nature.

With the establishment of the sequence of methylations on the way from porphyrinogen to corrin and the techniques necessary to isolate the metabolites containing up to 4 methyl groups, the stage is set for a determined attack on the problem of finding the remaining intermediates, on the assumption that they can be isolated free of enzyme. During the last decade's search for such compounds, the enzymology of corrin biosynthesis has been developed in parallel with the sequential isolation work. The method of choice for discovering the remaining intermediates may indeed require knowledge of all of the enzymes of the pathway. Extensive studies on PBG deaminase and uro'gen III synthase, the enzymes which forge the type III macrocycle are described in Chapter 1.

11. The methyl transferases

Purification of the post uro'gen III enzymes has begun recently and the first two of these have been described. The first of these, S-Adenosylmethionine uro'gen III methyl transferase (SUMT), has been found to be non-specific in that it can use isomers of uro'gen III as substrate. Until recently it has been assumed that all natural porphinoids arise from uro'gen III[7], the biosynthetic precursor of heme, chlorophylls, cytochromes, and vitamin B_{12}. The discovery [69,70] of Factors S_1-S_4 derived from a uro'gen I template in B_{12}-producing organisms suggested that a new family based on type I porphinoids exists, although their biological role is not yet understood. A study was therefore undertaken to compare the products of the reaction catalyzed by uro'gen III-SAM-methyltransferase (SUMT) [75] the first methylating enzyme of the B_{12} pathway, on uro'gen I (43, Figure 25) with those of its normal substrate, uro'gen III (44). SUMT was first isolated from *P. shermanii* by Muller [68] and a 50-fold purification achieved.

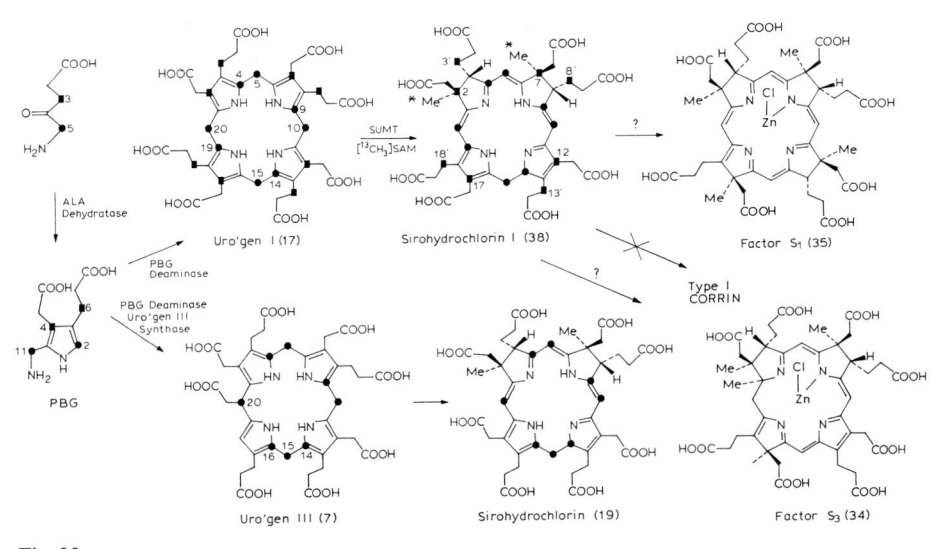

Fig. 25

A purified preparation of SUMT [75], which uses S-adenosylmethionine (SAM) and uro'gen III as substrates to produce sirohydrochlorin was incubated with SAM and uro'gen I. FAB-MS and UV-VIS spectroscopy of the major product (60%) of this reaction indicated that two methyl groups had been added to the substrate and that the initially formed product was oxidized during isolation to the observed isobacteriochlorin chromophore [76]. [5-^{13}C]-Aminolevulinic acid (ALA; •) (Figure 25), was incubated with ALA dehydratase from *Rhodopseudomonas spheroides* to give [2,11-^{13}C$_2$]-porphobilinogen (PBG; •), which was converted to enriched uro'gen I (•) using cloned PBG deaminase from *Escherichia coli* [77]. Oxidation to uro-I (I$_2$), esterification (methanol/H$_2$SO$_4$), and purification of the methyl ester by silica gel HPLC [78], gave a sample of uro I octamethyl ester containing less than 0.4% of the uro III isomer. A sample (4 mg) of this ester was saponified, reduced back to uro'gen I with sodium amalgam, and incubated with SUMT [75] and [^{13}CH$_3$]-SAM. Purification of the esterified, dimethylated product by Overpressure-Thin Layer Chromatography (O.P.L.C.) gave 490 µg of ^{13}C enriched dimethyl-isobacteriochlorin octamethyl ester, which exhibited 4 meso ^{1}H resonances (as carbon-coupled doublets) at δ 6.79 (H-5, $^1J_{CH}$ 155 Hz), 7.47 (H-10, $^1J_{CH}$ 152 Hz), 7.64 (H-20, $^1J_{CH}$ 155 Hz), and 8.84 (H-15, $^1J_{CH}$ 152 Hz) together with signals at δ 1.61 and 1.67, confirming the presence of the two SAM derived methyl groups. The signals for the protons at C-10 (δ 7.47) and C-20 (δ 7.64) were assigned from selective ^{1}H-^{13}C decoupling experiments, in close agreement with the corresponding data [41] for sirohydrochlorin.

The ^{13}C NMR spectrum (Figure 26) consists of four pairs of adjacent (•; Figure 25)

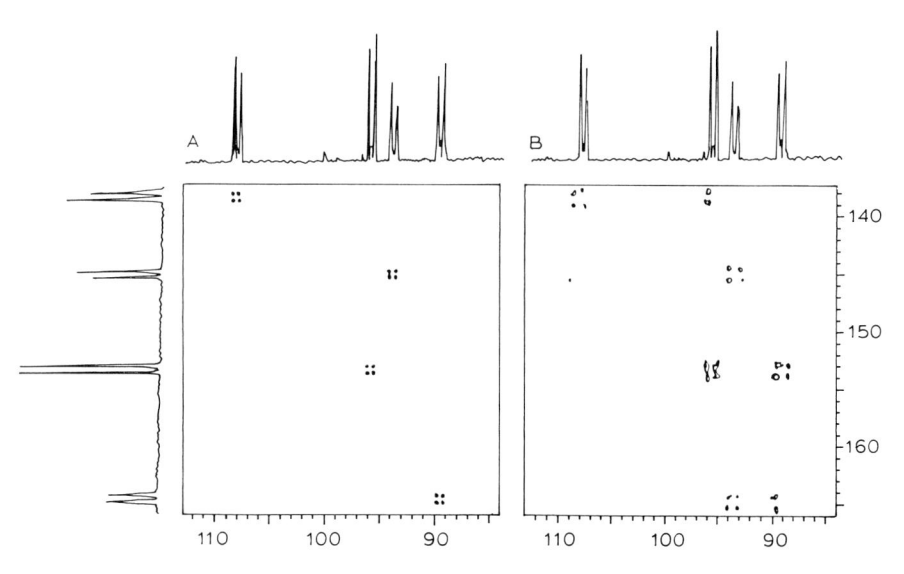

Fig. 26. ^{13}C COSY short range (A) and long range (B) spectra (lower right quadrant) of sirohydrochlorin I methyl ester (39) derived from [5-^{13}C]ALA.

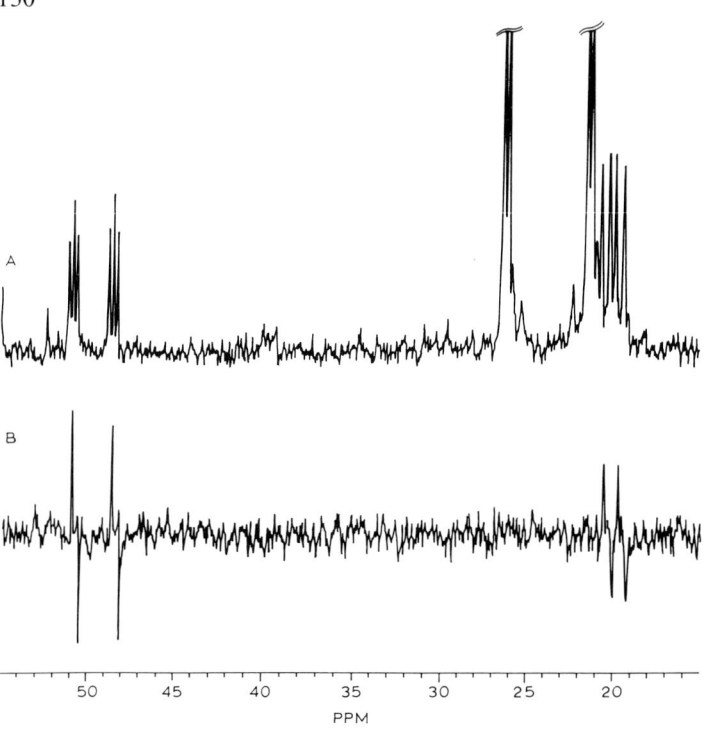

Fig. 27. (A) ^{13}C NMR and (B) inadequate spectra of sirohydrochlorin I methyl ester (40) derived from [^{13}C]ALA and [^{13}CH$_3$]SAM.
Compound (38) (methyl ester) has max (rel. intensity) 275 (0.455), 362 sh (0.855), 377 (1.217), 514 (0.519), 547 (0.251), 584 (0.378), 643 nm (0.065) (CH$_2$Cl$_2$) identical with the spectrum of sirohydrochlorin methyl ester (23) measured in CH$_2$Cl$_2$. The ^{1}H NMR and FAB-MS spectra of (38) were virtually identical with those of (23). CD of (38): max (ε) 270 (+3.5), 296 (−3.0), 350 (+8.5), 400 (+5.5), 512 (−1.0), 546 (−2.0), 583 (−3.3), 638 nm (−0.5). CD of (23): 264 (+2.5), 290 (−4.5), 351 (+2.6), 396 (+4.0), 510 (−1.6), 545 (−3.5), 584 (−5.9), 635 nm (−0.5) (CH$_2$Cl$_2$).

^{13}C resonances (J$_{CC}$ 71–77 Hz) in contrast to sirohydrochlorin (39) which is labelled from [5-^{13}C]-ALA in the 14, 15, and 16 positions (●), producing a triplet pattern for C-15 (Figure 25). Spectral assignment was accomplished by selective ^{1}H decoupling, which indicated that peaks at δ 89.8 and 108.6 were due to C-5 and C-15 respectively. A carbon COSY experiment was then performed, optimized for 80 Hz couplings. Correlations appearing in the lower right quadrant (Figure 26A) show that the low field (δ 108.6) meso carbon (C-15) is coupled to the high field (δ 138.3) pyrrole carbon (C-14) and vice versa, while the carbons of intermediate shift value (C-10, C-20) of each type correlate directly. The carbon COSY experiment was then optimized for long range coupling constants (Figure 26B) indicating that C-9 (δ 153.3) was coupled to C-5 and that C-19 (δ 145.0) was coupled to C-15. The signal assignments and coupling constants observed confirm that the new isolate is a type-I structural isomer of (19) and is accordingly named sirohydrochlorin I.

The positions of methylation in sirohydrochlorin I were revealed by repeating the enzymatic synthesis using uro'gen I derived from [3-^{13}C]-ALA ( ) and [^{13}CH$_3$]-SAM (*) as substrates. The ^{13}C NMR spectrum of the methyl ester (40) of (38) labelled in this mode (Figure 27A) revealed doublets centred at δ 20.4 and 19.5 for the two SAM derived methyl groups (*) attached to C-2 (δ 48.5) and C-7 (δ 50.8) as confirmed by INADEQUATE spectroscopy [79] which clearly identifies the 2 pairs of coupled nuclei (Figure 27B). The absolute stereochemistry of sirohydrochlorin I (38) was shown to correspond to that of natural sirohydrochlorin by comparison of the appropriate CD spectra (see legend Fig. 27).

Since small amount of uro'gen I and its tetramethyl metabolites, Factors S$_1$ [70] and S$_3$ [69] are found in *P. shermanii*, the in vitro reaction described above most probably represents the first two methylations on the way to these type I corphinoids in vivo. The possibility was examined that a complete corrinoid structure based on a type I template could be prepared using uro'gen I or reduced sirohydrochlorin I (38) labelled with ^{13}C, as substrates with the cell-free system capable of converting uro'gen III to cobyrinic acid. However, no ^{13}C-enrichment was observed in the spectra of the small amount of normal (endogenous) cobyrinic acid produced in these experiments which would necessarily act as carrier for a 'type-I' cobyrinic acid, suggesting that Nature, although capable of inserting at least 4 methyl groups into Uro'gen I by C-alkylation is unable to effect the key step of ring contraction of a corphinoid to a corrin using the type-I pattern of acetate and propionate side chains. This result, although negative, is in accord with a suggestion [80] concerning the requirement for two adjacent acetate side chains in pre-corrinoids, one of which (at C-2) can participate in lactone formation to the C-20 meso position (see Figure 17, precorrin 6b) a postulate supported indirectly by recent experiments [53] with ^{18}O labelled precursors, while the second acetate function (at C-18) is used as an auxiliary to control the necessary activation of C-15 for C-methylation. In any event, the methylase SUMT from *P. denitrificans* (and by analogy from *P. shermanii*) has been shown to be capable of transforming uro'gen I (normally formed in small amounts in vivo) by two successive methylations with absolute stereochemical fidelity to a type-I isomer of

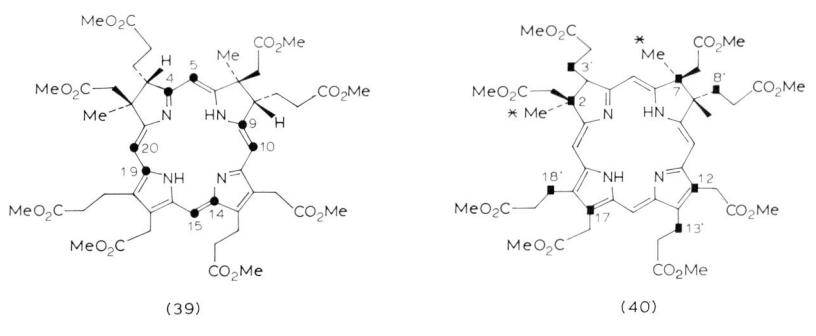

Fig. 28

sirohydrochlorin. The possible biochemical conversion of sirohydrochlorin I to factors S_1 and S_3 as outlined in Figure 25 and the biological and evolutionary roles of these remarkable type-I metabolites are being investigated. A similar enzyme has recently been found in *E. coli*, which requires the synthesis of sirohydrochlorin and thence siroheme from uro'gen III to form the prosthetic group of sulfite reductase. The gene *cysG*, [81] part of the genetic machinery for cysteine biosynthesis was found to encode a methyl transferase (M-1) [82,83] of close homology to SUMT. The specificities of these enzymes differ however since M-1 is capable (at high concentration) of performing 2 C-methylations not only on uro'gens I and III but, as distinct from SUMT, the II and IV isomers [84]. In addition prolonged incubation of uro'gens I and III with M-1 and SAM catalyzes a third C-methylation, this time on ring C to form (41), the first example of a 'natural' pyrrocorphin [85]. The lack of regiospecificity of M-1 may be of synthetic utility for the preparation of permethylated B_{12} intermediates from unnatural substrates. Although *E. coli* does not synthesize B_{12}, the requirement for siroheme ensures that M-1 is expressed. A second methyl transferase [54] which introduces the C-20 methyl into the substrate precorrin-2 → precorrin-3 has recently been isolated and purified [86].

12. Biosynthesis of the nucleotide loop and coenzyme B_{12}

The steps from cobyrinic acid to cobalamin (Figure 4) involve amidation to cobyric acid (9), attachment of 1-amino-2-propanol to ring D propionate to give cobinamide (10) and stepwise attachment of the nucleotide from ribose and dimethylbenzimidazole. A cell-free system capable of cobinamide synthesis via amidation by aminopropanol and amidation of the propionate on ring A has been described [87]. A remarkable, regiospecific non-enzymatic self assembly of the nucleotide segment to ring D propionate has been disclosed [80].

The origin of the aminopropanol group was shown to be threonine [88]. The phos-

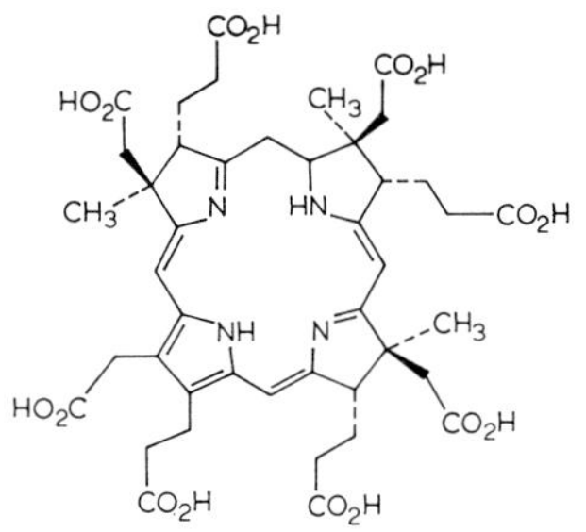

Trimethylpyrrocorphin (41)

Fig. 29

phorylation of cobinamide to cobinamide phosphate is mediated by ATP and is followed by attachment of GDP (from GTP). Addition of the lower base to GDP-cobinamide involves a novel reaction of nicotinate mononucleotide (Nicotinate-ribose-phosphate) with inversion of the α-glycosidic linkage to β- [87–89]:

Nicotinate-R-P+DMBI $\rightarrow$ nicotinate+DMBI-R-P

The 5'-nucleotide that reacts with GDP cobinamide to form cobalamin phosphate is again a rather unusual step [90,91]:

GDP-cobinamide+DMBI-R-P $\rightarrow$ Cobalamin-P+GMP

in which the 3'-OH of the sugar displaces GMP from GDP-cobinamide. The final step involves dephosphorylation to cobalamin by a non-specific phosphatase [92,93]. Cobalamin undergoes Co β-5'-adenosylation by cob(I)alamin adenosyltransferase to produce coenzyme B_{12}. The enzyme has been purified [94] and the stereochemistry of the displacement shown to involve inversion at C-5' using stereospecifically deuterated ATP [95].

Genetic mapping of the loci of the B_{12}-synthesizing enzymes has been reported for *Pseudomonas denitrificans* [96]. This complements a most interesting study on the genetics of *Salmonella typhimurium* which cannot make B_{12} when grown aerobically [97]. A mutant requiring methionine, cobinamide or cyanocobalamin when grown anaerobically produces B_{12} de novo thus leading to the isolation of other mutants blocked in B_{12} synthesis including one which cannot make Factor II required for siroheme production. All of the cobalamin mutations are close together on the chromosome. In *P. denitrificans* the genes encoding B_{12} synthesizing enzymes are scattered widely over the chromosome although *hems* C-D are contiguous.

13. Evolutionary aspects of B_{12} biosynthesis

13.1 The C_5 pathway

Until quite recently it had been assumed that the Shemin pathway (glycine-succinate) to ALA was ubiquitous in bacterial production of porphyrins and corrins. However it is now clear that in many archaebacteria (e.g. *Methanobacterium thermoautotrophicum* [98], *Clostridium thermoaceticum* [99,100]) the C_5 pathway from glutamate is followed. The same pathway has been suggested for the synthesis of an unusual corrinoid from *Sporomusa ovata* [101]. Phylogenetically the C_5 route is conserved in higher plants and it appears from recent work that *hem*A of *E. coli* (and perhaps of *S. typhimurium*) in fact catalyzes the glutamate $\rightarrow$ ALA conversion, i.e. the C_5 pathway is much more common than had been realised [102]. Although *E. coli* does not appear to synthesize B_{12} the enzyme SUMT is expressed as part of the genetic machinery for making siroheme. In *C. thermoaceticum* it has been shown [100] that the B_{12} produced by this thermophilic archaebacterium is synthesized with ALA produced from glutamate.

13.2 Chemical methods

Eschenmoser has speculated that corrinoids resembling B_{12} could have arisen by prebiotic polymerization of hydrogen cyanide and has developed an impressive array of chemical models [71,80] to support this hypothesis including ring contraction of porphyrinoids to acetyl corrins, deacetylation and C-acetylation chemistry all of which mimic the corresponding biochemical sequences. Corrin may therefore have existed 4×10^9 years ago, that is before the origin of life [102] or the genetic code [103] on earth.

Indeed B_{12} is found in primitive anaerobes and requires no oxidative process in its biogenesis unlike the routes to heme and chlorophyll which are oxidative. A recent review is concerned with this topic [80].

13.3 Why type III?

If B_{12} indeed was the first natural substance requiring uro'gen III as a precursor, the question arises 'why type III?' Since the chemical synthesis of the uro'gen mixture from PBG under acidic conditions leads to the statistical ratio of uro'gens [I 12.5%; II 25%; III 50%; IV 12.5%] containing a preponderance of uro'gen III, natural selection of the most abundant isomer could be the simple answer. However, it has been suggested [80] that this unique juxtaposition of two adjoining acetate and side chains in the type III isomer (which does not obtain in the symmetrical uro'gen I) may be responsible for a self assembly mechanism requiring these functions to hold the molecular scaffolding in place via lactone and ketal formations as portrayed in Figure 17. These ideas can be tested with ^{18}O labelling and by studying the possible biotransformation of types I, II and IV porphyrinogens to corrins.

With the mapping of B_{12} biosynthesis now under way, rapid progress can be expected in the discovery of the remaining intermediates post-precorrin-3 and the na-

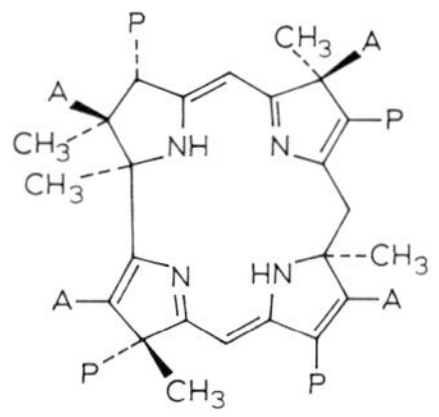

Fig. 30

ture of the enzymes which mediate the methyl transfers, decarboxylation, ring contraction and cobalt insertion. It is anticipated that the powerful combination of molecular biology and NMR spectroscopy which has been essential in solving the previous problems in B_{12} biosynthesis will again be vital to the solution of those enigmas still to be unravelled in the fascinating saga of the natural assembly of the corrin template.

This prophecy is already under fulfilment with the very recent description of the isolation and characterization of an intermediate between precorrin-3 and cobyrinic acid [104,105]. This intermediate, precorrin-6x (42), was found to accumulate in a crude cell free system from a culture of a recombinant strain of *Pseudomonas denitrificans* to which factor III (oxidised precorrin-3) had been added. Good quantitative conversion of precorrin-6x into hydrogenocobyrinic acid (cobalt-free cobyrinic acid) could be performed by addition of NADPH to this cell free system. Determination of the structure by NMR revealed that precorrin-6x had arisen by ring contraction, methylation at C-11, but not decarboxylation, and was at the dehydrocorrin oxidation level. Thus for conversion into cobyrinic acid, precorrin-6x has to undergo reduction with NADPH, methylation at C-5 and C-15, decarboxylation followed, presumably, by a [1,5] sigmatropic rearrangement, tautomerization and cobalt insertion. Since there is no enzymatic coversion of cobalt-free cobyrinic acid, it now seems that two distinct pathways for corrin biosynthesis exist. The first inserts cobalt 'early' into precorrin-3 (anaerobic route) while the second, aerobic pathway leads to the cobalt-free corrins via 6x.

Although precorrin-6x is obviously relevant to the elucidation of B_{12} biosynthesis, we will have to wait for the isolation of the purified enzyme system to confirm that the formation of precorrin-6x is not a result of non-specific transformation within the crude cell free system used in this work and whether it is converted to B_{12} by the *P. Shermanii* system.

References

1 Dolphin, D. (Ed.) (1982) 'B_{12}' Vols. 1 and 2, John Wiley, New York.
2 Scott, A.I. (1990) Acc. Chem. Res. 23, 308–317.
3 Scott, A.I. (1986) Pure Appl. Chem. 58, 753–766.
4 Battersby, A.R. (1986) Acc. Chem. Res. 19, 147–152.
5 Leeper, F.J. (1985) Nat. Prod. Rep. 2, 19–47.
6 Leeper, F.J. (1985) Nat. Prod. Rep. 2, 561–568.
7 Leeper, F.J. (1987) Nat. Prod. Rep. 4, 441–469.
8 Leeper, F.J. (1989) Nat. Prod. Rep. 6, 171–203.
9 Bray, R.C., Shemin, D. (1958) Biochem. Biophys. Acta 30, 647–648.
10 Shemin, D., Corcoran, J.W., Rosenblum, C. and Miller, I.W. (1956) Science 124, 272.
11 Bray, R.C. and Shemin, D. (1963) J. Biol. Chem. 238, 1501–1508.
12 Shemin, D. and Bray, R.C. (1964) Ann. N.Y. Acad. Sci. 112, 615–621.

13 Scott, A.I., Townsend, C.A., Okada, K. Kajiwara, M., Whitman, P.J. and Cushley, R.J. (1972) J. Am. Chem. Soc. 94, 8267–8269.
14 Scott, A.I., Townsend, C.A., Okada, K., Kajiwara, M. and Cushley, R.J. (1972) J. Am. Chem. Soc. 94, 8269–8271.
15 Brown, C.E., Katz, J.J. and Shemin, D. (1971) Proc. Nat. Acad. Sci., USA 68, 1083–1088.
16 Battersby, A.R., Ihara, M., McDonald, E. and Stephenson, J.R. (1973) J. Chem. Soc. Chem. Commun. 404–405.
17 Scott, A.I., Townsend, C.A. and Cushley, R.J. (1973) J. Am. Chem. Soc. 95, 5759–5761.
18 Doddrell, D. and Allerhand, A. (1971) Proc. Natl. Acad. Sci. USA 68, 1083–1088.
19 Battersby, A.R., Ihara, M., McDonald, E., Stephenson, J.R. and Golding, B. (1974) J. Chem. Soc. Chem. Commun. 458–459.
20 Scott, A.I., Yagen, B. and Lee, E., (1973) J. Am. Chem. Soc. 95, 5761–5762.
21 Dauner, H. and Muller, G. (1975) Hoppe-Seyler's Z. Physiol. Chem. 356, 1353–1358.
22 Scott, A.I., Kajiwara, M., Takahashi, T., Armitage, I.M., Demou, P. and Petrocine, D. (1976) J. Chem. Soc. Chem. Commun. 544–546.
23 Hensen, O.D., Hill. H.A.O., Thornton, J., Turner, A.M. and Williams, R.J.P. (1976) Royal Soc. London Phil. Trans. 273, 353–357.
24 Imfield, M., Townsend, C.A. and Arigoni, D. (1976) J. Chem. Soc. Chem. Commun. 541–542.
25 Battersby, A.R., Hollenstein, R., McDonald, E. and Williams, D.C. (1976) J. Chem. Soc. Chem. Commun. 543–544.
26 Arigoni, D. (1978) Ciba Foundation Symposium No. 60, 'Molecular Interactions and Activity in Proteins', pp. 243–248, Exerpta Medica, Amsterdam.
27 Scott, A.I., Townsend, C.A., Okada, K. and Kajiwara, M. (1974) J. Am. Chem. Soc. 96, 8054–8069.
28 Scott, A.I., Townsend, C.A., Okada, K., Kajiwara, M., Cushley, R.J. and Whitman, P.J. (1974) J. Am. Chem. Soc. 96, 8069–8080.
29 Scott, A.I., Yagen, B., Georgopapadakou, N., Ho, K.S., Kliose, S., Lee, E., Lee, S.L., Temme, G.H., Townsend, C.A. and Armitage, I.M. (1975) J. Am. Chem. Soc. 97, 2548–2550.
30 Battersby, A.R., Ihara, M., McDonald, E., Satoh, F. and Williams, D.C. (1975) J. Chem. Soc. Chem. Comm. 436–437.
31 Battersby, A.R., McDonald, E., Hollenstein, R., Ihara, M., Satoh, F. and Williams, D.C. (1977) J. Chem. Soc. Perkin, I. 166–178.
32 Hensens, O.D., Hill, H.A.O., McClelland, C.E. and Williams, R.J.P. (1982) in: 'B_{12}', (D. Dolphin, Ed.) Vol. 1,463–500, Wiley, New York.
33 Anton, D.L., Hogenkamp, H.P.C., Walker, T.E. and Matwiyoff, N.A. (1982) Biochemistry 21, 2372–2378.
34 Ernst, L. (1981) Liebigs Ann. Chem. 376–386.
35 Battersby, A.R., Edington, C., Fookes, C.J.R. and Hook, J.M. (1982) J. Chem. Soc. Perkin, I. 2265–2277.
36 Ernst, L. (1984) J. Chem. Soc. Chem. Commun. 2267–2270.
37 Kurumaya, K. and Kajiwara, M. (1989) Chem. Pharm. Bull 37, 9–12.
38 Summers, M.F., Marzilli, L.G. and Bax, A. (1986) J. Am. Chem. Soc. 108, 4285–4294.
39 Siegel, L.M., Murphy, M.J. and Kamin, H. (1973) J. Biol. Chem. 248, 251–264.
40 Scott, A.I., Irwin, A.J. and Siegel, L.J.N. (1978) J. Am. Chem. Soc. 100, 316–318.
41 Scott, A.I., Irwin, A.J., Siegel, L.M., Lewis, M. and Shoolery, J.N. (1978) J. Am. Chem. Soc. 100, 7987–7994.
42a Deeg, R., Kriemler, H-P., Bergmann, K-H. and Muller, G. (1977) Hoppe-Seyler's Z. Physiol. Chem. 358, 339–352.
42b Imfield, M., Arigoni, D., Deeg, R. and Müller, G. (1979) in: Vitamin B_{12} (B. Zagalak and W. Friedrich, Eds.), pp. 315–318, de Gruyter, New York.

43 Bergmann, K-H., Deeg, R., Gneuss, K.D., Kriemler, H-P. and Muller, G. (1977) Hoppe-Seyler's Z. Physiol. Chem. 358, 1315–1323.
44 Nussbaumer, C., Imfield, M., Worner, G., Muller, G. and Arigoni, D. (1981) Proc. Nat. Acad. Sci. U.S. 78, 9–10.
45 Battersby, A.R., Bushnell, M.J., Jones, C., Lewis, N.G. and Pfenninger, A. (1981) Proc. Nat. Acad. Sci., U.S. 78, 13–15.
46 Mombelli, L., Nussbaumer, C., Weber, H., Muller, G. and Arigoni, D. (1981) Proc. Nat. Acad. Sci. U.S. 78, 11–12.
47 Muller, G., Gneuss, K.D., Kriemler, H-P., Scott, A. I. and Irwin, A.J. (1979) J. Am. Chem. Soc. 101, 3655–3657.
48 Muller, G., Gneuss, K.D., Kriemler, H-P., Irwin, A.J. and Scott, A.I. (1981) Tetrahedron (Supp.) 37, 81–90.
49 Battersby, A.R., Frobel, K., Hammerschmidt, F. and Jones, C. (1982) J. Chem. Soc. Chem. Comm. 455–457.
50 Uzar, H.C., Battersby, A.R., Carpenter, T.A. and Leeper, F.J. (1987) J. Chem. Soc. Perkin I. 1689–1696.
51 Scott, A.I., Mackenzie, N.E., Santander, P.J. Fagerness, P., Muller, G., Schneider, E., Sedlmeier, R. and Worner, G. (1984) Bioorg. Chem. 13, 356–362.
52 Scott, A.I., Williams, H.J., Stolowich, N.J., Karuso, P., Gonzalez, M.D., Muller, G., Hlineny, K., Savvidis, E., Schneider, E., Traub-Eberhard, U. and Wirth, G. (1989) J. Am. Chem. Soc. 111, 1897–1900.
53 Kurumaya, K., Okasaki, T. and Kajiwara, M. (1989) Chem. Pharm. Bull. 37, 1151–1154.
54 Blanche, F., Handa, S., Thibaut, D., Gibson, C.L., Leeper, F.J. and Battersby, A.R. (1988) J. Chem. Soc. Chem. Comm. 1117–1119.
55 Battersby, A.R., Deutscher, K.R. and Martinoni, B. (1983) J. Chem. Soc. Chem. Comm. 698–700.
56 Rasetti, V., Pfaltz, A., Kratky, C. and Eschenmoser, A. (1981) Proc. Natl. Acad. Sci., USA. 78, 16–19.
57 Muller, G., private communication.
58 Kulka, J., Nussbaumer, C. and Arigoni, D. (1990) J. Chem. Soc. Chem. Comm. 1512–1514.
59 Dresov., B., Ernst, L., Grotjahn, L. and Koppenhagen, V.B. (1981) Angew. Chem. Int. Ed. Engl. 20, 1048–1049.
60 Scott, A.I., Georgopapadakou, N.E. and Irwin, A.J. unpublished.
61 Scott, A.I., Kajiwara, M. and Santander, P.J. (1987) Proc. Nat. Acad. Sci. U.S. 84, 6616–6618.
62 Battersby, A.R., Edington, C. and Fookes, C.J.R., (1984) J. Chem. Soc. Chem. Comm. 527–530.
63 Battersby, A.R. and Sheng, Z.C. (1982) J. Chem. Soc. Chem. Comm. 1393–1394.
64 Johnson, A.P., Wehrle, P., Fletcher, R. and Eschenmoser, A., (1968) Angew. Chem. Int. Ed. Engl. 7, 623–625.
65 Muller, P.M. (1973) Dissertation No. 5735 E.T.H.
66 Muller, P.M., Farooq, S., Hardegger, B., Salmond, W.S. and Eschenmoser, A. (1973) Angew. Chem., Int. Ed. Engl. 12, 914–916.
67 Scott, A.I. (1978) Acc. Chem. Res. 11, 29–36.
68 Muller, G. (1979) in: Vitamin B_{12}, (B. Zagalak and W. Friedrich, Eds.) pp. 279–291, de Gruyter, New York.
69 Muller, G., Schmiedl, J., Schneider, E., Sedlmeier, R., Worner, G., Scott, A.I., Williams, H.J., Santander, P.J., Stolowich, P.E., Fagerness, P. and Mackenzie, N.E. (1986) J. Am. Chem. Soc. 108, 7875–7877.
70 Muller, G., Schmiedl, J., Savidis, L., Wirth, G., Scott, A.I., Santander, P.J., Williams, H.J., Stolowich, N.J. and Kriemler, H-P. (1987) J. Am. Chem. Soc. 109, 6902–6904.
71 Eschenmoser, A. (1986) Ann. N.Y. Acad. Sci. 471, 108–129.

72 Pfaltz, A., Livingston, D., Jaun, B., Diekert, G., Thauer, R.K. and Eschenmoser, A. (1985) Helv. Chim. Acta 68, 1338–1358.
73 Leumann, C., Hilpert, K., Schreiber, J. and Eschenmoser, A. (1983) J. Chem. Soc. Chem. Commun. 1404–1407.
74 Scott, A.I., Townsend, C.A., Okada, K. and Kajiwara, M. (1973) Trans. N.Y. Acad. Sci. 35, 72–79.
75 Blanche, F., Debussche, L., Thibaut, D., Crouzet, J. and Cameron, B. (1989) J. Bacteriol. 171, 4222–4231.
76 Reference ommitted (see legend Fig. 27)
77 Scott, A.I., Stolowich, N.J., Williams, M.J., Gonzalez, M.D., Roessner, C.A., Grant, S.K. and Pichon, C. (1988) J. Am. Chem. Soc. 110, 5898–5900.
78 Gonzalez, M.D., Grant, S.K., Williams, H.J. and Scott, A.I. (1988) J. Chromatogr. 437, 311–315.
79 Bax, A., Freeman, R. and Kempsell, S. (1980) J. Am. Chem. Soc. 102, 4849.
80 Eschenmoser, A. (1988) Angew. Chem., Int. Ed. Engl. 27, 5–39.
81 Cole, J.A., Newman, B.M. and White, P. (1980) J. Gen. Microbiol. 120, 475–483.
82 Warren, M.J., Stolowich, N.J., Santander, P.J., Roessner, C.A., Sowa, B.A. and Scott, A.I. (1990) FEBS Lett. 261, 76–80.
83 Warren, M.J., Roessner, C.A., Santander, P.J. and Scott, A.I. (1990) Biochem. J. 265, 725–729.
84 Warren, M.J., Gonzalez, M.D., Williams, H.J., Stolowich, N.J. and Scott, A.I. (1990) J. Am. Chem. Soc. 112, 5343–5345.
85 Scott, A.I., Warren, M.J. Roessner, C.A., Stolowich, N.J. and Santander, P.J. (1990) J. Chem. Soc. Chem. Commun. 593–597.
86 Blanche, F. et al, private communication; see also reference 54.
87 Ford, S.M. (1985) Biochem. Biophys. Acta 841, 306–317.
88 Krasna, A.J. Rosenblum, C. and Sprinson, D.B. (1957) J. Biol. Chem. 225, 745–750.
89 Friedmann, H.C. and Harris, D.L. (1965) J. Biol. Chem. 240, 406–412.
90 Friedmann, H.C. (1965) J. Biol. Chem. 240, 413–418.
91 Fyfe, J.A. and Friedmann, H.C. (1969) J. Biol. Chem. 244, 1659–1666.
92 Renz, P. (1968) Hoppe-Zeyler's J. Physiol. Chem. 349, 979–983.
93 Schneider, F. and Friedmann, H.C. (1972) Arch. Biochem. Biophys. 152, 488–495.
94 Sato, K., Nakashima, T. and Shimizu, S. (1984) J. Nutr. Sci. Vitaminol. 30, 405–408.
95 Parry, R.J., Ostrauder, J.M. and Arzou, I.Y. (1985) J. Am. Chem. Soc. 107, 2190–2191.
96 Cameron, B., Briggs, K., Pridmore, S., Brefort, G. and Crouzet, J. (1989) J. Bacteriol. 171, 547–557.
97 Jeter, R.M., Olivera, B.M. and Roth, J.R. (1984) J. Bacteriol. 159, 206–213.
98 Zeikus, J.G. (1983) Adv. Microb. Physiol. 24, 215–299.
99 Stern, J.R. and Bambers, G. (1966) Biochemistry 5, 1113–1118.
100 Oh-hama, T., Stolowich, N.J. and Scott, A.I. (1988) FEBS Lett. 228, 89–93.
101 Stupperich, E. and Eisinger, H.J. (1989) Arch. Microbiol. 151, 372–377.
102 Decker, K. Jungermann, K. and Thauer, R.K. (1970) Angew. Chem., Int. Ed. Engl. 9, 153–162.
103 Eigen, M., Lindemann, B.F., Tietze, M., Winkler-Oswatitsch, R., Dress, A. and von Haeseler, A. (1989) Science 244, 673–679.
104 Thibaut, D., Debussche, L. and Blanche, F. (1990) Proc. Natl. Acad. Sci. USA. 87, 8795–8799.
105 Thibaut, D., Blanche, F., Debussche, L., Leeper, F.J. and Battersby, A.R. (1990) Proc. Natl. Acad. Sci. USA. 87, 8800–8804.

CHAPTER 4

Biochemistry of coenzyme F430, a nickel porphinoid involved in methanogenesis

HERBERT C. FRIEDMANN[2], ALBRECHT KLEIN[1] and RUDOLF K. THAUER[1]

[1]*Fachbereich Biologie, Philipps-Universitat, D-3550 Marburg, F.R.G. and* [2]*Department of Biochemistry and Molecular Biology, The University of Chicago, Chicago IL 60637, U.S.A.*

1. Introduction

In 1978 Gunsalus and Wolfe [1] isolated from methanogenic bacteria – anaerobic archaebacteria – a yellow low-molecular weight substance which they named factor 430 because of its absorption maximum at 430 nm. One year later Schonheit et al. [2] found that methanogenic bacteria require nickel for growth. The same group soon thereafter discovered this transition metal to be required for the synthesis of F430. The substance was shown to contain nickel [3,4]. Whitman and Wolfe [5] at almost the same time made corresponding observations. A number of biosynthetic experiments soon indicated the new nickel-containing compound to be in fact a macrocyclic tetrapyrrole [6,7,8,9]. The structure of the nickel porphinoid (Fig. 1) was unravelled in the laboratory of Eschenmoser and Pfaltz at the ETH Zurich [10,11,12,13,14]. As to function, Ellefson, Whitman and Wolfe [15] discovered that F430 is the prosthetic group of methyl coenzyme M reductase, the enzyme that catalyzes the final step in methane formation [16]. F430 thus has coenzyme function. Until now coenzyme F430 has been found only in methanogenic bacteria, in which it is always present [17].

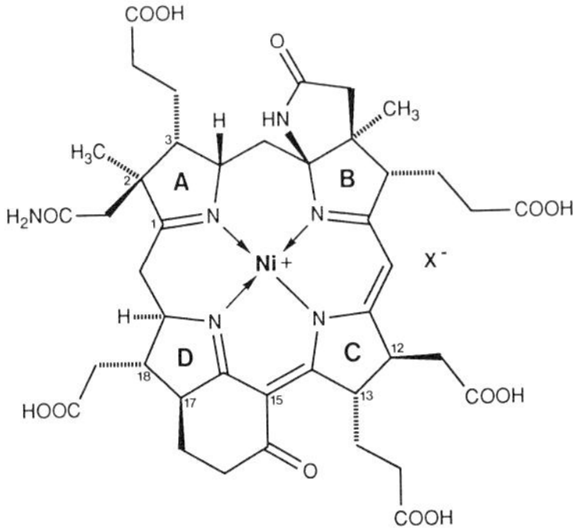

Fig. 1. Structure of coenzyme F430 [10–14]. The configuration at C(17), C(18) and C(19) has only recently been determined [146].

The primary structure of methyl coenzyme M reductase, which contains three different subunits, has been determined via analysis of the encoding genes [18,19, 20,21,22,23]. With a knowledge of the structure of coenzyme F430 and of the amino acid sequence of the protein subunits the groundwork has been laid for further structural analysis of methyl coenzyme M reductase. Insight into the three-dimensional structure of the enzyme is a prerequisite for a detailed understanding of the role of the novel tetrapyrrole in the catalysis of methane formation.

2. Structural relations to other tetrapyrroles

Coenzyme F430 (Fig 1) is the most recently discovered cyclic tetrapyrrole with a known biological function. The elucidation of the structure of this yellow nonfluorescent substance greatly increased the structural variations shown by naturally occurring macrocyclic tetrapyrroles [24,25], all of which are derived from uroporphyrinogen III (Fig. 2A). The chemical peculiarities of coenzyme F430 are seen vividly upon comparison of its structure to the structures of some tetrapyrroles that may be more familiar to most readers.

The presence of a ligand nickel atom is the first striking feature of coenzyme F430. Nickel brings to five the number of different metals found in the center of biological cyclic tetrapyrroles. Up to the discovery of coenzyme F430 the different metals known to occur were iron in hemes and in siroheme, magnesium in chlorophylls, cobalt in corrinoids, and copper in turacin, the pigment of turaco bird feathers [26].

Additional interesting features are shown by the side chains of coenzyme F430 and the state of reduction of its pyrrole ring system.

With respect to the side chains in hemes, chlorophylls and all bile pigments with the exception of bactobilin (=urobiliverdin, related to uroporphyrinogen I rather than III [27,28]) the eight side chains correspond to, or are derived from, those in protoporphyrinogen IX (Fig. 2B). In each case, six of the uroporphyrinogen III side chains

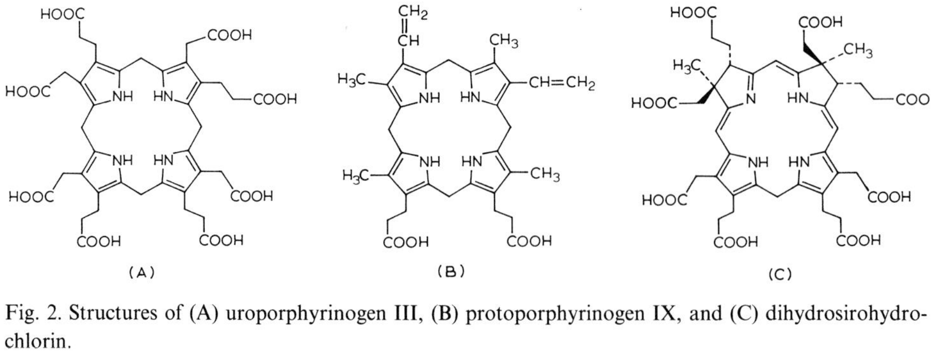

Fig. 2. Structures of (A) uroporphyrinogen III, (B) protoporphyrinogen IX, and (C) dihydrosirohydrochlorin.

have been shortened by direct decarboxylation (4 acetic acid groups → 4 methyl groups), or by oxidative decarboxylation (2 propionic acid groups → 2 vinyl groups). No decarboxylations have, however, occurred in turacin, in siroheme (in distinction to heme), in vitamin B_{12} (with the exception of one acetic acid chain), and in coenzyme F430. Thus turacin (copper-uroporphyrin III) and siroheme contain all the eight side chains found in uroporphyrinogen III; B_{12} contains all of the seven remaining side chains in the amidated form, resulting in three acetamide and four propionamide groups; in coenzyme F430, finally, only two of the side chains are amidated. From the viewpoint of amidation, coenzyme F430 is thus positioned between siroheme and B_{12}. Coenzyme F430, however, is a pentaacid, not a hexaacid: the propionic acid substituent on the 'inverted' ring D gives rise to a six-membered carbocyclic ring formed by attachment of its carboxyl carbon to the bridge carbon between rings C and D. (It is instructive to note that the six-membered carbocyclic ring of coenzyme F430 is quite different from the five-membered carbocyclic ring attached to ring C in chlorophyll). Furthermore, one of the two amidated side chains of coenzyme F430 has become cyclized as well: instead of an acetamide side chain on ring B a lactam ring whose nitrogen is joined to a ring B α-carbon is present.

The periphery of coenzyme F430 shows some further interesting features in addition to the limited number of amidations and to the two 'new' cyclic structures: there are two carbon-linked methyl groups on rings A and B. These two methyl groups occur at precisely the same position and with the same stereochemistry as those in dihydrosirohydrochlorin (Fig. 2C), the compound from which siroheme and B_{12} are derived [29].

With respect to the state of reduction of the pyrrole ring system one encounters the interesting fact, possibly related to its anaerobic evolution and function, that coen-

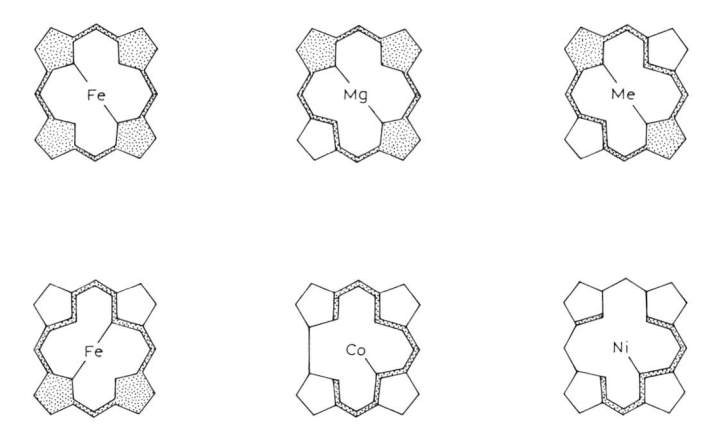

Fig. 3. Degree of unsaturation and conjugation in various macrocyclic tetrapyrroles. Shaded areas show double bonds or resonating conjugated bond systems. From left to right (upper line): hemes, plant type chlorophylls, bacteriochlorophylls a and b; (lower line) siroheme, corrinoids, coenzyme F430 [24].

zyme F430 is by far the most highly reduced or saturated amongst all known tetrapyrroles [24,25] (Fig. 3). Coenzyme F430 has only 5 double bonds while 6 double bonds occur in the ring structure of B_{12}, 8 in the porphyrinogens, 9 in siroheme, 10 in chlorophylls, and 11 in uro-and protoporphyrins and in heme. Two pairs of the double bonds in coenzyme F430 are conjugated, but they are separated by two single bonds and so do not form a larger conjugated bond system. The yellow colour of coenzyme F430, in contrast to the red colour of porphyrins and of corrinoids, is no doubt related to this lower degree of conjugation.

Since the discovery of coenzyme F430 a second nickel-containing tetrapyrrole, tunichlorin, has been isolated. This compound, a nickel porphinoid whose side chains are related to those of protoporphyrin IX, was obtained from the tunicate, *Tridemnum solidum*, where it occurs in traces ($10^{-5}\%$) [30]. No function has yet been discovered for this substance.

The substituents that occur in tunichlorin at the β-pyrrole positions, along with the absence of extra carbon-linked methyl groups, relate it structurally not to coenzyme F430 but to chlorophyll, along with the nickel [31,32,33,34] and vanadium petroporphyrins [31,35] found in oil shale. It is thought that the petroporphyrins originated from chlorophylls via replacement of magnesium by nickel, or vanadium, and by geochemical modification of the macrocyclic ring system. From this viewpoint it is of interest that coenzyme F430 and corrinoids, in contrast to chlorophylls, cannot be demetallated without cleavage of the macrocycle. This is yet another circumstance that excludes them as petroporphyrin precursors.

The biosynthetic and structural relationships among the different macrocyclic tetrapyrroles are depicted in Fig. 4.

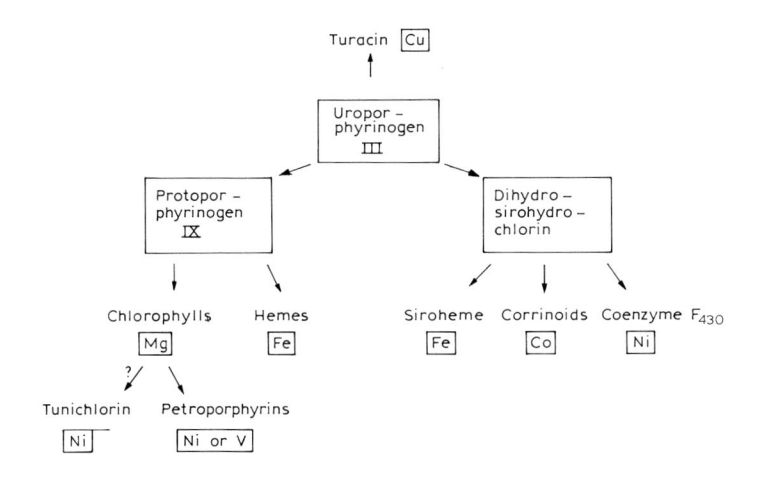

Fig. 4. Structural and biosynthetic relationships (genealogy) of different macrocyclic tetrapyrroles [36].

3. Biosynthesis from glutamate via uroporphyrinogen III and dihydrosirohydrochlorin

A molecular weight of approximately 1,000, the presence of strongly bound nickel, and the absence of amino acids and sugars early suggested coenzyme F430 to be a nickel-tetrapyrrole [3,36]. This suggestion was confirmed by the first biosynthetic experiments which demonstrated that ^{14}C-labelled 5-aminolaevulinic acid was incorporated into the compound, and that the specific radioactivity of the product was exactly 8 times that of the 5-aminolaevulinic acid used [7]. These experiments were followed by labelling studies with L-[methyl-^{14}C]methionine and L-[methyl-^{3}H]methionine which established the presence of two methyl groups derived from methionine [8]. At this point NMR-spectroscopic analysis was performed with various differentially labelled coenzyme F430 preparations isolated from *Methanobacterium thermoautotrophicum* cells grown in the presence of 5-amino[2-^{13}C]laevulinic acid, 5-amino[3-^{13}C]laevulinic acid, 5-amino[4-^{13}C]laevulinic acid, 5-amino[5-^{13}C]laevulinic acid, or L-[methyl-^{13}C]methionine [10]. The results of these and other studies led to the structure given in Fig. 1 [10,11,12,13,14b].

This structure, which shows a close analogy of the side chains to those of uroporphyrinogen III (Fig. 2A), and the presence of the two carbon-bound methyl groups that, as mentioned above, are identical to those in dihydrosirohydrochlorin (Fig. 2C), strongly indicated that a step in coenzyme F430 biosynthesis must be the conversion of uroporphyrinogen III to dihydrosirohydrochlorin, a change also exhibited in the course of B_{12} formation [37a–c]. The occurrence of this conversion was substantiated by the subsequent observations that coenzyme F430 is formed both from uroporphyrinogen III in growing methanogenic bacteria [38a], and from sirohydrochlorin by cell extracts of these bacteria [38b].

More detailed inspection of the structure of coenzyme F430 suggests that dihydrosirohydrochlorin is converted to coenzyme F430 via the following steps (not necessarily in this order) [39]: (i) insertion of nickel; (ii) amidation of the ring A and B acetate groups; (iii) reduction of 2 double bonds; (iv) cyclization of the acetamide of ring B; (v) cyclization of the propionic acid group of ring D (Fig. 5).

In the above reaction sequence only the formation of 15,17^3-seco-F430–17^3-acid (Fig. 5) and its conversion to coenzyme F430 have been demonstrated experimentally [39]. Seco-F430 contains an uncycled propionic acid group at ring D. Before cyclization can occur this group almost certainly must be activated, possibly via the CoA-thioester.

As to the early steps, 5-aminolaevulinic acid is made in methanogens from glutamate rather than from succinyl-CoA and glycine [40]. The involvement of a tRNA^{Glu} has been demonstrated [41,42]. Thus the pathway of 5-aminolaevulinic acid formation for coenzyme F430 biosynthesis is the same as in plants and in many eubacteria [43a–d].

4. Properties of free coenzyme F430 including its redox behaviour

Coenzyme F430 is present in methanogenic bacteria both in the free state and bound to methyl coenzyme M reductase [44]. The bound form is released only upon denaturation of the enzyme. Originally it was thought that the two forms differ from each other [45]. It has since been shown, however, that the free and the bound coenzyme F430 (after release from the protein) are chemically identical [11,46,47]. The original designation of factor 430 was changed to coenzyme F430 when it was recognized that the structure shown in Fig. 1 represents the prosthetic group of methyl coenzyme M reductase.

Evidence is available that coenzyme F430 participates in methane formation only when bound to methyl coenzyme M reductase. The free form apparently functions only as 'precursor' of the bound form [44]. Despite this fact the properties of free coenzyme F430 are of considerable interest in relation to its biological role.

Free coenzyme F430 is yellow and non-fluorescent, and has absorption maxima at 430 nm (ε_{430}=23,100 cm^{-1} M^{-1}) [4,10] and at 274 nm (ε_{274}=20,000 cm^{-1} M^{-1}) [10]. The pentaacid (Fig. 1) has a molecular mass of 905 Da (^{58}Ni). The nickel is present in a divalent state (Ni(II)).

Two striking aspects of coenzyme F430 structure, already mentioned above, are the presence both of nickel and of the most reduced bond system known for a cyclic tetrapyrrole. The fact that these two structural peculiarities are found in the same molecule results in special chemical properties (for review see 14). One consequence of the highly reduced ring system is that coenzyme F430 is the most puckered of all cyclic tetrapyrroles. This puckering is enhanced by the insertion of nickel which brings about an unusual deformation of the macrocycle via contraction of the Ni-nitrogen bonds in the equatorial plane [25,48,49].

An important property of coenzyme F430 is its epimerization to 13-epi-F430 and 12,13-di-epi-F430 [12,17,50–52]. These successive changes already occur at considerable rates at room temperature in aqueous media. One has to bear this epimerization in mind when preparations of coenzyme F430 are studied with respect to their physical and biological behaviour.

In aprotic solutions (tetrahydrofuran or dimethylformamide), the pentamethylester of coenzyme F430 (F430M) can undergo reduction without affecting the π-chromophore system. Ni(II) complexed to the coenzyme F430 ligand system is reduced to Ni(I) as evidenced by electron spin resonance spectroscopy (Fig. 6) [53]. Reduction of nickel is also associated with a change in the UV-vis absorption spectrum, resulting in a strong absorption band at 382 nm (ε_{382}=29,600 cm^{-1} M^{-1}) and a weaker band at 754 nm (ε_{754}=2500 cm^{-1} M^{-1}) (Fig. 7).

Jaun and Pfaltz [53] measured the redox potential of the pentamethylester of coenzyme F430 (F430M) in dimethylformamide. Cyclic voltammograms of coenzyme F430M showed a single one-electron wave at -1.32 V versus the ferricenium/ferrocene couple (0.815 V), and at 0.84 V versus the 0.1 M calomel electrode (0.335

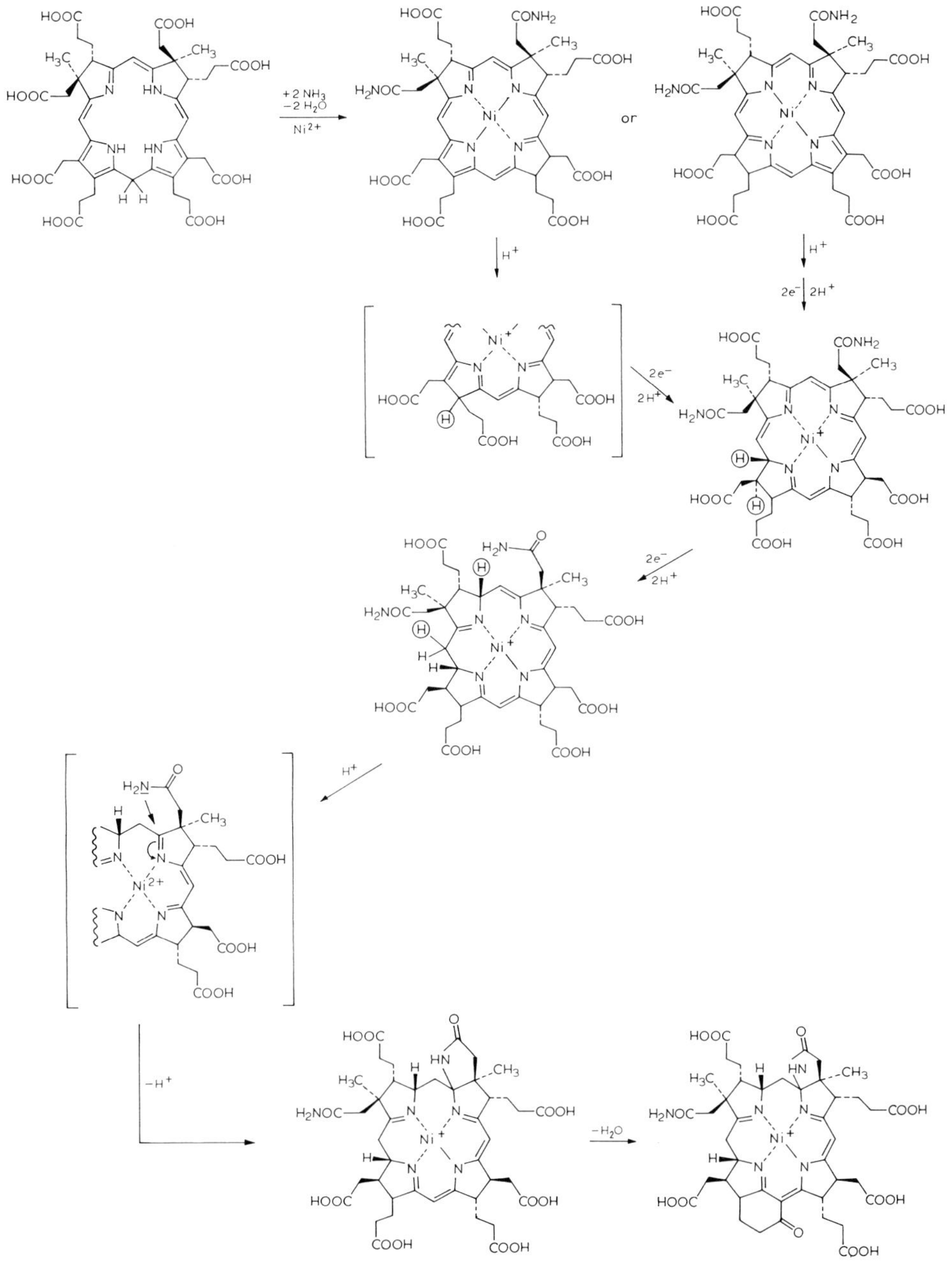

Fig. 5. Proposed steps in the biosynthesis of coenzyme F430 from dihydrosirohydrochlorin. The structure immediately preceding coenzyme F430 is 15,17^3-seco-F430–17^3-acid [39]. For correct configuration at C(17), C(18) and C(19) see Fig. 1.

V). The redox potential versus the hydrogen electrode (0 V) is thus −0.504 V. (For studies on the reductive chemistry of other nickel hydroporphyrins see [54]).

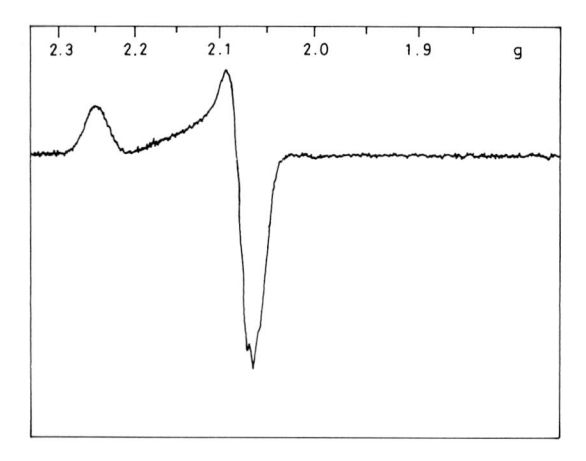

Fig. 6. The ESR spectrum of the nickel(I) form of F430M in frozen tetrahydrofuran at 88 °K [14,53]. F430M=pentamethylester of coenzyme F430. For correct configuration at C(17), C(18) and C(19) see Fig. 1.

Recently it was found that coenzyme F430 in H_2O catalyzes the reduction of methyl chloride to methane ($E^{o'}=+0.438$ V) with Ti(III) citrate ($E^{o'}=-0.48$ V) [55]. (The same reaction is mediated by aquocobalamin and other corrinoids but only at very much lower rates [56]). A mechanism has been proposed involving reduction of coenzyme F430 with Ti(III) citrate to the Ni(I) form which in turn reduces the alkylhalide to methane [55]. Indeed, reduced F430M in dimethylformamide has been shown to react with methyliodide to yield methane [57].

Finally, it must be mentioned that coenzyme F430 (Ni(II)) is susceptible to autoxidation with O_2 in the π-chromophore system: the blue 12,13-didehydro-F430 (F560) is formed from coenzyme F430 [12,58]. Cell extracts of methanogenic bacteria contain an enzyme activity catalyzing the regeneration of coenzyme F430 from the didehydro form by reduction [59].

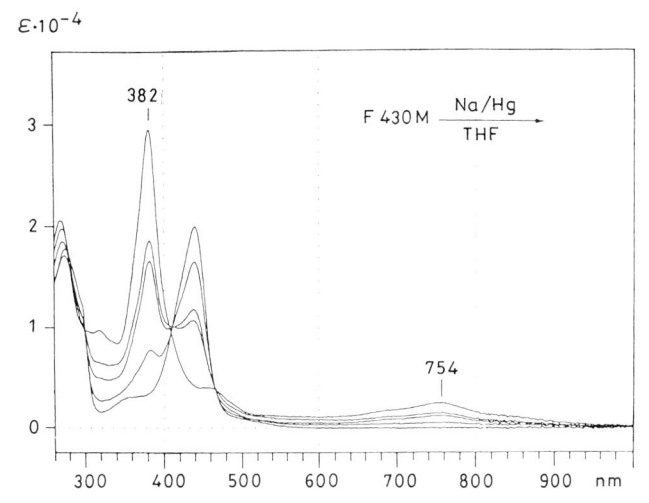

Fig. 7. UV/vis spectrum of F430M at different degrees of reduction to F430M$_{red}$ with sodium amalgam in tetrahydrofuran at 22 °C [14,53]. F430M=pentamethylester of coenzyme F430.

5. Function of coenzyme F430 as prosthetic group of methyl coenzyme M reductase in methanogenesis

Only two years after its discovery in methanogenic bacteria, coenzyme F430 was recognized as the chromophore of methyl coenzyme M reductase [15], a key enzyme of methane formation [16]. Every year 10^9 tons of methane are generated by the action of methanogenic bacteria [36]. In vivo the reduction of methyl coenzyme M to methane is coupled to energy conservation via the chemiosmotic mechanism (for recent literature on the mechanism of energy conservation in methanogenic bacteria see 60,61). These considerations are sufficient comments on the practical and theoretical importance of the novel porphinoid compound.

Methyl coenzyme M reductase is a hexamer of three different subunits with an α_2, β_2, γ_2 composition and an apparent molecular mass of 300 kDa. The subunits have apparent masses of 66 kDa, 48 kDa and 37.5 kDa, respectively [16,62,63]. Each molecule of the holoenzyme contains two molecules of tightly but not covalently bound coenzyme F430. There is evidence that coenzyme F430 is probably associated with the α-subunits [64]. Methyl coenzyme M reductase has absorption maxima at 278 nm and at 420 nm, with a shoulder at 445 nm. The A_{278}/A_{420} ratio is 6. The extinction coefficient at 420 nm, 22,000 cm^{-1} M^{-1}, is practically the same as that of the free coenzyme at 430 nm [16,62]. At present the shift in the absorption maximum and the generation of the 445 nm absorbance shoulder are not understood. It has been postulated that the absorbance changes are due to the formation of a Schiff base between the carbonyl group of the carbocyclic ring of coenzyme F430 and an amino group of

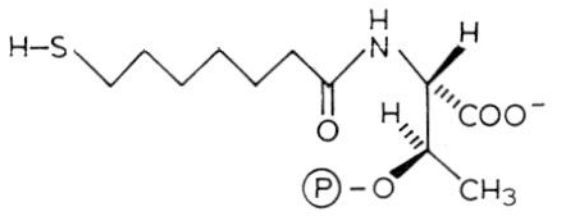

Fig. 8. Structures of methyl coenzyme M (CH_3-S-CoM) and of 7-mercaptoheptanoylthreonine phosphate (H-S-HTP) [16].

the α-subunit. However, attempts to reduce the putative base and thus to link the coenzyme covalently to the peptide chain were unsuccessful.

The function of methyl coenzyme M reductase was long thought to be the catalytic reduction of methyl coenzyme M (CH_3-S-CoM) by an unknown reductant to yield methane and to regenerate coenzyme M (H-S-CoM) (for structure see Fig. 8):

(a) CH_3-S-CoM + 2[H] ────→ CH_4 + H-S-CoM

Recently the reductant was found to be a novel coenzyme present in methanogens, 7-mercaptoheptanoylthreonine phosphate (H-S-HTP) (Fig. 8) [16]. Methane is formed by reaction of this thiol compound with methyl coenzyme M, but instead of free coenzyme M, as in equation (a), the heterodisulfide of H-S-CoM and H-S-HTP is produced [16,62,63]:

(b) CH_3-S-CoM + H-S-HTP ────→ CH_4 + CoM-S-S-HTP

Thus the regeneration of coenzyme M must occur in a reaction subsequent to the one that produces methane. Coenzyme M is re-formed as the result of a further reduction catalyzed by the enzyme CoM-S-S-HTP reductase, which is quite distinct from methyl coenzyme M reductase [65]:

(c) CoM-S-S-HTP + 2[H] ────→ H-S-CoM + H-S-HTP

Reduced viologen dyes can be used as artificial reductants in this reaction, but the biological electron donor has not yet been identified. With methylene blue as electron acceptor the enzyme also mediates the reverse reaction [66].

The question of what is the role of coenzyme F430 as prosthetic group in the reaction catalyzed by methyl-coenzyme M reductase needs to be addressed. A definitive answer cannot yet be given, one reason being that the purified enzyme shows, at the most, only 5% of the specific activity expected from the in vivo enzyme concentration and from the observed in vivo methane formation rates [62,63]. The enzyme from most methanogenic bacteria is actually completely inactive in the isolated form and becomes active only in the presence of additional proteins, a reductant and ATP [67]. It is hypothesized that the activation occurs by reduction of the enzyme-bound coenzyme F430. There is evidence from electron spin resonance spectroscopy that methyl coenzyme reductase is probably active only when coenzyme F430 is present in the Ni(I) form [68,69]. In this context it becomes relevant to note that methyl coenzyme reductase is rapidly inactivated by low levels of chloroform [16], a potent oxidant of free coenzyme F430 (Ni(I)) [55].

In the literature on the subject of methane formation one often finds reference to various components required for the final steps of methane formation. These were named by the letters of the alphabet before being further characterized. Component A is a protein fraction containing several enzymes, amongst them those required for the above-mentioned activation of methyl coenzyme M reductase [67]. Component B is a heat-stable factor now known to be H-S-HTP (Fig. 8), and component C turned out to be methyl coenzyme M reductase [16].

It remains to be stressed that methyl coenzyme M reductase is highly substrate-specific both for methyl coenzyme M and for 7-mercaptoheptanoylthreonine phosphate. Neither the coenzyme M analogues 3(methylthio)propane sulfonate or 1(methylthio)methane sulfonate nor the H-S-HTP analogues 8-mercaptooctanoyl-threonine phosphate or 6-mercaptohexanoylthreonine phosphate are active. In fact, these compounds are potent reversible inhibitors as are for example the further coenzyme M analogues 2-bromoethanesulfonate, and 3-bromopropanesulfonate. The latter compound has an apparent K_i of 0.05 μM compared to the apparent K_m for methyl coenzyme M and H-S-HTP of 4 mM and 75 μM, respectively (for literature see 62,63)

To understand the mechanism of methyl coenzyme M reduction it is necessary to know the axial ligands of nickel in coenzyme F430 in the active enzyme [70]. More extensive knowledge of the binding of coenzyme F430 to the protein and knowledge of the quarternary structure of the enzyme will help to answer this question. The elucidation of the primary structure of the three subunits of methyl coenzyme M reductase, described in the next section, is an important step in this direction.

6. Comparative analysis of genes encoding methyl coenzyme M reductase

As described above methyl coenzyme M reductase is so far recognized to comprise three pairs of subunits. Based on this observation, two groups have undertaken comparative analyses of the structural genes encoding subunits α, β and γ. They followed complementary approaches using either antibodies against the purified subunits [71] or oligonucleotides derived from partial polypeptide sequences [21] in order to detect

150

SUBUNIT α

```
MADKLFINALKKKFEESPEEKKTTFYTLGGWKQSERKTEFVNAGKEVAAKRGIPQYNPDI
            K  F    G  Q  RKE *    K *A  RG*   YNP

GTPLGQRVLMPYQVSTTDTYVEGDDLHFVNNAAMQQMWDDIRRTVIVGLNHAHAVIEKRL
G PLGQR * PY *S TD    E DDLH**NN A*QQ WDDIRRT *VG** AH  *E* L

GKEVTPETITHYLETVNHAMPGAAVVQEHMVETHPALVADSYVKVFTGNDEIADEIDPAF
GKEVTPETI   Y*  *NH** G*AVVQE MVE HP L  D Y **FTG*D *A E**  *

VIDINKQFPEDQAETLKAEVGDGIWQVVRIPTIVSRTCDGATTSRWSAMQIGMSMISAYK
*IDINK F   QA  *K  *G    Q   *P**V R  DG*  SRW*AMQIGMS  I*AY

QAAGEAATGDFAYAAKHAEVIHMGTYLPVRRARGENEPGGVPFGYLADICQSSRVNYEDP
  AGEAA *D **A*KHA ** *G  L  RRARG NE GG* FG L D* Q**RV    DP

VRVSLDVVATGAMLYDQIWLGSYMSGGVGFTQYATAAYTDNILDDFTYFGKEYVEDKYGL
 L*VV*  G   LYDQIWLG  YMSGGVGFTQYATA YTD*ILD* *Y*  *Y*   KY

CEAPNNMDTVLDVATEVTFYGLEQYEEYPALLEDQFGGSQRAAVVAAAAGCSTAFATGNA
  *  *  D*A*E  T  Y ** Y*  *P  *ED FGGSQRA **AAAAG    A  GN

QTGLSGWYLSMYLHKEQHSRLGFYGYDLQDQCGASNVFSIRGDEGLPLELRGPNYPNYAM
*  G* GWYLS  LHKE   RLGF*G*DLQDQCGA N    DE  P ELRGPNYPNYAM

NVGHQGEYAGISQAPHAARGDAFVFNPLVKIAFADDNLVFDFTNVRGEFAKGALREFEPA
NVGH  G YAGI QA H *RGDAF   L*K* FAD  L F**   R EF**GA*REF

GERALITPAK
GER ** PA
```

SUBUNIT β

```
MAKFEDKVDLYDDRGNLVEEQVPLEALSPLRNPAIKSIVQGIKRTVAVNLEGIENALKTA
        D * **D *G  *   *V * ***P  N  I  **   KR*VAVNL  I*  L  *

KVGGPACKIMGRELDLDIVGNAESIAAAAKEMIQVTEDDDTNVELLGGGKRALVQVPSAR
  GG * *G *  *V *A* IA     ** V  DDT *   GK       P R

FDVAAEYSAAPLVTATAFVQAIINEFDVSMYDANMVKAAVLGRYPQSVEYMGANIATMLD
 *A**  *    * A * **      *** V**** G YPQ ** G***  *L

IPQKLEGPGYALRNIMVNHVVAATLKNTLQAAALSTILEQTAMFEMGDAVGAFERMHLLG
  P    EG G**LRNI  NH  A *    **A A  S** E ** FEMG A*G FER  LLG

LAYQGMNADNLVFDLVKANGKEGTVGSVIADLVERALEDGVIKVEKELTDYKVYGTDDLA
LAYQG*NA*NL* *  *K N*    GT*G*V*   *V *A    G*I V*K    * Y   D*

MWNAYAAAGLMAATMVNQGAARAAQGVSSTLLYYNDLIEFETGLPSVDFGKVEGTAVGFS
 WNA  AA G *AA  V   GA*RAAQ **S  *LY ND**E E GLP  D*G*   GTAVGFS

FFSHSIYGGGGPGIFNGNHIVTRHSKGFAIPCVAAAMALDAGTQMFSPEATSGLIKEVFS
FFSHSIYGGGGP*FNGNH*VTRHS*GFAIP V AA*  DAGTQMFS  E  TS*L*  V*

QVDEFREPLKYVVEAAAEIKNEI
   EFREP*K  V
```

SUBUNIT γ

```
MAQYYPGTTKVAQNRRNFCNPEYELEKLREISDEDVVKILGHRAPGEEYPSVHPPLEEMD
MAQ*YP  T *A NRR      LEKLREI DED*  **GHR PG  Y * HPPL E*

EPEDAIREMVEPIDGAKAGDRVRYIQFTDSMYFAPAQPYVRSRAYLCRYRGADAGTLSGR
  P    ** V    GA  G R*RYIQF DSMY AP* PY RS      RG D GTLSGR

QIIETRERDLEKISKELLETEFFDPARSGVRGKSVHGHSLRLDEDGMMFDMLRRQIYNKD
Q**E  RE *** *K   *TE  D A  G*RG *VHGHS*RL*E*G*MFD L R

TGRVEMVKNQIGDELDEPVDLGEPLDEETLMEKTTIYRVDGEAYRDDVEAVEIMQRIHVL
  G *  K*Q**    D  *D*G P*        TTIYR  D   RD *  *E * RI

RSQGGFNLE
R*   G*
```

the genes in genomic libraries. The total sequences of the methyl CoM reductase genes from *Methanosarcina barkeri, Methanococcus vannielii, Methanobacterium thermoautotrophicum* (Marburg), *Methanococcus voltae,* and *Methanothermus fervidus* were established and could be compared [19,20,21,22,23,72,73].

The most obvious result of the comparative analyses is the high degree of conservation of the derived polypeptide sequences amongst the different organisms. Weil et al. [20] noted the highest conservation in all subunits amongst the two thermophilic species. In general it transpired that the molecular mass of the smallest subunit, γ, had been overestimated on the basis of the electrophoretic mobilities of the denatured polypeptides. The reason for this atypical behaviour is unknown. It has been noted, however, that this anomaly is most likely not due to a modification, since heterologous expression in *Escherichia coli* yielded the same apparent mass on denaturing gels [22]. More interesting is the general 'consensus' sequence among the primary structures of the subunits as derived from the DNA sequences (Fig. 9). It shows areas of complete amino acid sequence conservation and/or substitution by functionally homologous amino acids at the respective positions in all five organisms, which may be considered as parts of functionally essential domains determining, for instance, cofactor binding, nucleation of protein folding, or subunit interaction. Since the enzyme has been recognized to be membrane associated [74,75], yet the protein is not hydrophobic in general, hydrophobic areas within subunits are also of interest. Only few such regions are found in the α and β subunits centering around alanine-rich conserved areas. Clearly, folding of the polypeptides can generate additional hydrophobic surface areas which might interact with the cell membrane.

In general, no definite conclusions relating primary structure of the polypeptides to their functions can be drawn at present. It must be hoped that assays for functions of the individual subunits can be established, which could be helped by the heterologous expression of the cloned genes. The experimental suggestion that F430 binds to the α subunit [64] could be the starting point for such investigations. Clearly the eventual goal must be crystallization of the proteins and the establishment of their three-dimensional structure.

In the course of the analysis of the transcription unit, in which the three genes encoding the known methyl coenzyme M reductase subunits are arranged, two more genes have been found in all analysed cases (Fig. 10). They are interspersed between the genes coding for subunits β and γ and have different degrees of homology among the compared species, the smaller one being more heterogeneous both in size and in

Fig. 9. Amino acid sequences of the subunits of methyl coenzyme M reductase from *Methanobacterium thermoautotrophicum* (Marburg), as derived from the nucleotide sequence, and consensus sequences of conserved positions in the homologous subunits from four other methanogens. The following amino acids were considered to be functionally equivalent: A/S, S/T, I/V/L/M, F/Y, K/R, D/E/N/Q, A/G. Positions occupied by such pairs or groups in all 5 polypeptides are indicated by asterisks. The data were compiled from references [20] and [23].

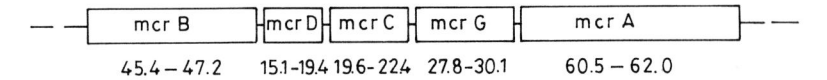

```
— —[  mcr B  ]—[mcrD]—[mcr C]—[ mcr G ]—[   mcr A   ]—  —
    45.4 – 47.2   15.1-19.4 19.6-22.4  27.8-30.1    60.5 – 62.0
```

Fig. 10. Common arrangement of the methyl coenzyme M reductase gene clusters in methanogenic bacteria. Genes *mcrA*, *mcrB*, and *mcrG* code for subunits α, β and γ of the enzyme, respectively. The function of the *mcrC* and *mcrD* gene products are so far unknown. The lower and upper limits of the molecular masses (in kD) of each type of the encoded polypeptides, as derived from the five known nucleotide sequences of *Methanobacterium thermoautotrophicum*, *Methanothermus fervidus*, *Methanococcus voltae*, *Methanococcus vannielii* and *Methanosarcina barkeri*, are shown below the genes.

its derived polypeptide sequence. From the structure of their translation signals, their codon usage and from attempts to detect the gene products in cell extracts by immuno-precipitation, it appears most likely that these polypeptides are not produced in a 1:1 stoichiometry compared with the α, β and γ subunits. Their role remains to be elucidated. Even though all partial reactions of the methyl coenzyme M reduction have been shown to occur in vitro in the apparent absence of these small polypeptides [63], their coordinate expression with the main subunits of methyl coenzyme M reductase suggests that they play a role in the function or assembly of the enzyme. This problem merits further attention.

Acknowledgement

This work was supported by grants from the Deutsche Forschungsgemeinschaft and by the Fonds der Chemischen Industrie.

References

1 Gunsalus, R.P. and Wolfe, R.S. (1978) FEMS Microbiol. Lett. 3, 191–193.
2 Schonheit, P., Moll, J. and Thauer, R.K. (1979) Arch. Microbiol. 123, 105–107.
3 Diekert, G., Klee, B. and Thauer, R.K. (1980) Arch. Microbiol. 124, 103–106.
4 Diekert, G., Weber, B. and Thauer, R.K. (1980) Arch. Microbiol. 127, 273–278.
5 Whitman, W.B. and Wolfe, R.S. (1980) Biochem. Biophys. Res. Commun. 92, 1196–1201.
6 Diekert, G., Gilles, H., Jaenchen, R. and Thauer, R.K. (1980) Arch. Microbiol. 128, 256–262.
7 Diekert, G., Jaenchen, R. and Thauer, R.K. (1980) FEBS Lett. 119, 118–120.
8 Jaenchen, R., Diekert, G. and Thauer, R.K. (1981) FEBS Lett. 130, 133–136.
9 Jaenchen, R., Gilles, H. and Thauer, R.K. (1981) FEMS Microbiol. Lett. 12, 167–170.
10 Pfaltz, A., Jaun, B., Fassler, A., Eschenmoser, A., Jaenchen, R., Gilles, H., Diekert, G. and Thauer, R.K. (1982) Helv. Chim. Acta 65, 828–865.
11 Livingston, D.A., Pfaltz, A., Schreiber, J., Eschenmoser, A., Ankel-Fuchs, D., Moll, J., Jaenchen, R. and Thauer, R.K. (1984) Helv. Chim. Acta 67, 334–351.
12 Pfaltz, A., Livingston, D.A., Jaun, B., Diekert, G., Thauer, R.K. and Eschenmoser, A. (1985) Helv. Chim. Acta 68, 1338–1358.
13 Fassler, A., Kobelt, A., Pfaltz, A., Eschenmoser, A., Bladon, C., Battersby, A.R. and Thauer, R.K. (1985) Helv. Chim. Acta 68, 2287–2298.

14a Pfaltz, A. (1988) In: The Bioinorganic Chemistry of Nickel (J.R. Lancaster, Jr., Ed.) pp. 275–298, VCH Publishers, New York.
14b Färber, G., Keller, W., Kratky, C., Juan, B., Pfaltz, A., Spinner, C., and Kobelt, A. and Eschenmoser, A (1991) Helv. Chim. Acta, 74, 697–716.
15 Ellefson, W.L., Whitman, W.B. and Wolfe, R.S. (1982) Proc. Natl. Acad. Sci. USA 79, 3707–3710.
16 Rouviere, P.E. and Wolfe, R.S. (1988) J. Biol. Chem. 263, 7913–7916.
17 Diekert, G., Konheiser, U., Piechulla, K. and Thauer, R.K. (1981) J. Bacteriol. 148, 459–464.
18 Allmansberger, R., Bollschweiler, C., Konheiser, U., Muller, B., Muth, E., Pasti, G. and Klein, A. (1986) System. Appl. Microbiol. 7, 13–17.
19 Bokranz, M. and Klein, A. (1987) Nucleic Acids Res. 15, 4350–4351.
20 Weil, C.F., Cram, D.S., Sherf, B.A. and Reeve, J.N. (1988) J. Bacteriol. 4718–4726.
21 Cram, D.S., Sherf, B.A., Libby, R.T., Mattaliano, R.J., Ramachandran, K.L. and Reeve, J.N. (1987) Proc. Natl. Acad. Sci. USA 84, 3992–3996.
22 Bokranz, M., Baumner, G., Allmansberger, R., Ankel-Fuchs, D. and Klein, A. (1988) J. Bacteriol. 170, 568–577.
23 Klein, A., Allmansberger, R., Bokranz, M., Knaub, S., Muller, B. and Muth, E. (1988) Mol. Gen. Genet. 213, 409–420.
24 Eschenmoser, A. (1988) Angew. Chemie, Intl. Ed. 27, 5–39.
25 Krautler, B. (1987) Chimia 41, 277–292.
26 Blumberg, W.E. and Peisach, J. (1965) J. Biol. Chem. 240, 870–876.
27 Brumm, P.J., Fried, J. and Friedmann, H.C. (1983) Proc. Natl. Acad. Sci. USA 80, 3943–3947.
28 Valasinas, A., Diaz, L., Frydman, B. and Friedmann, H.C. (1985) J. Org. Chem. 50, 2398–2400.
29 Battersby, A.R., Frobel, K., Hammerschmidt, F. and Jones, C. (1982) J. Chem. Soc., Chem. Commun. 455–457.
30 Bible, K.C., Buytendorp, M., Zierath, P.D. and Rinehart, K.L. (1988) Proc. Natl. Acad. Sci. USA 85, 4582–4586.
31 Treibs, A. (1936) Angew. Chem. 49, 682–686.
32 Habermehl, G.G., Springer, G. and Frank, M.H. (1984) Naturwissenschaften 71, 261–263.
33 Storm, C.B., Krane, J., Skjetne, T., Telnaes, N., Branthaver, J.F. and Baker, E.W. (1984) Science 223, 1075–1076.
34 Ocampo, R., Callot, H.J., Albrecht, P. and Kintzinger, J.P. (1984) Tetrahedron Lett. 25, 2589–2592.
35 Ekstrom, A., Fookes, C.J.R., Hambley, T., Loeh, H.J., Miller, S.A. and Taylor, J.C. (1983) Nature 306, 173–174.
36 Thauer, R.K. (1985) Biol. Chem. Hoppe-Seyler 366, 103–122.
37a Scott, A.I., Williams, H.J., Stolowich, N.J., Karuso, P. and Gonzalez, M.D. (1989) J. Am. Chem. Soc. 111, 1897–1900.
37b Scott, A.J., Williams, H.J., Stolowich, N.J., Karuso, P., Gonzalez, M.D., Blanche, F., Thibaut, D., Müller, G., Savvidis, E. and Hlinency, K. (1989) J. Chem. Soc. Commun. 522–524.
37c Blanche, F., Debussche, L., Thibaut, D., Crouzet, J. and Cameron, B. (1989) J. Bacteriol. 171, 4222–4231.
38a Gilles, H. and Thauer, R.K. (1983) Eur. J. Biochem. 135, 109–112.
38b Mucha, H., Keller, E., Weber, H., Lingens, F. and Trosch, W. (1985) FEBS Lett. 190, 169–171.
39 Pfaltz, A., Kobelt, A., Huster, R. and Thauer, R.K. (1987) Eur. J. Biochem. 170, 459–467.
40 Gilles, H., Jaenchen, R. and Thauer, R.K. (1983) Arch. Microbiol. 135, 237–240.
41 Friedmann, H.C. and Thauer, R.K. (1986) FEBS Lett. 207, 84–88.
42 Friedmann, H.C. Thauer, R.K., Gough, S.P. and Kannangara, C.G. (1987) Carlsberg Res. Commun. 52, 363–371.
43a Avissar, Y.J., Ormerod, J.G. and Beale, S.I. (1989) Arch. Microbiol. 151, 513–519.
43b Li, J.-M., Brathwaite, O., Cosloy, S.D. and Russell, C.S. (1989) J. Bacteriol. 171, 2547–2552.
43c Avissar, Y.J. and Beale, S.I. (1989) J. Bacteriol. 171, 2919–2924.

43d Rieble, S., Ormerod, J.G. and Beale, S.I. (1989) J. Bacteriol. 171, 3782–3787.
44 Ankel-Fuchs, D., Jaenchen, R., Gebhardt, N.A. and Thauer, R.K. (1984) Arch. Microbiol. 139, 332–337.
45 Keltjens, J.T., Whitman, W.B., Caerteling, C.G., van Kooten, A.M., Wolfe, R.S. and Vogels, G.D. (1982) Biochem. Biophys. Res. Commun. 108, 495–503.
46 Hausinger, R.P., Orme-Johnson, W.H. and Walsh, C. (1984) Biochemistry 23, 801–804.
47 Huster, R., Gilles, H. and Thauer, R.K. (1985) Eur. J. Biochem. 148, 107–111.
48 Eschenmoser, A. (1986) Ann. N.Y. Acad. Sci., 471, 108–129.
49 Kratky, C., Waditschatka, R., Angst, C., Johansen, J.E., Plaquevent, J.C., Schreiber, J. and Eschenmoser A. (1985) Helv. Chim. Acta 68, 1312–1337.
50 Shiemke, A.K., Hamilton C.L. and Scott, R.A. (1988) J. Biol. Chem. 263, 5611–5616.
51 Shiemke, A.K., Scott, R.A. and Shelnutt, J.A. (1988) J. Am. Chem. Soc. 110, 1645–1646.
52 Shelnutt, J.A. (1987) J. Am. Chem. Soc. 109, 4169–4173.
53a Jaun, B. and Pfaltz, A. (1986) J. Chem. Soc., Chem. Commun. 1327–1329.
53b Jaun, B. (1990) Helv. Chim. Acta 73, 2209–2217.
54 Stolzenberg, A.M. and Stershic, M.T. (1988) J. Am. Chem. Soc. 110, 6391–6402.
55 Krone, U.E., Laufer, K., Thauer, R.K. and Hogenkamp, H.P.C. (1989) Biochemistry
56 Krone, U.E., Thauer, R.K. and Hogenkamp, H.P.C. (1989) Biochemistry, 28, 4908–4914.
57 Jaun, B. and Pfaltz, A. (1988) J. Chem. Soc., Chem. Commun. 293–294.
58 Thauer, R.K., Diekert, G. and Schonheit, P. (1980) Trends Biochem. Sci. 5, 304–306.
59 Keltjens, J.T., Hermans, J.M.H., Rijsdijk, G.J.F.A., van der Drift, C. and Vogels, G.D. (1988) Antonie v. Leeuwenhoek 54, 207–220.
60 Muller, V., Winner, C. and Gottschalk, G. (1989) Eur. J. Biochem.
61 Kaesler, B. and Schonheit, P. (1989) Eur. J. Biochem. in press.
62 Ellermann, J., Hedderich, R., Bocher, R. and Thauer, R.K. (1988) Eur. J. Biochem. 172, 669–677.
63 Ellermann, J., Rospert, S., Thauer, R.K., Bokranz, M., Klein, A., Voges, M. and Berkessel, A. (1990) Eur. J. Biochem. 184, 63–68.
64 Hartzell, P.L. and Wolfe, R.S. (1986) Proc. Natl. Acad. Sci. USA 83, 6726–6730.
65a Hedderich, R. and Thauer, R.K. (1988) FEBS Lett. 234, 223–227.
65b Hedderich, R., Berkessel, A. and Thauer, R.K. (1990) Eur. J. Biochem. 193, 255–261.
66 Hedderich, R., Berkessel, A. and Thauer, R.K. (1989) FEBS Lett.
67 Rouviere, P.E., Bobik, T.A. and Wolfe, R.S. (1988) J. Bacteriol. 170, 3946–3952.
68 Albracht, S.P.J., Ankel-Fuchs, D., Van der Zwaan, J.W., Fontijn, R.D. and Thauer, R.K. (1986) Biochim. Biophys. Acta 870, 50–57.
69 Albracht, S.P.J., Ankel-Fuchs, D., Bocher, R., Ellermann, J., Moll, J., Van der Zwaan, J.W. and Thauer, R.K. (1988) Biochim. Biophys. Acta, 955, 86–102.
70 Cheesman, M.R., Ankel-Fuchs, D., Thauer, R.K. and Thompson, A.J. (1989) Biochem. J. 260, 613–616.
71 Konheiser, U., Pasti, G., Bollschweiler, C. and Klein, A. (1984) Mol. Gen. Genet. 198, 146–152.
72 Allmansberger, R., Bokranz, M., Krockel, L., Schallenberg, J. and Klein, A. (1989) Can. J. Microbiol. 35, 52–57.
73 Weil, C.F., Sherf, B.A. and Reeve, J.N. (1989) Can. J. Microbiol. 35, 101–108.
74 Ossmer, R., Mund, T., Hartzell, P.L., Konheiser, U., Kohring, G.W., Klein, A., Wolfe, R.S., Gottschalk, G. and Mayer, F. (1986) Proc. Natl. Acad. Sci. USA 83, 5789–5792.
75 Aldrich, H.C., Beimborn, D.B., Bokranz, M. and Schonheit, P. (1987) Arch. Microbiol. 147, 190–194.

CHAPTER 5

Biochemistry and regulation of photosynthetic pigment formation in plants and algae

SAMUEL I. BEALE and JON D. WEINSTEIN

Division of Biology and Medicine, Brown University, Providence, RI 02912, and Department of Biological Sciences, Clemson University, Clemson, SC 29634, U.S.A.

1. The variety and functions of plant and algal tetrapyrroles

A great diversity of naturally occurring tetrapyrroles is found among plants and algae. These species share with other organisms the need for heme-containing cytochromes and oxidases, and in addition, employ tetrapyrroles as pigments for the photosynthetic processes of trapping light energy and converting it to chemical energy.

The tetrapyrrole pigments that are characteristic of plant and algal species fall into two structural groups: Mg-containing closed-macrocycle chlorophylls and their structural relatives, and open-macrocycle bilins. In this chapter the primary focus is on the biosynthesis of chlorophylls and bilins, along with a limited discussion of plant and algal hemes.

1.1. Introduction to the branched tetrapyrrole biosynthetic pathway

Fig. 1 illustrates the biosynthetic relationships among the major groups of tetrapyrrole pigments. The earliest well-characterized precursor that is committed to the tetrapyrrole pathway is ALA. A major branch point occurs at protoporphyrin IX, the last common intermediate leading to both the chlorophylls and the other major products. Another important branch point occurs at protoheme, which is the last common intermediate leading to both other hemes and the phycobilins (including the phytochrome chromophore). A third branch point occurs at uroporphyrinogen III, the last common intermediate that leads to siroheme and the corrinoids. (See Chapter 3).

1.2. Chlorophylls and bacteriochlorophylls

Tetrapyrrole pigments that are grouped under the general classification of chlorophylls all contain Mg as the centrally chelated metal, and all contain a fifth, so called isocyclic, ring (Fig. 2). The macrocycle oxidation state may be that of a porphyrin (as in the chlorophylls *c*), dihydroporphyrin (chlorophylls *a* and *b*, and bacteriochloro-

```
                                                    Other Hemes
                                                   ↗
                                        Protoheme
                                      ↗            ↘
ALA → → Uroporphyrinogen III → → Protoporphyrin IX    Biliverdin IXα
                    ↙         ↘         ↓                    ↓
                                   Mg-Protoporphyrin IX   ┌─────────────────┐
                                         ↓                │ Phytochromobilin│
              Siroheme      Vitamin B₁₂  ↓                │ Phycocyanobilin │
                                         ↓                │ Phycoerythrobilin│
                                         ↓                │ Other Phycobilins│
                                                          └─────────────────┘
                                    Protochlorophyllide
                    Chlorophylls c ↙      ↓
                                      Chlorophyllide a
                                          ↓
                                  ↙   Chlorophyll a
                   Chlorophyll b ↙      ↙        ↘
                                       ↙          ↘
                        Bacteriochlorophylls c,d,e    Bacteriochlorophylls, a,b,g
```

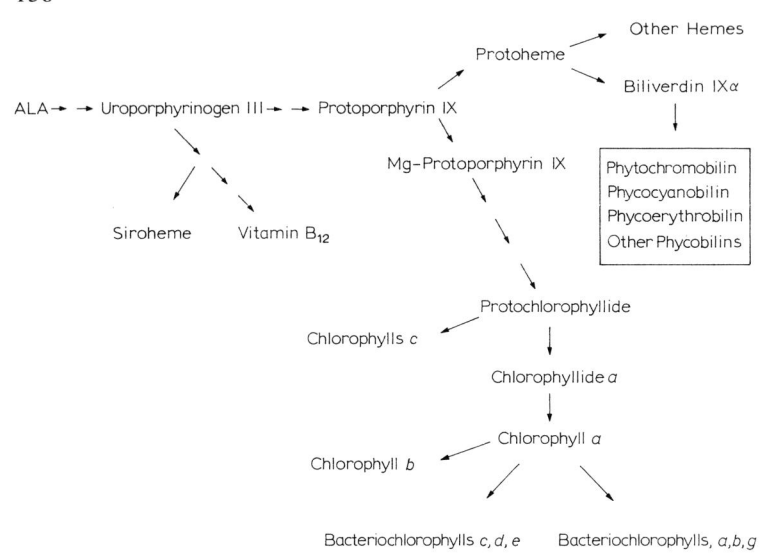

Fig. 1. Outline of the tetrapyrrole biosynthetic pathway, illustrating the end products characteristic of photosynthetic organisms, and their biosynthetic relationships.

phylls c, d, and e) or tetrahydroporphyrin (bacteriochlorophylls a, b, and g). (The bacteriochlorophylls are covered in Chapter 6). In all cases the macrocycle is aromatic and contains a complete conjugated double bond system.

1.3. Hemes

Plant and algal cells contain all three common heme types found in nonphotosynthetic organisms: protoheme (heme b) is a constituent of respiratory cytochromes, photosynthetic electron transport chain components, peroxidases, plant microsomal cytochrome P-450 [1] and other oxidative enzymes; heme a is found in plants as the prosthetic group of mitochondrial cytochrome c oxidase [2]; heme c is found in cytochromes that function in the electron transport chains of both mitochondria (cytochrome c) and plastids (cytochrome f).

1.4. Phycobilins and the phytochrome chromophore

In plants and algae, unlike the situation in most other organisms, bilins are functional tetrapyrrole end products. Cyanobacteria, red algae, and cryptophytes employ bilins as primary light-harvesting photosynthetic pigments, and higher plants and some green algae contain another bilin as the chromophore of the important photomorphogenetic pigment phytochrome. Thus in many photosynthetic species, the bilins must be included in discussions of tetrapyrrole biosynthesis.

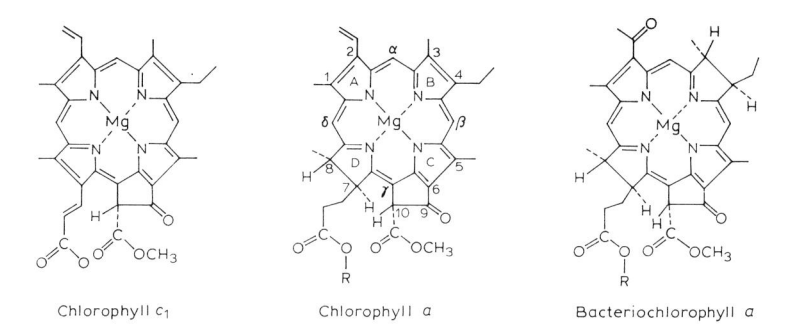

Fig. 2. Three 'chlorophyll' pigments having different degrees of reduction of the macrocyclic ring: chlorophyll c_1 (a porphyrin); chlorophyll a (a dihydroporphyrin); bacteriochlorophyll a (a tetrahydroporphyrin). R is usually phytyl in chlorophyll a, and can be either phytyl or geranylgeranyl in bacteriochlorophyll a. Chlorophyll a is illustrated with the conventional designations for the pyrrole rings by capital letters, pyrrole substituent positions by numbers, and meso positions by lower case Greek letters.

Phycobilins are open-chain tetrapyrroles that function as photosynthetic light-harvesting pigments when covalently linked to specific proteins (Fig. 3). Two major classes of phycobiliproteins are phycocyanins and phycoerythrins, which are respectively colored blue and red. These pigments are largely responsible for the characteristic colors of the organisms which contain them. The different colors arise from slightly different phycobilin chromophores. In bluegreen and red algae, phycobiliproteins form functional aggregates, called phycobilisomes, which decorate the external surface of thylakoid membranes and mediate efficient light absorption and excitation transfer to the photosynthetic reaction centers. The cryptomonad algae also contain phycobiliproteins and utilize them for light harvesting, but in these organisms the phycobiliproteins occur within the inner loculi of the thylakoid membranes. Phycocyanobilin may occur alone as the only phycobilin in some algal species, or it may occur together with other phycobilins, but in all cases, it is the centrally important chromophore necessary for utilization of light energy absorbed by other phycobilin pigments.

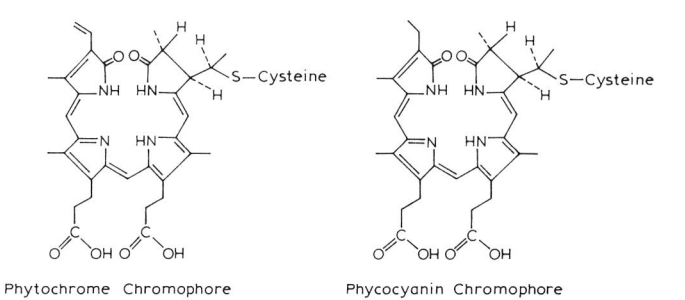

Fig. 3. The chromophores of phytochrome and phycocyanin illustrated with covalent thioether bond linkage to the apoproteins.

The photomorphogenetic pigment phytochrome is a biliprotein whose chromophore structure closely resembles the phycobilins, and whose biosynthesis may share features with the latter. Phytochrome is ubiquitous in higher plants and also occurs in some algae.

1.5. Other tetrapyrroles found in phototrophic organisms

Plant cells contain siroheme as the prosthetic group of nitrite and sulfite reductases [3–5]. Many prokaryotic photosynthetic species are also capable of synthesizing corrinoids. The biosynthesis of these uroporphyrinogen-derived molecules will not be specifically addressed in this article.

There are still other tetrapyrroles found in photosynthetic species, which are more restricted in their distribution. Among these are the dinoflagellate luciferin, which is a bilin that has been proposed to be derived from chlorophyll [6].

2. The biosynthetic route

The tetrapyrrole biosynthetic pathway can be considered to begin from two roots, each giving rise to the precursor ALA by a different mechanism. This is followed by a common trunk leading from ALA to protoporphyrin IX. The pathway then splits into three branches which give rise to hemes (see Chapter 2), bilins, and chlorophylls (Fig. 1). There is also the important branch from the trunk leading to siroheme and the corrinoids (see Chapter 3).

Many, but not all steps of tetrapyrrole synthesis in plants and algae, are similar or identical to those occurring in photosynthetic bacteria. Several of the enzymes were first detected, and some have been more thoroughly characterized, from photosynthetic bacterial sources. Although the focus of this chapter is on the biosynthetic pathway in plants and algae, information derived from the photosynthetic bacteria will be discussed where relevant.

2.1. ALA formation

The earliest well-characterized committed tetrapyrrole precursor is ALA. Two distinct mechanisms exist in photosynthetic species for the diversion of general metabolic intermediates into the tetrapyrrole biosynthetic pathway by transformation to ALA. In yeast and animal mitochondria and in some bacteria ALA, is formed by condensation of glycine and succinyl-CoA (Fig. 4). In other bacteria and in plants and algae (including the prokaryotic cyanobacteria), ALA is formed from the intact carbon skeleton of glutamic acid, in a process requiring three enzymatic reactions and tRNA^{Glu} (Fig. 5).

2.1.1. ALA biosynthesis from glycine and succinyl-CoA
In vitro ALA formation was first reported in extracts of anaerobically grown cells of

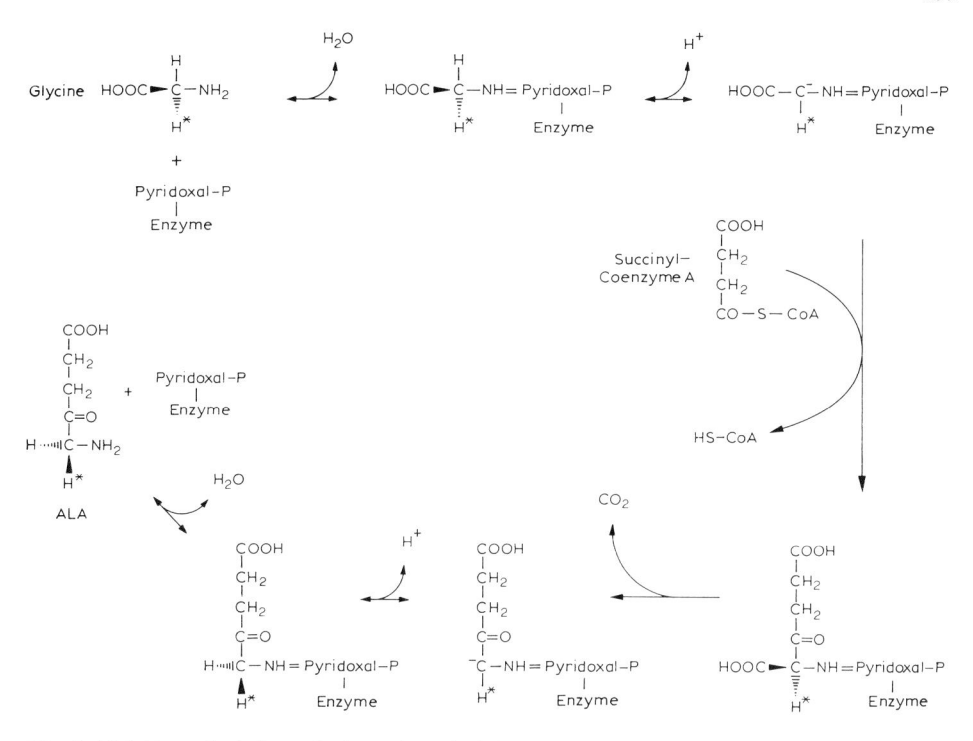

Fig. 4. ALA biosynthesis from glycine and succinyl-CoA.

the facultatively aerobic photosynthetic bacterium *Rhodobacter spheroides*, which synthesize large quantities of bacteriochlorophyll as well as lesser amounts of heme and corrinoids [7, 8]. ALA is formed in these cells by the condensation of succinyl-CoA and glycine, mediated by the pyridoxal phosphate-requiring enzyme, ALA synthase (succinyl-CoA:glycine C-succinyltransferase (decarboxylating) EC 2.3.1.37). In the reaction, the carboxyl carbon of glycine is lost as CO_2 and the remainder is incorporated into the ALA.

ALA synthase was purified over 1600 fold from *R. spheroides* [9]. The native enzyme has a molecular weight of 80,000 and consists of two subunits of 41,000–45,000 molecular weight. A native molecular weight of 61,000–65,000 was derived for ALA synthase from *Rhodopseudomonas palustris* [10]. The ALA synthase reaction mechanism was studied with the *R. spheroides* enzyme. During the reaction, the glycine 2-H atom having the *pro-R* configuration is specifically removed and the *pro-2S* H atom occupies the *S* position at C-5 of ALA [11, 12]. In the absence of succinyl-CoA, the enzyme catalyzes the exchange of one of the C-2 hydrogen atoms of glycine with the medium [13], and in the absence of either substrate, the enzyme catalyzes exchange of one of the C-5 hydrogen atoms of ALA with the medium [14]. The powerful allosteric inhibitor heme has little effect on these exchange reaction with glycine but does affect the exchange reaction with ALA (see Chapter 1).

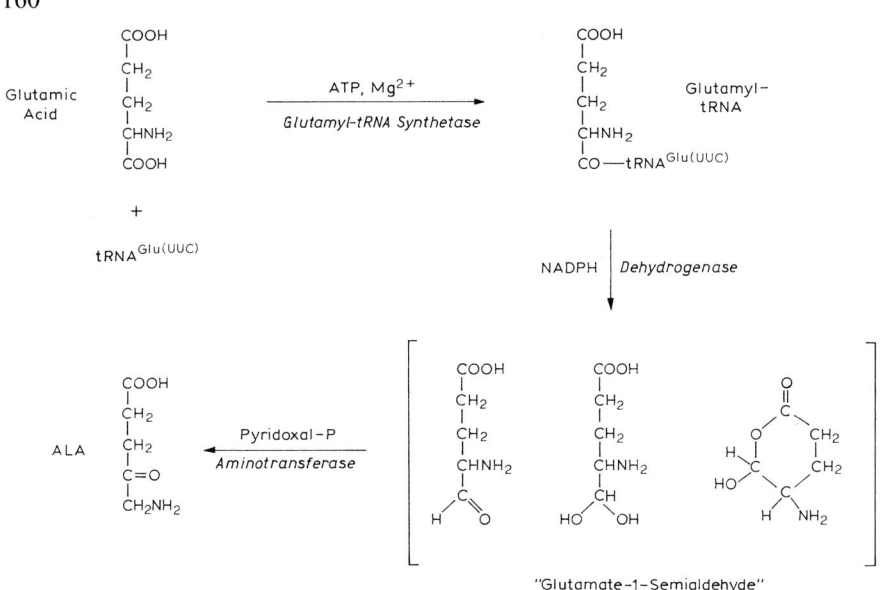

Fig. 5. ALA biosynthesis from glutamate. The structures of three proposed forms of the dehydrogenase reaction product are shown: glutamate-1-semialdehyde; the hydrated form of the semialdehyde; and the cyclized internal ester between the γ-carboxyl and hydrated aldehyde functional groups.

The presence of ALA synthase in plants and algae has remained doubtful. One early brief report of activity in spinach leaf extracts [15] was never confirmed. Certain atypical plant cells, such as nongreening soybean callus cultures [16–19], and greening peels of cold-stored potatoes [20–22] have been reported to contain ALA synthase. Indirect evidence supporting the existence of ALA synthase was reported in dark-grown barley [23], some algae [24–27], and one moss [28]. Operation of ALA synthase was inferred from the relative rates of in vivo label incorporation into ALA from specifically ^{14}C-labeled exogenous presumptive substrates. ALA synthase activity was reported to be present in extracts of certain pigment mutants of the unicellular green alga *Scenedesmus obliquus*, but the reports were too brief to allow evaluation of the nature of the reaction and product [29, 30]. On the other hand, it has been determined that the unicellular red alga *Cyanidium caldarium* [31] and etiolated maize epicotyl sections [2] synthesize all cellular hemes, including the heme *a* prosthetic group of mitochondrial cytochrome oxidase, from ALA that is generated exclusively from glutamate.

ALA synthase activity was found in extracts of the green photosynthetic phytoflagellate *Euglena gracilis* [32]. An examination of the physical and kinetic properties of the enzyme [33], and the regulation of its activity [34, 35] led to the conclusion that in *Euglena*, ALA synthase functions exclusively to provide precursors for nonplastid tetrapyrroles. This conclusion was subsequently proven directly by measuring incorporation of labeled precursors into isolated tetrapyrroles specific to plastids and mitochondria [36]. Under all growth conditions examined, heme a of mitochondrial cytochrome c oxidase was formed from glycine, even while in cells growing under conditions permitting chlorophyll formation, the plastid pigments were formed exclusively from glutamate. Thus unlike higher plants and other algae examined, *Euglena* has both ALA-forming pathways, each separately compartmented and responsible for synthesizing distinct pools of precursors for different classes of tetrapyrrole end products.

2.1.2. ALA biosynthesis from glutamate

2.1.2.1. In vivo evidence for ALA and tetrapyrrole formation from glutamate. When ALA dehydratase inhibitors are administered to greening etiolated plant tissues, ALA accumulates. Radioactivity from ^{14}C-labeled exogenous compounds is incorporated into the accumulated ALA. In greening etiolated plant tissues, glycine and succinate were relatively inefficient contributors of label to ALA, whereas the five-carbon compounds glutamate, α-ketoglutarate, and glutamine, were much better contributors [22, 37, 38]. Based on the uniform degree of incorporation of all of the carbons of glutamate, and the carbon-to-carbon correspondence of label position in precursor and product molecules, a number of hypothetical routes were proposed for the transformation of the intact carbon skeleton of glutamate or α-ketoglutarate into ALA [38, 39]. Preferential and specific transfer of label from glutamate or α-ketoglutarate to ALA formed in vivo, in the presence of levulinic acid, has been found in a variety of cell types, including cyanobacteria [26, 40, 41] red [42] and green [26] algae, and many higher plant tissues [37–39]. However, a contrary result was obtained with the green alga *S. obliquus*, where glycine and succinate were more efficiently incorporated than glutamate [24, 25]. Recently, ALA formation from glutamate has been found to occur in many groups of bacteria (see below).

Evidence supporting the physiological role of a five-carbon pathway in providing ALA for photosynthetic pigments was first provided by Castelfranco and Jones [43]. They demonstrated that both protoheme and chlorophyll were labeled most efficiently by five-carbon precursors in greening barley. Similarly, ^{14}C-labeled α-ketoglutarate was superior to glycine or succinate in contributing label to phycocyanobilin in growing cultures of *Anacystis nidulans* [44]. Based on the relative abilities of C_1- or C_2-labeled acetate to contribute label to chlorophyll and glutamate in *Synechococcus* 6301, McKie et al. [45] concluded that the five-carbon pathway operates exclusively in tetrapyrrole precursor formation. Finally, Oh-hama et al. [46] performed ^{13}C-NMR analysis on chlorophyll formed from ^{13}C-labeled glycine or glutamate in *S. obliquus*. Glycine was shown to contribute label only to the methoxyl group adjacent to the isocyclic ring, while glutamate contributed label in a manner that was consistent with the exclusive operation of the five-carbon pathway in ALA formation. This result was confirmed in maize [47], and similar results were obtained with the purple sulfur photosynthetic bacterium *Chromatium* [48] and the green sulfur bacterium *Prosthecochloris* [49, 50].

It is now generally accepted that a five-carbon pathway is the route of synthesis for ALA destined for all cellular tetrapyrroles in plants and most algae. By measuring the relative ability of ^{14}C-labeled glycine and glutamate to contribute radioactivity to heme a, the prosthetic group of mitochondrial cytochrome oxidase, it was determined that glutamate is the source of the carbon for this heme in etiolated maize epicotyl sections [2] and in *Cyanidium* cells [31]. Other cellular hemes also appear to be made from glutamate in plants. Peroxidase is excreted into the medium by peanut cell suspension cultures. In these cultures the heme moiety was more efficiently labeled with five-carbon ALA precursors than with ALA synthase precursors [51]. Likewise plant microsomal cytochrome *P-450* heme appears to be made from glutamate, as deduced from the inhibition of its biosynthesis by gabaculine (3-amino-2,3-dihydrobenzoic acid), a potent, specific inhibitor of the aminotransferase step in the conversion of glutamate to ALA (see below) [1].

2.1.2.2. Mechanism of ALA formation from glutamate. Much of the earlier work on ALA formation from five-carbon precursors by intact chloroplasts and unpurified cell extracts has been summarized previously [52]. Particulate-free cell extracts capable of converting glutamate to ALA have been obtained from barley [53], *Euglena* [54], the green algae *Chlamydomonas* [55], *Scenedesmus* [56], and *Chlorella* [57], the red alga *Cyanidium* [58], the cyanobacteria *Synechocystis* [59, 60] and *Synechococcus* [59], the prochlorophyte *Prochlorothrix* [59], and the photosynthetic bacterium *Chlorobium* [61]. Similar or identical reaction mechanisms appear to operate in all cases, and reaction components from some heterologous sources can be mixed to reconstitute activity in fractionated systems. Recently, the five-carbon ALA biosynthetic route has been found to occur in several groups of nonphotosynthetic bacteria (see below).

A minimum of three enzyme reactions is required for transformation of glutamate to ALA. In the first, glutamate is ligated to tRNA in a reaction identical or very similar to the charging reaction in protein biosynthesis. Like aminoacyl-tRNA formation in general, this reaction requires ATP and Mg^{2+}. Next, the tRNA-bound glutamate is converted to a reduced form in a reaction that requires a reduced pyridine nucleotide. The product of this reduction has been characterized as glutamate-1-semialdehyde [63] or its hydrated hemiacetal form [63]. However, some of the reported properties of the product are inconsistent with the presence of an α-aminoaldehyde group, and other structures have been proposed [64]. Finally, the positions of the nitrogen and oxo atoms of the reduced five-carbon intermediate are interchanged to form ALA.

Consistent with the above scheme, the ALA-forming systems extracted from barley [65], *Chlorella* [66], and *Synechocystis* [67] have each been separated into four macromolecular components, all of which must be present to catalyze in vitro ALA formation from glutamate. Three of these are enzymes and the fourth is a low molecular weight RNA.

2.1.2.2.1. $tRNA^{Glu}$. In vitro ALA-forming activity in extracts from barley plastids [68] and whole cells of *Chlamydomonas reinhardtii* [69], *S. obliquus* [56], and *Chlorella vulgaris* [70] was blocked by preincubation of the extracts with RNase A. Addition of RNase inhibitor plus low molecular weight RNA from the same species restored activity. RNA was also found to be required for ALA formation from glutamate in extracts of cyanobacteria and a prochlorophyte [59], and the green photosynthetic bacterium *Chlorobium* [61]. In the barley chloroplast system, the required RNA was purified, sequenced, and characterized as tRNA^{Glu}, bearing the UUC glutamate anticodon [71]. Two of the anticodon bases bear modifications: the first U is 5-methylaminomethyl-2-thiouridine and the second is pseudouridine. Although the extent of the modifications in the anticodon region is greater than previously reported for any tRNA^{Glu}, this species is the only tRNA^{Glu} found in barley chloroplasts [72], and thus it must take part in both protein synthesis and ALA formation. Two other glutamate-accepting tRNAs in barley chloroplasts were found to carry glutamine anticodons [72], and it was concluded that in barley chloroplasts, as in *B. subtilis* [73], glutaminyl-tRNA is formed by ligation of tRNA^{Gln} with glutamate, and subsequent amidation of the γ-carbon of glutamate to form the glutaminyl moiety while the molecule is bound to the tRNA.

In several plant and algal species examined, the tRNA required for ALA formation was found to contain the UUC glutamate anticodon. This was determined by affinity purification using an affinity ligand directed against the UUC glutamate anticodon [74]. In the prokaryotic species *Synechocystis*, the tRNA required for ALA formation was first purified by affinity chromatography directed against the UUC anticodon. The purified UUC anticodon-bearing tRNA was then further fractionated into two components by mixed mode (anion exchange and hydrophobic interaction) HPLC [75]. Each homogeneous tRNA was tested for the ability to be charged with glutamate and to participate in ALA formation and protein synthesis in extracts derived from *Synechocystis*. The results indicated that the same species of tRNA can participate in both processes. In vitro, the two tRNAs differ functionally only in their relative abilities to be charged with glutamate by *Synechocystis* extracts. Once charged, they both participate equally well in both ALA formation and protein synthesis. The two *Synechocystis* tRNA^{Glu}s were reported to have the same nucleotide sequence and to differ by some unspecified base modification [60].

2.1.2.2.2. Glutamyl-tRNA. Activity of barley tRNA^{Glu} in ALA formation required the presence of the 3'-terminal CCA, suggesting that the tRNA functions as a glutamate acceptor in the ALA-forming system [71]. Glutamyl-tRNA served as a substrate for ALA formation in *Chlamydomonas* extracts [76]. However, the efficiency of incorporation of label from ^{14}C-glutamyl-tRNA into ALA was low, and the extracts to which the ^{14}C-glutamyl-tRNA was added still had the ability to form ALA from glutamate. More direct proof for the intermediacy of glutamyl-tRNA was obtained with fractionated *Chlorella* extract. Glutamyl-tRNA, but not free glutamate, served as precursor to ALA in a partially reconstituted system lacking one fraction. The missing enzyme fraction had glutamyl-tRNA synthetase activity [77].

2.1.2.2.3. Glutamyl-tRNA synthetase. A single glutamyl-tRNA synthetase was detected in barley chloroplasts [65]. The synthetase was purified by immunoaffinity chromatography. The purified 54,000 molecular weight enzyme was capable of linking glutamate to the tRNA^{Glu} and both tRNA^{Gln} species present in chloroplasts. The aminoacylation reaction required ATP. The presence of this enzyme was required for ALA synthesis from glutamate in the presence of other enzyme fractions. The glutamyl-tRNA synthetase in *Chlorella* extracts that participates in ALA formation was separated from the other enzyme components by serial affinity chromatography on Blue-Sepharose and ADP-agarose [66]. In the absence of this enzyme fraction, ALA was formed from glutamyl-tRNA, but not from free glutamate [77]. The *Chlorella* enzyme that is required for the reaction to proceed starting from free glutamate has a molecular weight of 73,000, while the *Synechocystis* enzyme has a molecular weight of 63,000 [67].

2.1.2.2.4. Glutamate-1-semialdehyde. Glutamate-1-semialdehyde was chemically synthesized and used as a substrate for the in vitro enzyme system [78, 79]. The conversion required only the enzyme system plus glutamate-1-semialdehyde, and was inhibited by the aminotransferase inhibitors, aminooxyacetate and cycloserine. These results were questioned by Kah and Dörnemann [80], who were unable to synthesize glutamate-1-semialdehyde, and by Meisch and Maus [81], who synthesized glutamate-1-semialdehyde and found it to be extremely unstable in solution. They also concluded on theoretical grounds that the method published by Kannangara and Gough [78] could not have yielded glutamate-1-semialdehyde. In response to the above criticisms, a more stable derivative of glutamate-1-semialdehyde, the diethyl acetal, was synthesized and its structure confirmed by mass spectrometry and by carbon and proton magnetic resonance [62]. The diethyl acetal of glutamate-1-semialdehyde could be hydrolyzed in dilute acid to a compound that was converted to ALA by the soluble barley chloroplast enzyme system. The hydrolyzed product was chromatographically indistinguishable from glutamate-1-semialdehyde prepared by the previous method [78]. Other investigators have reported the conversion of glutamate-1-semialdehyde to ALA in in vitro preparations [55, 82, 83]. In these cases the glutamate-1-semialdehyde was supplied by Kannangara and co-workers. Material identical to chemically synthesized glutamate-1-semialdehyde was reported to accumulate in greening barley leaves [85] and extracts of *Scenedesmus* cells [56] when treated with gabaculine, a mechanism-based inhibitor of ω-aminotransferases that blocks chlorophyll synthesis.

Recently, Jordan and co-workers have reported that the compound synthesized by the procedure of Kannangara and Gough is not glutamate-1-semialdehyde, but the cyclic ester between the γ-carboxyl group and the hydrated aldehyde group [64]. The cyclic structure does not contain free aldehyde or carboxylic acid functions, and is more compatible with previously reported properties of the chemically synthesized product (stability in aqueous solution, heat stability) than the free α-aminoaldehyde. The cyclic compound, and not glutamate-1-semialdehyde, was reported to be the product of the dehydrogenase enzyme and the substrate of the barley aminotransferase [64].

2.1.2.2.5. Dehydrogenase. In unfractionated or reconstituted ALA-forming systems from algae or barley chloroplasts, a reduced pyridine nucleotide is required for activity. In *Chlorella* extracts, NADH is about half as effective as NADPH [57], in *Euglena* extracts the two reduced pyridine nucleotides are about equally effective [54], and in *Synechocystis* extracts NADPH is more effective at low concentrations but becomes inhibitory above 1 mM, and NADH is more effective at 5 mM [59]. The ALA-forming system from *Chlorobium vibrioforme* has a nearly exclusive preference for NADPH [67].

For technical reasons, the dehydrogenase activity is usually measured in coupled enzyme assays where the substrate is generated in vitro from glutamate plus tRNA, and/or the product is converted in situ to ALA. The isolated dehydrogenase reaction, conversion of glutamyl-tRNA to a product identical to chemically synthesized glutamate-1-semialdehyde, was reported in *Chlamydomonas* extracts when the conversion of the dehydrogenase product to ALA was prevented by separation of the aminotransferase enzyme from the other two enzymes by affinity chromatography prior to incubation [76]. In this preparation, glutamyl-tRNA synthetase activity was not physically separated from the dehydrogenase.

The enzyme component that utilizes the pyridine nucleotide cofactor was physically separated from the other enzyme components by affinity chromatography [66], and this enzyme was shown to be active in ALA formation from glutamyl-tRNA in the absence of glutamyl-tRNA synthetase [77]. Barley dehydrogenase was also separated from glutamyl-tRNA synthetase by immunoaffinity chromatography and chromatography on salicylate-Sepharose, and found to retain activity when recombined with the glutamyl-tRNA synthetase fraction [65]. Earlier, aminotransferase activity was physically separated from the other enzymes in barley extracts by Blue-Sepharose affinity chromatography. The product formed by incubation of the other enzymes with glutamate, ATP, Mg^{2+} and NADPH copurified with chemically synthesized glutamate-1-semialdehyde [85].

2.1.2.2.6. Aminotransferase. An enzyme capable of converting chemically synthesized glutamate-1-semialdehyde to ALA was purified from extracts of barley chloroplasts [78, 85] and *Chlamydomonas* cells [55] by affinity chromatography. The transamination reaction requires no added substrate or cofactor other than glutamate-1- semialdehyde. Barley aminotransferase has a native molecular weight of 80,000 [86, 87] and is inhibited by very low concentrations of gabaculine [63]. Gabaculine-resistant mutants of *Chlamydomonas* contain elevated levels of aminotransferase activity [88]. However, gabaculine-resistant mutants of *Synechococcus* 6301 were recently reported to contain a variant aminotransferase that is relatively insensitive to gabaculine [89]. Native aminotransferase from *Synechocystis* has a molecular weight of 99,000 [67]. Purified aminotransferase from both barley and *Synechococcus* has a molecular weight of 46,000 on denaturing SDS-PAGE [90].

The question of whether the migration of the amino group in the conversion of glutamate to ALA involves an intramolecular or intermolecular transfer was examined in *Chlamydomonas* extracts by the use of ^{13}C- and ^{15}N-labeled glutamate. When the heavy isotope labels were present on separate substrate molecules, a significant proportion of the ALA product molecules contained two heavy atoms, suggesting that the conversion occurs by intermolecular nitrogen transfer [91]. However, the isotopic stability of ALA in the incubations was not measured, leaving open the possibility that the observed redistribution of heavy isotopes among product molecules occurred after the initial formation of ALA.

In most reports on in vitro transformation of glutamate to ALA, pyridoxal phosphate was included in all media used for cell homogenization, enzyme fractionation, and incubation, even though dependence upon added pyridoxal phosphate for activity could not be demonstrated [55, 78, 85]. The evidence for the involvement of pyridoxal phosphate as a cofactor for the aminotransferase step consisted of the observation that in vitro ALA formation was inhibited by compounds such as aminooxyacetate and gabaculine, which are considered to be pyridoxal antagonists [57, 84]. However, when cell disruption and enzyme fractionation were carried out in the absence of added pyridoxal phosphate, and the relative proportion of the enzymes in the reconstituted ALA-forming assay was adjusted so that the aminotransferase was rate limiting, strong pyridoxal phosphate dependence of ALA formation could be directly demonstrated with the *Chlorella* enzyme [83].

2.1.3. Phylogenetic distribution of the two ALA-forming pathways
The in vivo and in vitro results discussed earlier indicate that in higher plants, green algae, red algae, and cyanobacteria, all cellular tetrapyrroles are formed from glutamate. Some photosynthetic bacteria have the glutamate pathway while others have the glycine pathway. *Euglena* is uniquely able to form plastid tetrapyrroles from glutamate while simultaneously forming mitochondrial hemes from glycine and succinate.

Recently, all of the photosynthetic bacterial groups were surveyed for the mode of ALA formation by in vitro measurement of label transfer to ALA from [1-^{14}C]glutamate and [2-^{14}C]glycine, and by measurement of the effect of RNase on the label transfer [92]. The results indicated that the pathway from glutamate is widely distributed among the bacterial groups and is probably the more primitive and evolutionarily earlier pathway. Genera containing the glutamate pathway include green sulfur bacteria (*Chlorobium*), green nonsulfur bacteria (*Chloroflexus*), Gram positive green bacteria (*Heliospirillum*), purple sulfur bacteria (*Chromatium*), and *Desulfovibrio* (which is not photosynthetic but may be closely related to photosynthetic groups). The glycine pathway was found only in the purple nonsulfur bacteria (*R. spheroides*, *Rhodospirillum rubrum*). Several strains of bacteria have been reported to form bacteriochlorophyll *a* under aerobic growth conditions. In all cases where the route of ALA formation has been determined in these species, the ALA synthase route has been found [93, 94]. It should also be noted that the five-carbon ALA biosynthetic pathway is not restricted to chlorophyll-containing organisms. The operation of this pathway has been demonstrated in vivo and in extracts of *Clostridium thermoaceticum* [95], *Methanobacterium thermoautotrophicum* [96, 97], *E. coli* [83, 98–100] and *B. subtilis* [100].

2.2. The Pathway from ALA to protoporphyrin IX

The steps leading from ALA to protoporphyrin, and the enzymes catalyzing the reactions, in plants and algae are generally identical or very similar to those in nonphotosynthetic species. An important consideration in plants and eukaryotic algae is the intracellular distribution of the enzymes. Isolated chloroplasts are capable of synthesizing protochlorophyllide from exogenous glutamate [101, 102] thus they must contain all of the enzymes catalyzing the reactions leading from glutamate to porphyrins. It is not yet clear how many of the biosynthetic steps leading to hemes may also occur in other cellular regions, and, for those that do, whether the properties of the nonplastid enzymes differ from those of the plastids. Knowledge gained in this area will shed light on the degree of cellular dependence on the plastids for tetrapyrrole biosynthesis and have implications for the functions of the plastid metabolism and its regulation, especially in nonphotosynthetic tissues.

2.2.1. ALA dehydratase
Formation of PBG by asymmetric condensation of two ALA molecules is catalyzed by the enzyme ALA dehydratase (EC 4.2.1.24). Plant and algal ALA dehydratase has been studied extensively. The enzyme from *Chlorella regularis* has a native molecular weight of 316,000, a pH optimum of 8.5, and a K_m for ALA of 0.5 mM [103]. Similar values were reported for the enzyme derived from radish cotyledons [104] spinach leaves [105], and tobacco tissue culture [106]. Spinach leaf ALA dehydratase is a hexamer of about 300,000 molecular weight [107]. The enzyme from *R. spheroides* is also a hexamer [108]. These results are in apparent contrast to the octameric structure of the animal enzyme. Another apparent difference among the enzymes from different sources is the metal requirement. Whereas the animal enzyme requires Zn^{2+} for activity, the plant and bacterial enzymes require Mg^{2+} [103, 104, 107, 109, 110]. The similarity of plant and bacterial ALA dehydratases suggests that the plant enzyme is of the prokaryotic type and located within the plastids. Most of the ALA dehydratase activity in greening radish cotyledons is associated with the plastid stroma, but a portion is bound to the thylakoid membranes [111]. The significance of this intraplastid distribution is unknown. Whether green plant cells also contain an additional, animal-like ALA dehydratase in a nonplastid region is not known. One report local-

izes ALA dehydratase in the plastids of pea leaves and *Arum* spadices [112]. The latter tissue is nongreening. However, it is interesting that ALA dehydratase from nongreening soybean callus cells was reported to share with the animal cytoplasmic enzyme a requirement for Zn^{2+}, rather than Mg^{2+} [113].

In single turnover experiments, [114] established that in the reaction catalyzed by ALA dehydratase from *R. spheroides*, the first bound ALA molecule is the one that contributes the propionic acid side chain of the product. In the formation of PBG, the removal of hydrogen to form the aromatic pyrrole ring must occur on the enzyme, as is indicated by the stereospecific retention of one of the two hydrogen atoms derived from the C_5 hydrogens of ALA [115].

Levulinic acid and dioxoheptanoic acid act as specific, competitive inhibitors of ALA dehydratase, and when administered to greening plant tissues or algal cells, cause the accumulation of ALA [116, 117]. The use of these inhibitors has facilitated the determination of the route of ALA formation and the role of ALA availability in regulating the rate of pigment formation in plants and algae.

2.2.2. PBG deaminase

PBG deaminase condenses four PBG molecules to form the first tetrapyrrole, uroporphyrinogen. The initial product of enzymic catalysis is the linear tetrapyrrole hydroxymethylbilane, which, in the absence of uroporphyrinogen III synthase, spontaneously cyclizes to form uroporphyrinogen I. Biosynthesis of the biologically relevant isomer, uroporphyrinogen III, requires the presence of uroporphyrinogen III synthase during or immediately after release of the initial tetrapyrrole product of PBG deaminase.

PBG deaminase was purified 200-fold from *Euglena* [118]. The native enzyme, a monomer of 41,000 molecular weight, does not require metal ions for activity. This appears to conflict with a report of the isolation of a pteridine compound from *Euglena* cells which stimulates *Euglena* PBG deaminase activity in vitro [119, 120]. PBG deaminase from wheat germ and spinach leaves also has a molecular weight of about 40,000 [121]. The molecular weight of the enzyme from *Chlorella regularis* was 35,000 [122] and that of *R. spheroides* 36,000 [123], while the molecular weight of PBG deaminase from *R. palustris* was reported to be 74,000 [124]. PBG deaminase activity from dark grown *Euglena* cells has been resolved into two components having molecular weights of 40,000 and 20,000 [125].

cDNA coding for PBG deaminase has been cloned from light-grown *Euglena* [126]. The cDNA codes for a 51,744 molecular weight peptide comprising a mature protein of 36,927 molecular weight plus a 139 amino acid N-terminal extension. The N-terminal extension has properties similar to those of transit peptides for nuclear-encoded chloroplast proteins.

PBG deaminases from *R. spheroides* and *Euglena* were used to establish that the order of assembly of the four PBG units is ABCD, as they appear in uroporphyrinogen [127, 128]. Covalently enzyme-bound mono- through tetrapyrrole intermediates are formed [129, 130] and then hydroxymethylbilane is released [131, 132].

The nascent mono- through tetra-pyrrole enzyme-complexes have been shown to be bound via a dipyrrole cofactor [133, 134]. In PBG deaminase from *E. coli*, the dipyrrole cofactor appears to attach to a cysteine group on the apoenzyme during formation of the protein, and remains permanently attached to the enzyme while the link between the cofactor and the nascent pyrrole chain is severed after the tetrapyrrole stage is reached. PBG deaminase from plant sources also contains the dipyrrole cofactor [135].

Although PBG deaminase is a cytoplasmic enzyme in animal cells, it has been localized within the plastids in green and etiolated pea leaves and in *Arum* spadices (a nongreening tissue) [112], and has been shown to be associated with the stroma of developing pea leaf chloroplasts [136]. Earlier work had also suggested that a major proportion of activity in spinach leaves is associated with the chloroplasts [137]. Dark-grown *Euglena* cells appear to contain particulate form of PBG deaminase in addition to a soluble form [125].

2.2.3. Uroporphyrinogen III synthase
Uroporphyrinogen III synthase has been purified from *Euglena* [138]. The native enzyme is a monomer of 31,000 molecular weight and contains no reversibly-bound cofactors or metal ions. The molecular weight value is similar to that of the rat liver enzyme [139] but contrasts with the value of 62,000 reported for the wheat germ enzyme [121]. The insensitivity of the *Euglena* enzyme to diethyl pyrocarbonate [138] suggests that it contains no essential histidine groups, in contrast to the human enzyme [140].

PBG deaminase and uroporphyrinogen III synthase may form a complex that facilitates transfer of hydroxymethylbilane between the two enzymes. The presence of *Euglena* uroporphyrinogen III synthase influences the K_m of *Euglena* deaminase for PBG [141]. The sedimentation velocity of wheat germ deaminase is also influenced by the presence of wheat germ uroporphyrinogen III synthase [121]. The presence of *R. spheroides* uroporphyrinogen III synthase was reported to facilitate release of the tetrapyrrole product from PBG deaminase [142].

2.2.4. Uroporphyrinogen decarboxylase
Uroporphyrinogen decarboxylase catalyzes the decarboxylation of all four of the acetate residues on uroporphyrinogen to yield coproporphyrinogen, which contains methyls in their place. There have been relatively few studies of this enzyme from photosynthetic organisms. Uroporphyrinogen decarboxylase activity was measured in leaf extracts of several plants and purified 72 fold from tobacco leaves [143]. The highest activity was found in the soluble fraction of the leaf homogenate, and very little activity was found in the organelle fractions. No metal requirements were observed, and EDTA or other metal chelating agents enhanced activity. Uroporphyri-

nogen III was reported to be a much better substrate than uroporphyrinogen I for the tobacco enzyme.

2.2.5. Coproporphyrinogen oxidase

Although the ability of isolated cucumber plastids to form protoporphyrin from exogenous ALA indicates that plastids must contain coproporphyrinogen oxidase activity [144, 145] the only reported extraction of this enzyme from higher plants indicated that in tobacco leaves coproporphyrinogen oxidase was associated with the mitochondria [146].

Euglena coproporphyrinogen oxidase accepts ring A monovinyl porphyrinogen more readily than ring B monovinyl porphyrinogen as a substrate, suggesting that oxidative decarboxylation of the ring A propionate of coproporphyrinogen occurs before that of ring B [147]. Although O_2 is the electron acceptor in aerobic systems, extracts of anaerobic *R. spheroides* can carry out the reaction anaerobically in the presence of ATP, oxidized pyridine nucleotide, and methionine [148]. These requirements are similar to those reported for anaerobic yeast extracts [149].

2.2.6. Protoporphyrinogen oxidase

Protoporphyrinogen oxidase activity has been detected in extracts of *R. spheroides* [150] and barley leaves [151, 152]. The enzyme was found in both the plastid and mitochondrial fractions of etiolated barley leaves [153]. The enzyme from the two organelles appeared to be identical. The approximately 210,000 molecular weight enzyme accepts both protoporphyrinogen and mesoporphyrinogen as substrate, in contrast to the rat liver enzyme which accepts only protoporphyrinogen. Although the K_m of 5 μM for protoporphyrinogen was similar to the reported value for mammalian protoporphyrinogen oxidase, the pH optimum of 5–6 differed markedly from the optimum of 7.5–8.5 reported for the mammalian and yeast enzymes. Purified barley protoporphyrinogen oxidase migrated as a single 36,000 molecular weight band on SDS-PAGE electrophoresis. Activity was stimulated by unsaturated fatty acids, suggesting that the plant enzyme may have a lipid requirement for activity [154].

The electron acceptor in the reaction catalyzed by extracts of aerobically-grown cells is O_2. However, nitrite or fumarate can function as the electron acceptor in the reaction in extracts of anaerobically grown *E. coli*, and presumably also in anaerobic photosynthetic bacteria.

2.3. The Fe branch

2.3.1. Ferrochelatase and protoheme formation

Ferrochelatase (protoheme ferrolyase, EC 4.99.1.1), the enzyme responsible for insertion of Fe into the porphyrin macrocycle, is described elsewhere in this volume.

Membrane-bound ferrochelatase from the photosynthetic bacterium, *R. spheroides*, has been purified and characterized [155, 156]. The enzyme is a single polypeptide of molecular weight 115,000. The apparent K_m values for protoporphyrin and Fe^{2+} are 18 and 22 μM, respectively. Although the enzyme has not been purified to homogeneity from any plant source, its activity has been measured in cell-free extracts of plant material [157–161] and it has been partially purified from etiolated barley [157]. Unlike preparations from bacteria and animals, in the barley preparation, 50% of the activity appeared to be in the soluble fraction after a 20,000 × g centrifugation. Ferrochelatase activity co-purified with zinc chelatase activity through ammonium sulfate fractionation and gel filtration chromatography on Sephadex G-150. The estimated molecular weight, by gel filtration, was 55,000–65,000. Mesoporphyrin, hematoporphyrin, and deuteroporphyrin were all better substrates than protoporphyrin in the ferrochelatase assay. Ferrochelatase activity has also been characterized from spinach chloroplasts that were stripped of their outer envelope membranes [161]. The preparation was free of mitochondrial contamination, as judged by phase contrast microscopy. The K_m for Fe^{2+} in the presence of 25 μM protoporphyrin or mesoporphyrin was 8 μM or 36 μM, respectively. The K_m values for protoporphyrin and mesoporphyrin were 0.2 μM and 0.4 μM, respectively (at non-saturating Fe^{2+} concentration). Like most ferrochelatase preparations from other sources, the enzyme was membrane bound.

Because chloroplasts and mitochondria have independent needs for hemes, it is logical to assume that each organelle has its own independent machinery for heme biosynthesis. A more detailed discussion of this premise will be presented in the section on regulation. Porra and Lascelles [159] prepared chloroplasts, proplastids and mitochondria from a variety of plant tissues by differential centrifugation. Spinach chloroplasts had ferrochelatase activity, as did chloroplasts and proplastids from greening bean cotyledons and oat seedlings. Microscopic examination of the above preparations indicated little contamination by mitochondria. Because the specific activity did not change upon washing, the authors concluded that the activity was not due to mitochondrial contamination. Mitochondria from the above tissues also had ferrochelatase activity, but the preparations were obviously contaminated with chloroplast fragments. Mitochondria, slightly contaminated with leucoplasts, were prepared from potato tubers. These mitochondria had ferrochelatase activity and cytochrome oxidase activity. When the mitochondria were washed, the ratios of the two activities remained the same. No ferrochelatase activity was detected in the leucoplast fraction. The authors concluded that ferrochelatase is localized in both mitochondria and chloroplasts of plant cells.

The subcellular localization of ferrochelatase in both chloroplasts and mitochondria of plant cells was put on a more firm foundation by the studies of Little and Jones [158]. These investigators prepared washed etioplasts and washed mitochondrial fractions from etiolated barley by differential centrifugation. These fractions were then subjected to sucrose-density gradient centrifugation. The density gradient-purified etioplasts had ferrochelatase activity and no detectable cytochrome oxidase activity. Upon density gradient centrifugation, the washed mitochondria yielded two bands of ferrochelatase activity, one coinciding with cytochrome oxidase and succinate dehydrogenase activity and the other having a density corresponding to the etioplasts. Even with density gradient centrifugation, it was not possible to completely free the mitochondria from contamination with chloroplast fragments. Ferrochelatase in the washed etioplast preparation had optimum activity at pH 7.3, while the activity in the gradient-purified mitochondrial fraction had an optimum at pH 8.0. The ferrochelatase activity from both organelles was inhibited 50% or more by Fe-, Mg-, or Zn-protoporphyrin at concentrations between 3.0 and 7.0 μM.

Brown et al. [162] measured ferrochelatase activity in extracts of *Cyanidium*, by using Co^{2+} and deuteroporphyrin IX in place of the physiological substrates Fe^{2+} and protoporphyrin IX. In wild-type cells, which do not normally form pigment in the dark, the in vitro ferrochelatase level increased several fold within 72 h after the cells were transferred from darkness to light. The increase in ferrochelatase activity paralleled the accumulation of phycocyanin, which suggests that the level of this enzyme may be an important rate-controlling factor in phycobilin synthesis. The intracellular localization of the measured ferrochelatase activity was not reported.

N-methyl porphyrins apparently inhibit plant ferrochelatase, as they do the animal enzyme. Administration of N-methyl proto- or mesoporphyrin to growing *Cyanidium* cells caused protoporphyrin accumulation and inhibited phycobilin formation [163,164]. This result supported other evidence that heme is a precursor to the phycobilins.

2.3.2. Bilin formation

2.3.2.1. Heme as a phycobilin precursor. The participation of heme in phycobilin formation is suggested by similarities of the phycobilins to the tetrapyrrole macrocycle ring-opening reaction products appearing in animal heme catabolism. Also, mechanistic studies have indicated the necessity of the central Fe atom in the ring-opening reaction.

Direct evidence for the participation of heme was reported by Brown et al. [165] and Schuster et al. [166], who showed that exogenous ^{14}C-heme could contribute label to phycocyanobilin in greening *Cyanidium* cells. The specificity of heme incorporation was indicated by the fact that unlabeled chlorophyll was formed simultaneously with the labeled phycocyanobilin when greening cells were incubated with ^{14}C-heme. Further evidence supporting a role for heme was provided by the observation that nonradioactive heme was able to decrease the incorporation of ^{14}C-labeled ALA into phycocyanobilin [165].

2.3.2.2. Biliverdin as a phycobilin precursor. Biliverdin was reported to accumulate in the culture medium of *Cyanidium* cells grown in the dark with exogenous ALA

[167]. Although this result was interpreted as evidence that biliverdin is a precursor of phycocyanobilin, the biliverdin accumulation could also have resulted from degradation of excess heme that might have been formed as a result of ALA administration.

Beale and Cornejo [168] found that the phycocyanin chromophore became labeled when purified ^{14}C-biliverdin IXα was administered to *C. caldarium* cells growing in the dark in the presence of N-methyl mesoporphyrin IX to block endogenous heme formation. The strain of cells used in these experiments was capable of forming phycocyanin in the dark. Dark growth was used to eliminate possible phototoxic effects of administered N-methyl mesoporphyrin or biliverdin. Cellular protoheme remained unlabeled during the incubations with ^{14}C-biliverdin, indicating that the conversion of biliverdin to phycocyanobilin was direct, rather than via degradation of the administered labeled compound and subsequent reutilization of the ^{14}C. The ability of exogenous ^{14}C-biliverdin to label the phycocyanin chromophore in vivo was confirmed [162, 169].

2.3.2.3. Algal heme oxygenase. Several key features of the heme oxygenase reaction occur in phycobilin formation in vivo. When intact *Cyanidium* cells were allowed to form phycocyanin in the presence of [^{14}C-5]ALA, ^{14}CO and ^{14}C-phycocyanobilin were formed in equimolar amounts [170]. Also, each of the lactam oxygen atoms of phycocyanobilin is derived from a different O_2 molecule [171, 172].

Beale and Cornejo [173] detected heme oxygenase activity in extracts of *Cyanidium*. Originally, it was necessary to utilize mesoheme in place of the physiological substrate protoheme. The advantage of mesoheme is that the reaction product mesobiliverdin is more stable than biliverdin in the cell extract, and thus accumulates to detectable levels during the incubation. Using optimized incubation conditions, conversion of protoheme to biliverdin was also detected in these cell extracts [175].

Cyanidium heme oxygenase differs from the animal cell-derived microsomal system in that it is soluble, with virtually all of the activity appearing in the high-speed supernatant. This finding is not too surprising in view of the fact that the reaction is thought to occur in the plastids, and also presumably occurs in prokaryotic bluegreen algae. Neither plastids nor prokaryotes have microsomes.

Like the animal system, *Cyanidium* heme oxygenase requires reduced pyridine nucleotide (NADPH is about twice as effective as NADH), as well as O_2. In addition to the reduced pyridine nucleotide, isoascorbate or another moderately strong reductant stimulates the reaction in unpurified cell extracts, and is required in more purified preparations. Perhaps the isoascorbate serves to reduce heme to the ferrous state, or otherwise supplies reducing power in the incompletely purified system. Like the microsomal enzyme [175], the algal heme oxygenase is powerfully inhibited by Sn-protoporphyrin IX.

Algal heme oxygenase was fractionated into three required protein components, having molecular weights of 22,000, 37,000, and 38,000 [174]. The small 22,000 molecular weight protein contains an Fe/S cluster, as is the case with other nonmicrosomal monooxygenases, and can be replaced by commercial ferredoxin [174]. The 37,000 molecular weight protein fraction binds NADPH and has ferredoxin-linked NADPH-cytochrome c reductase activity. The 38,000 molecular weight fraction is inactivated by diethyl pyrocarbonate and the inactivation is blocked by heme, indicating that this fraction is the one that binds heme. The proposed roles of the three *Cyanidium* protein fractions that reconstitute heme oxygenase activity are illustrated in Fig. 6.

2.3.2.4. Biliverdin reduction to phycocyanobilin. All of the phycobilins that have been described (Fig. 7) contain four more hydrogen atoms than biliverdin. Beale and Cornejo [176] measured enzymatic conversion of biliverdin to free phycocyanobilin in cell-free extracts of *Cyanidium*. In addition to biliverdin IXα, the reaction required a reduced pyridine nucleotide, NADPH being more effective than NADH. Activity was retained in the high-speed supernatant fraction and eluted with the protein fraction on gel filtration. Product identification was by comparative absorption spectroscopy, reverse-phase HPLC, and chemical derivatization. Products of the reaction included both the Z- and E-ethylidine isomers of phycocyanobilin. At early reaction times, the less stable Z isomer was the predominant reaction product [176]. The time courses for the appearance of the two isomers suggests a precursor-product relationship between the Z and E forms. Preliminary evidence indicates that enzymatic isomerization is catalyzed by cell extract in the presence of reduced glutathione [177]. Several other cis-trans isomerases have been reported to require reduced glutathione for activity [178, 179].

Interestingly, both phycocyanobilin ethylidine isomers are also formed upon methanolytic cleavage of the phycocyanin chromophore from the protein moiety [180], but the Z isomer, being less stable [181], isomerizes rapidly at the elevated temperatures employed for methanolysis, and the equilibrium isomer ratio strongly favors the E form.

α-Hydroxymesobiliverdin IXα was reported to be excreted, in addition to biliver-

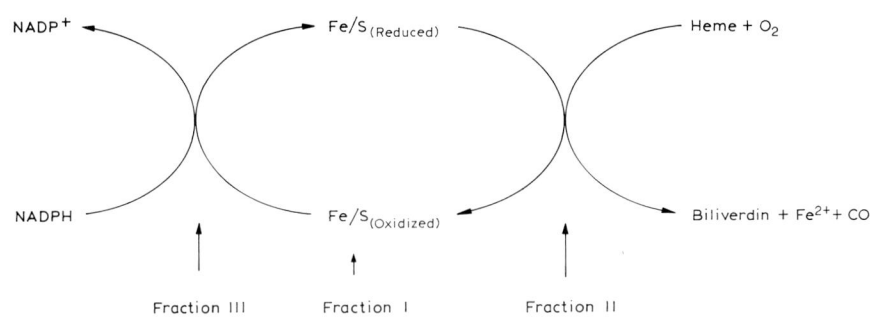

Fig. 6. Proposed roles for the three protein components of the algal heme oxygenase system from *C. caldarium*. In this system, Fraction III accepts electrons from NADPH and reduces Fraction I, which is a ferredoxin-like protein. Electrons are passed from Fraction I to Fraction II, which binds heme and catalyzes the macrocycle ring opening reaction.

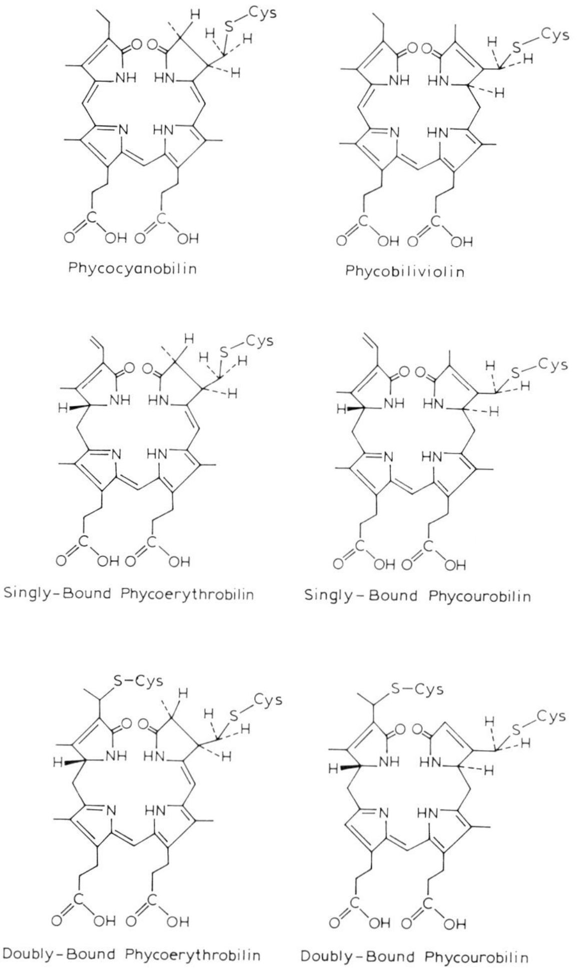

Fig. 7. Structures of phycobilin chromophores and a hypothetical biosynthetic relationship. Although the chromophores are illustrated in their protein-bound form, biosynthesis proceeds at least as far as phycocyanobilin in the unbound form (see text).

din and free phycocyanobilin, when ALA is administered to *Cyanidium* cells in the dark [182]. Although the authors suggested that the compound might arise as a degradation product of phycocyanobilin, another interpretation is possible: If the initial two-electron reduction step, acting on biliverdin, is reduction of the pyrrole ring, then the resulting product will contain a vinyl group that is no longer conjugated to the rest of the tetrapyrrole double bond system (Fig. 8). This molecule, with its relatively unstable vinyl group, could be the activated substrate for ligation to the apoprotein. In the absence of a suitable acceptor apoprotein, the vinyl double bond would shift toward the pyrrole ring, to form the ethylidine and re-establish conjugation. This scenario would also explain the generation of free phycocyanobilin isomers having both *E*- and *Z*-ethylidine configurations in the in vitro conversion of biliverdin to phycocyanobilin [176]. An alternative fate of the reactive vinyl-containing intermediate, in the absence of apoprotein, would be for water to add across the vinyl group, forming α-hydroxymesobiliverdin. A recent report indicates that a chemically synthesized bilin, containing a reduced pyrrole ring and a vinyl group, is very unstable and rapidly isomerizes to Z-ethylidine phycocyanobilin in neutral solution [183].

Preliminary work on the enzyme system from *Cyanidium* that catalyzes the reductive conversion of biliverdin to phycocyanobilin has revealed that there are four enzyme components which are separable by differential ammonium sulfate precipitation, affinity chromatography, and gel filtration [177]. Two of these, a ferredoxin-linked NADPH-cytochrome c reductase and a low molecular weight Fe/S protein,

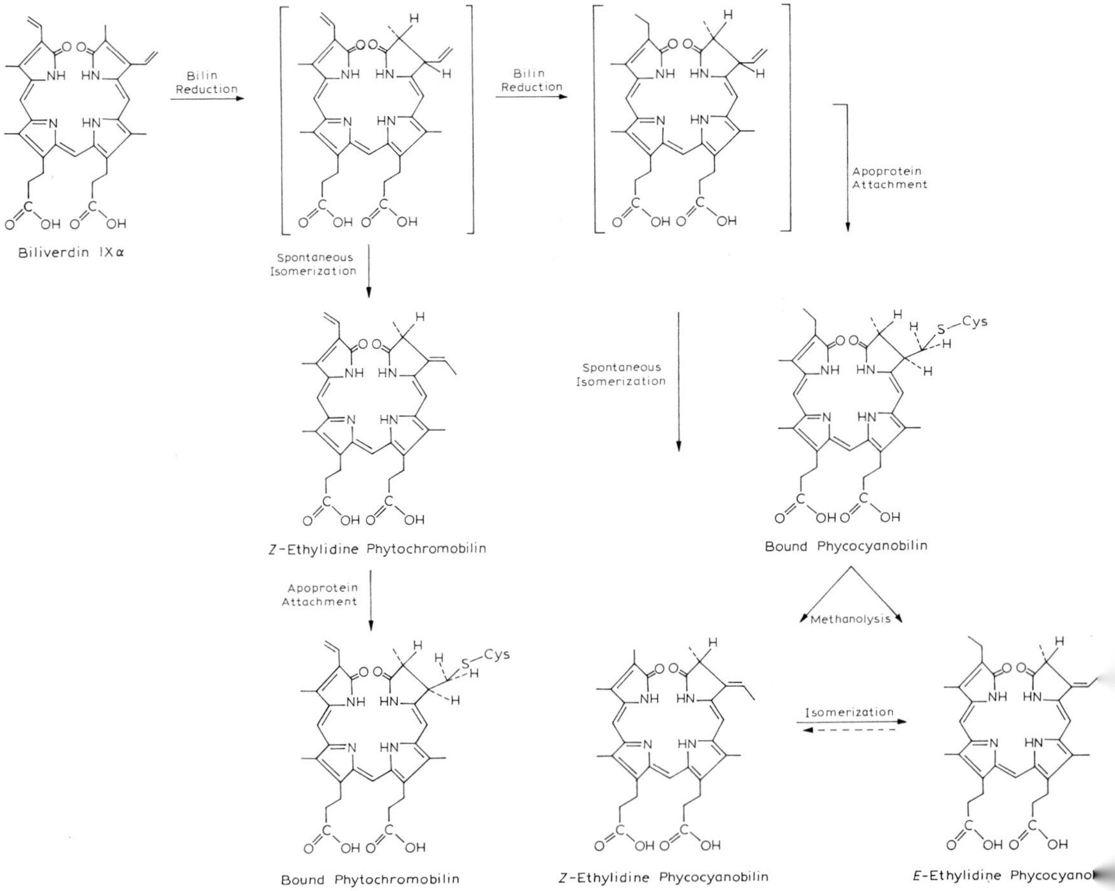

Fig. 8. Proposed biosynthetic pathway for phycocyanobilin, and structures of the unbound pigment forms derived from precursors and the protein-bound form.

appear to be identical to those of the *Cyanidium* heme oxygenase system. The third component appears to be a bilin dehydrogenase which transfers electrons from the reduced Fe/S protein to biliverdin, to form a reduced bilin whose structure has not been determined. The fourth component is an isomerase that transforms the product of the bilin reductase to phycocyanobilin having the Z-ethylidine configuration. This transformation proceeds in the absence of reduced pyridine nucleotide. A hypothetical model for the transformation of biliverdin to phycocyanobilin that is consistent with the above results is presented in Fig. 8.

2.3.2.5. Biosynthesis of other phycobilins. Phycocyanobilin and phycoerythrobilin were the first phycobiliprotein chromophores to be chemically characterized [184]. Other chromophores with mesobiliverdin, urobilin, and violin-like spectral characteristics have been detected in some phycobiliproteins, where they coexist with the two major phycobilins [185, 186]. Several of these chromophores have been characterized structurally (Fig. 7) [187–189].

All of the algae that contain other phycobilins also contain phycocyanobilin, but there are some species that contain only phycocyanobilin. This distribution suggests the possibility that phycocyanobilin may be a precursor to other phycobilins. However, it should be noted that in the absence of phycocyanobilin, excitation energy probably cannot be transferred from the other chromophores to chlorophyll. Therefore the distribution of the phycobilin types among algal species may be reflection of the functional dependence on phycocyanobilin, rather than indicating a biosynthetic relationship.

The possibility exists that some phycobilins are formed from protein-linked precursors. However, when *Cyanidium* cells (even mutant cells that do not accumulate phycocyanin) are incubated with exogenous ALA, they excrete a blue pigment that is identical to the free chromophore that is cleaved from phycocyanin by methanolysis [190, 191]. In this instance, it is unlikely that the excreted bilin is derived from cleavage of phycocyanin. This observation, and ability of cell-free extracts of *Cyanidium* to catalyze free phycocyanobilin formation from biliverdin, strongly suggests that, in vivo, at least this phycobilin can be synthesized as the free chromophore, with an ethylidine group present, and that a free bilin is the precursor of the bound pigment.

Although all four of the phycobilin chromophores thus far described are isomeric [187], they may not be derived one from another. Instead, all might be derived independently from biliverdin or from a precursor with a reduced pyrrole ring, the different products being determined by the site of the second two-electron reduction step.

2.3.2.6. Biosynthetic relationship of the phytochrome chromophore to phycobilins. The structure of the phytochrome chromophore resembles very closely those of the phycobilins [204]. Phytochrome is present in higher and lower plants and at least some algae, including several green algae and the rhodophyte *Porphyra tenera* [205]. Elich and Lagarias [206] obtained evidence that the phytochrome chromophore, like phycocyanobilin, can be synthesized in vivo from exogenously supplied biliverdin. These workers found that phytochrome levels are substantially reduced in oat seedlings germinated in the presence of gabaculine, a specific inhibitor of ALA formation via the five- carbon pathway. If either ALA or biliverdin was administered to the gabaculine- grown seedlings, there was a rapid increase in spectrophotometrically-detected phytochrome. Finally, ^{14}C-labeled biliverdin was shown to be specifically incorporated into the phytochrome chromophore in oat leaves, establishing a clear biosynthetic link between this chromophore and the phycobilins [203].

It is interesting to note that a hypothetical partially reduced intermediate between biliverdin and the phycobilins, having the reduced pyrrole ring but still retaining both the exo-vinyl group and a fully conjugated ring system, is chemically equivalent to the free form of the phytochrome chromophore (Fig. 7). Thus it could be proposed that the phytochrome chromophore originated in nature as a precursor to the phycobilin chromophores.

2.3.3. Ligation of phycobilins to apoproteins

The phycobilins exist, in their photosynthetically functional state, as covalently attached chromophores of the phycobiliproteins. A clearer picture of the bilin-to-apoprotein attachment modes is beginning to emerge from work with defined bilipeptides derived from the larger phycobiliproteins by proteolytic cleavage. The most frequently encountered linkage is a thioether bond between an apoprotein cysteine and the 3' position (the α-carbon of what originated as the ring-B vinyl group of protoheme). It is noteworthy that the linkage of heme to the apoprotein of cytochrome c is also by attachment of both, or occasionally one, of the tetrapyrrole vinyl groups to cysteinyl residues [192]. Stereochemical investigations [193–196] have revealed that the absolute configuration about the 3' carbon that is attached to cysteine sulfur in the phycobiliproteins examined, as well as both of the chiral centers in the attached ring, are all R. Spectral differences among the individual chromophores have been interpreted to indicate that some phycocyanobilin [197] and phycoerythrobilin [198] groups are linked to the apoprotein by the 18' position (the α-carbon of what originated as the ring-A vinyl group of protoheme) instead of the 3' position. Also, in some cases, phycoerythrobilin [199] and phycourobilin [200] chromophores may be doubly-linked to the apoprotein, by thioether links at both the 3' and 18' positions. Some of these conclusions regarding different linkage modes have been challenged, and the data reinterpreted in terms of different conformations of the bilins on the biliproteins leading to different bilin *meso* bridge carbon configurations upon denaturation of the proteins [201]. However, in the case of the β subunit of a cryptomonad phycoerythrin, a doubly-linked bilin has been unambiguously established by isolation of peptides crosslinked by the bilin [202]. Finally, some evidence has been interpreted to indicate the existence of an ester bond between an apoprotein serine group and one of the propionate carboxyl groups of the bilin [188].

There is little direct information concerning the mechanism by which the bilin chromophore is ligated to the cysteine sulfhydryl group(s) of the apo-phycobiliproteins. The bilin could be activated for apoprotein ligation by reduction of the pyrrole ring, leaving a vinyl group that is not conjugated with the rest of the tetrapyrrole double bond system. The existence of such an intermediate is consistent with the production of α-hydroxymesobiliverdin [182] and both E- and Z-ethylidine phycocyanobilin isomers [176] in the absence of apoprotein acceptor molecules.

Although in vitro ligation of phycobilins to apo-phycobiliproteins has not been reported, Elich et al. [203] were able to measure incorporation of exogenous biliverdin and phycocyanobilin into phytochrome in etiolated oat leaves that had been prevented from synthesizing endogenous bilins by administration of the inhibitor of ALA formation, gabaculine. After incubation of the gabaculine-treated leaves with exogenous bilins, the phytochrome was isolated and found to undergo photoreversible spectral changes. In the case of phycocyanobilin administration, the isolated phytochrome had its absorption maxima shifted as a result of incorporation of a bilin other than the natural one, phytochromobilin.

2.4. The Mg branch

2.4.1. Chlorophyll a formation

2.4.1.1. Chelation of Mg. Demonstration of the Mg chelatase reaction in whole cells of *R. spheroides* was first achieved by Gorchein [207]. Protoporphyrin was incorporated into a mixture of lipids previously extracted from the bacteria and dispersed by sonication. This substrate mixture was administered to cells in the light under semi-anaerobic conditions (incubations were vigorously bubbled with N_2 and sealed with rubber stoppers). Aside from the cells and substrate mixture, the incubation contained only buffer and EGTA. Under these conditions, a typical 90-min incubation yielded approximately 20 nmol of Mg-protoporphyrin monomethyl ester. Independent formation of unesterified Mg-protoporphyrin could not be demonstrated. Incubation in the dark, aeration of the incubation mixture, or omission of the lipid resulted in drastically diminished (at least six-fold) recovery of product. Dialysis of the cells against distilled water for 16 h resulted in loss of half the activity, but this activity could be restored by inclusion of S-adenosyl-methionine or ATP plus methionine in the incubation mixture. Compounds which inhibit S-adenosyl-methionine formation also inhibited Mg-protoporphyrin monomethyl ester formation, although the unesterified form did not accumulate under conditions where formation of the methyl ester was prevented.

Attempts to demonstrate activity in cell-free extracts were unsuccessful even with gentle methods of cell breakage [208]. Spheroplasts prepared by incubation with lysozyme retained much of the original activity, but this activity was lost when the spheroplasts were lysed by osmotic shock. Activity was not restored by addition of hypothetical cofactors. Although activity was supported in the dark in an atmosphere containing 5% O_2, concentrations above 5% were inhibitory. Experiments with inhibitors of electron transport, uncouplers, and oligomycin suggested a need for

ATP, which was not related to the methylation step. The experimental results of Gorchein [207, 208] suggest requirements for membrane intactness, low O_2 tension, ATP, and coupling of Mg insertion to methylation.

Although it is tempting to assume that Mg chelation in green plants occurs by the same process, the mechanisms may be as different as they are for ALA formation. Early attempts to measure Mg chelatase activity in extracts of higher plants were not successful or definitive. In the first partially successful attempt, unpurified homogenates of etiolated wheat seedlings formed very limited and non-reproducible amounts of Mg-protoporphyrin, as measured by incorporation of $^{28}Mg^{2+}$ [209]. ALA or protoporphyrinogen, but not protoporphyrin, could serve as a substrate for this reaction. Incubation of etioplasts or developing chloroplasts from cucumber cotyledons with protoporphyrin and a mixture of cofactors (ATP, NAD, CoA, GSH, and methanol) resulted in the accumulation of a mixture of components termed MPE-equivalents [210]. These components included Mg-protoporphyrin, Mg-protoporphyrin monomethyl ester, and other components that were presumed to be intermediates between Mg-protoporphyrin monomethyl ester and protochlorophyllide. Although the individual components were not resolved, the collection of products was detected and quantified by the fluorescence emission spectra.

A similar preparation of developing chloroplasts from cucumber cotyledons was further refined, in which only one (or two) major products accumulated, and the requirements and conditions for the reaction were examined in more detail [145]. Incubation of this chloroplast preparation with protoporphyrin, glutamate, and a cofactor mixture (ATP, NAD, and GSH) resulted in the accumulation of Mg-protoporphyrin and its monomethyl ester. The products were identified by their fluorescence excitation and emission spectra and by their retention times on reverse-phase HPLC. ATP and glutamate (or α-ketoglutarate) were absolutely required, but the requirement for protoporphyrin could be met by in situ generation from precursors. Essentially similar results were obtained using a radioactive assay and separating the fully methylated products by HPLC [211]. However, in contrast to the situation with plastids from cucumbers, the major product formed in incubations with plastids from dark-grown wheat was Zn-protoporphyrin.

The perplexing requirement for glutamate in Mg-protoporphyrin formation was subsequently explained by demonstrating that glutamate provided an indirect source of extra ATP by serving as a substrate for ATP generation via oxidative phosphorylation occurring in mitochondria which contaminated the chloroplast preparation [212]. High concentrations of ATP could be substituted for glutamate in developing plastids isolated by differential centrifugation, and plastids that were purified by density gradient centrifugation could not use glutamate to enhance Mg-protoporphyrin formation. Concentrations of ATP as high as 10 mM did not saturate the reaction, and other nucleoside triphosphates could not substitute for ATP. AMP was strongly inhibitory; 50% inhibition was observed at 3.5 mM AMP in the presence of 10 mM ATP. Properties of the Mg chelatase from developing chloroplasts of cucumber cotyledons were further defined by investigations of substrate concentration dependence [213] and methylation state of the product [214]. Endogenous Mg^{2+} and Zn^{2+} were removed by repeated washings of the plastids in buffer containing 10 mM EDTA. Subsequent incubation of the washed plastids indicated an optimal Mg^{2+} concentration around 10 mM, with concentrations greater than 10 mM causing substantial inhibition. In plastids isolated from cotyledons that were greened for 20 h, the optimal ATP concentration was 10 mM, with slight inhibition occurring at higher concentrations. Half-maximal activity was achieved at 3.5 mM ATP and 3.5 µM protoporphyrin. Saturation occurred at 10 µM protoporphyrin. All of the concentration curves were sigmoidal in shape [213]. The sigmoidal shape may reflect cooperative catalytic behavior of the enzyme, or it may reflect the barrier presented by the chloroplast envelope (However, see the information below on localization of the activity.). The washed plastids were capable of forming Zn-protoporphyrin when 5 mM ZnCl2 was added to the preparations, although half of this activity was non-enzymic.

Breakage of the developing chloroplasts by lysis in hypotonic media resulted in a drastic decrease in activity to a level less than 0.2% (on a pmol product h^{-1} mg^{-1} protein basis) of that in intact plastids [216]. Removal of the stromal components by centrifugation followed by incubation of the membranes with substrates resulted in 2.5% of the activity of intact plastids. Under these conditions, protoporphyrin dependence was saturated at 0.1 µM, with half maximal activity occurring at approximately 0.025 µM. The optimum concentration of added ATP was 3.0 mM, when an ATP regenerating system was included. The concentration dependence curves for protoporphyrin and ATP were not sigmoidal.

Mg chelatase activity was sensitive to the mercurial reagents, *p*-chloromercuribenzoate (PCMB) and *p*-chloromercuribenzoyl sulfonate (PCMBS), the latter being less capable of permeating biological membranes [216]. Pretreatment of plastids with PCMBS, followed by removal of unbound PCMBS and recovery of the plastids by centrifugation through a pad of 45% Percoll, resulted in complete inhibition of Mg-chelatase activity upon subsequent incubation with substrates. A similar treatment had only minor effects on two PCMB-sensitive stromal enzymes. Intact chloroplasts formed equivalent amounts of Mg-protoporphyrin with 10 µM protoporphyrin or 6.0 mM ALA as substrate [101]. The PCMBS inhibition of Mg-protoporphyrin accumulation was equivalent with either substrate. Thus PCMBS was not acting via an affect on the entry of protoporphyrin into the plastid. It is difficult to explain these results in any way other than by postulating that the Mg chelatase is localized in the chloroplast envelope [216].

In the normal incubations, the major product was the unesterified form of Mg-protoporphyrin. When 1.0 mM S-adenosyl-methionine was included in the reaction mixture, the major product was Mg-protoporphyrin monomethyl ester, and formation of the ester was dependent upon exogenous S-adenosyl-methionine [214]. ATP plus methionine could not substitute for S-adenosyl-methionine. Thus in developing chloroplasts from cucumber cotyledons, Mg chelation and methylation are not obligatorily coupled as they are in *R. spheroides*. In addition, the sequence of formation was confirmed as:

Protoporphyrin → Mg-protoporphyrin → Mg-protoporphyrin monomethyl ester

which is consistent with the earlier proposal of Radmer and Bogorad [217].

2.4.1.2. Mg-protoporphyrin methyl transferase. The enzyme catalyzing the methylation of Mg-protoporphyrin, S-adenosyl-methionine:Mg-protoporphyrin IX methyl transferase (EC 2.1.1.11), has been characterized from a number of sources, including *R. spheroides* [218, 219], wheat [220–222], *E. gracilis* [223, 222], maize [217], and barley and other plants [224]. This enzyme is relatively easy to assay and its kinetic mechanism has been the object of numerous studies.

The enzyme from *R. spheroides* was most active with Mg-protoporphyrin, and very much less so with protoporphyrin, Ca-or Zn-protoporphyrin, and not active at all with protoporphyrinogen, or heme [218]. The K_m values for Mg-protoporphyrin and S-adenosyl-methionine were 40 μM and 55 μM, respectively. The enzyme was tightly bound to membranous chromatophores, and was not present under conditions where bacteriochlorophyll is not formed (high aeration, in the dark). The enzyme from wheat is soluble in 0.5 M sucrose and 1.0 M NaCl, and has K_m values of 22 and 44 μM for Mg-protoporphyrin and S-adenosyl-methionine, respectively [220]. The wheat enzyme activity did not differ significantly in extracts from etiolated and greening tissues [221]. Two forms of the enzyme were found in *E. gracilis*, one soluble (15 to 20% of the total) and the other bound to the chloroplast membranes. The bound form could be solubilized with Tween-80 [223]. It was proposed that the soluble portion of the enzyme had not yet become incorporated into the membrane. The K_m for Mg-protoporphyrin was in the range of 10 to 34 μM, and the K_m for S-adenosyl-methionine varied between 20 to 160 μM, depending upon the fraction tested and the presence or absence of detergent in the assay. Cells grown in the light contained two to three times more activity than dark-grown cells.

The methyl transferase is a bi-substrate enzyme, and the kinetic mechanism has been investigated by classical initial velocity studies [220] and by combining initial velocity studies with behavior on affinity columns [219, 222, 225]. The wheat enzyme has a ping-pong mechanism, with S-adenosyl-methionine binding first, followed by release of S-adenosyl-homocysteine before binding of the second substrate, Mg-protoporphyrin [220, 222, 225, 226]. This mechanism is consistent with the existence of a methylated enzyme intermediate. Treatment of the enzyme with [methyl-^{14}C]-S-adenosyl-methionine followed by exhaustive dialysis resulted in 60 times more radioactivity incorporation into the methyl transferase than into an equivalent amount of bovine serum albumen treated in a similar manner. More direct evidence for the methylated enzyme was not obtained [222]. In contrast to the wheat enzyme, the enzyme from *E. gracilis* has a random mechanism [222], and the *R. spheroides* enzyme has an ordered mechanism, with Mg-protoporphyrin binding first [219].

The use of the affinity columns to gain insight into the kinetic mechanism is quite interesting, and the behavior of the wheat and *E. gracilis* enzymes on affinity columns will be described in more detail [222]. In the case of the wheat enzyme, it was known that heme is an inhibitor that is competitive with Mg-protoporphyrin, and that there is a drastic decline in activity upon shifting from the optimum pH of 7.8 to pH 9 [220]. A column with covalently-coupled heme was prepared and cell-free homogenates were passed through the column under a variety of conditions. The methyl transferase activity did not bind to the column unless S-adenosyl-methionine was included in the application buffer. Thus the presence of S-adenosyl-methionine is required for binding to heme (or Mg-protoporphyrin). This behavior would be expected for an enzyme with a ping-pong mechanism or an ordered mechanism with S-adenosyl-methionine binding first. If the mechanism were ordered, it would be expected that the methyl transferase would be eluted simply by removing S-adenosyl-methionine from the elution buffer. However, removal of S-adenosyl-methionine was not sufficient for elution. Instead, elution required changing the pH. In contrast, the *E. gracilis* enzyme bound to columns having Mg-protoporphyrin or S-adenosyl-homocysteine ligands in the absence of the other substrate. This behavior is indicative of a random kinetic mechanism [222, 226, 227]. As would be expected, these affinity columns were also very useful for purification of the methyl transferases. Purification factors for this step ranged from 460- to 2000-fold [226].

2.4.1.3.Isocyclic ring formation. Isocyclic ring formation is a complex process in which one of the porphyrin side chains is converted to the fifth ring. In this series of reactions the methyl propionate side chain at position 6 of the macrocycle is joined to the γ- meso bridge of the metalloporphyrin ring, forming a five-membered ring between pyrrole ring C and the meso bridge (Fig. 9). The ring is substituted with a keto group on the β-carbon, and the carboxyl group remains methylated. Attachment to the γ- meso bridge carbon is through the α-carbon of the side-chain. Formation of the ring creates a new asymmetric center in the R configuration on position 10 of the new macrocycle. The compound formed is called Mg-2,4-divinylpheoporphyrin a_5, and it has also been referred to as divinyl-protochlorophyllide. Reduction of the vinyl group at position 4 of the macrocycle yields protochlorophyllide. Formation of Mg- 2, 4-divinylpheoporphyrin a_5 occurs in photosynthetic bacteria as well as in chloroplasts of eukaryotic organisms, but there is no assurance that the mechanisms are the same in the two very different classes of organisms. Information on the chemical nature of the intermediates has come from analysis of mutants which excrete or accumulate intermediates in the pathway beyond Mg-protoporphyrin. In other in vivo studies, the accumulation of putative intermediates has been caused by administration of inhibitors or flooding the biosynthetic system with precursors. In each case, a sequence of the steps has been inferred from the structures of the accumulated products and the chemistry required to proceed from Mg-protoporphyrin

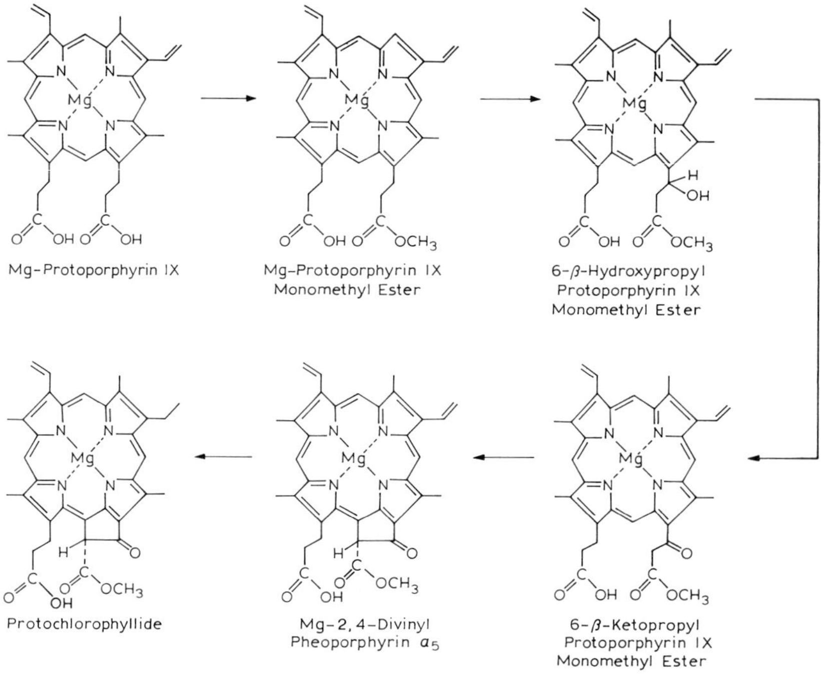

Fig. 9. Proposed steps in the conversion of Mg-protoporphyrin IX to protochlorophyllide. The step where reduction of the B-ring vinyl to ethyl occurs may vary with species and developmental conditions.

methyl ester to chlorophyllide. Those initial studies formed the basis of more recent investigations of the individual steps in cell-free systems.

Mg-divinylpheoporphyrin a_5 was identified in cultures of *R. spheroides* in which bacteriochlorophyll synthesis had been partially inhibited by treatment with 8-hydroxyquinoline [228, 229]. This compound also accumulates, or is excreted, in some bacteriochlorophyll-deficient mutants of *R. spheroides* [230, 231]. The Mg-divinylpheoporphyrin a_5 which accumulated in one of the above mutants was collected, purified, and administered to etioplast membranes from barley [232]. When NADPH was added, the Mg-divinylpheoporphyrin a_5 was photoconverted to chlorophyllide. Thus the Mg-divinylpheoporphyrin a_5 which accumulates in bacteriochlorophyll-deficient mutants of *R. spheroides* also served as a substrate for chlorophyllide biosynthesis in higher plants. The existence of Mg-divinylpheoporphyrin a_5 in the protochlorophyllide pool of etiolated cucumber cotyledons and other dark-grown higher plants was demonstrated by Rebeiz and co-workers [233, 234]. They also demonstrated that both Mg-divinylpheoporphyrin a_5 and protochlorophyllide (monovinyl) are photoconverted to divinyl- and monovinyl-chlorophyllides after a brief flash of light [235]. Identification of the compound as Mg-divinylpheoporphyrin a_5 was made on the basis of its fluorescence emission and excitation spectra at 77 °K and its behavior on thin layers of polyethylene. These properties were identical to those of a standard isolated from a *R. spheroides* mutant [234].

Granick [236] proposed that the formation of the isocyclic ring occurs via β-oxidation of the 6-methyl propionate group to a 6-methyl β-ketopropionate side chain. The α-methylene group would then attach to the γ-meso bridge of the porphyrin in an oxidative cyclization. In the fatty acid β-oxidation model, the β-keto group is introduced by desaturation of the α-carbon-β-carbon bond, followed by hydration of the double bond to form a β-hydroxy substituent. The β-hydroxy is then oxidized to form the β-keto group. It has been proposed that the presence of the methylated ester prevents the spontaneous decarboxylation of what would otherwise be a β-ketoacid [237]. Accumulated intermediates corresponding to the 6-methyl-, acrylate, β-hydroxy propionate, and β-ketopropionate derivatives of Mg-protoporphyrin were characterized in a series of chlorophyll-deficient mutants of *Chlorella* [238, 239].

Formation of protochlorophyllide from ^{3}H-labeled Mg-protoporphyrin monomethyl ester was demonstrated in crude extracts of etiolated wheat seedlings [240]. Incorporation of label required a complex cofactor mixture including ATP, coenzyme A, S-adenosyl-methionine, and inorganic phosphate. Activities were not reported, nor was the product distinguished as being the mono or divinyl form. A similar conversion was reported in developing cucumber chloroplasts [241].

Formation of the isocyclic ring from exogenous porphyrins by developing cucumber chloroplasts has been investigated by Castelfranco and co-workers. Preliminary studies in intact tissue suggested that the conversion required O_2 and Fe [242]. A preparation of developing chloroplasts similar to that used for Mg-chelatase studies (above) converted added Mg-protoporphyrin to protochlorophyllide in the dark when either S-adenosyl-methionine or ATP plus methionine were included in the incubation mixture [243]. The conversion did not occur in an atmosphere of N_2. It was also found that the Mg-protoporphyrin substrate was unstable in the light in air. Even in the absence of plastids, a 1-h incubation in strong light (90 mE m^{-2} s^{-1} PAR) caused almost complete disappearance of the Mg-protoporphyrin. The accumulated product was subsequently shown to be Mg-divinylpheoporphyrin a_5 by comparison of its fluorescence and chromatographic properties to published properties [244] and by detailed analysis of corrected fluorescence spectra at 77 °K, nuclear magnetic resonance, and secondary-ion mass spectroscopy of the highly purified product of the incubation mixtures [244].

The optimal requirements for Mg-divinylpheoporphyrin a_5 formation from Mg-protoporphyrin and Mg-protoporphyrin monomethyl ester in intact chloroplasts were also reported [244]. The optimal substrate concentration for either substrate was 10 μM, with activity declining at higher concentrations. The presence of 1 mM S-adenosyl-methionine stimulated activity with either substrate, although it was required only when Mg-protoporphyrin was the substrate. A separate experiment indicated that an active methyl esterase was present in the plastid preparation, and the stimulatory effect of the S-adenosyl-methionine could be a consequence of re-esterification of the de-esterified pigment. Although not absolutely required, the presence of either NADP or NADPH at 0.6 mM stimulated activity. Dependence on O_2 was investigated by incubating samples in air/N_2 and air/CO mixtures. When activity was plotted versus partial pressure of O_2 in the incubation, hyperbolic curves were obtained, with half-saturation occurring at 0.04 atm of O_2. Inhibitor studies did not support the involvement of a hemoprotein in a hydroxylation reaction to introduce the O_2, nor the involvement of a classical β-oxidation scheme. The location of the ring- forming enzyme(s) within the intact chloroplast was investigated by the use of permeant and non-permeant mercurial reagents [216]. Unlike Mg chelatase, the cyclase system in intact chloroplasts was insensitive to inhibition by PCMBS, but it was inhibited by PCMB, which is able to cross the membrane. Thus the cyclase system is probably localized within the chloroplast.

Also unlike Mg-chelatase, the cyclase system was active in broken chloroplasts [244]. This property was used to resolve the cyclase activity into two required enzymic fractions [246]. Developing chloroplasts were disrupted by sonication and fractionated into a high-speed supernatant and a membranous pellet. The enzymes in the supernatant were recovered by precipitation at 80% saturation of $(NH_4)_2SO_4$ followed by dialysis of the dissolved pellet. Reconstitution of cyclase activity required both membrane-bound and soluble protein fractions, plus Mg-porphyrin substrate, O_2, and reduced pyridine nucleotide. As with the intact system, activity with Mg-protoporphyrin methyl ester was stimulated by inclusion of S-adenosyl-methionine, and activity with Mg-protoporphyrin required S-adenosyl-methionine. The shape of the concentration curve for Mg-protoporphyrin methyl ester was hyperbolic to 10 μM, and activity declined at higher concentrations. Half-maximal activity was achieved with approximately 2 μM substrate. Unlike the intact plastid system, the reconstituted cyclase system had an absolute requirement for a reduced pyridine nucleotide. The curve of activity versus NADPH concentration was hyperbolic, the saturating concentration was approximately 5 mM, and half-maximal activity achieved at approximately 0.9 mM. The NADH concentration dependence curve was slightly sigmoidal at lower concentrations, with half-maximal activity occurring at approximately 1.8 mM and saturation at 5 mM. Activity with NADPH declined at concentrations greater than 5 mM, so that at higher concentrations, activity was equivalent with either cofactor. It must be noted that the activity measurements were for a 1-h incubation and were not initial rate measurements. Thus these values define the shape of the concentration dependence curves, but cannot be considered to be kinetic constants.

The reconstituted cyclase system was used to test a variety of inhibitors and possible intermediates [247–249]. The cyclase was inhibited by mercurial sulfhydryl reagents and the sulfhydryl alkylating agent, N-ethylmaleimide. Both the membranous and soluble fractions of the cyclase system were susceptible to inhibition by the latter reagent. The sulfhydryl-containing compounds, dithiothreitol and mercaptoethanol, were also inhibitory, suggesting the presence of a sensitive disulfide. The Mg-porphyrin specificity of the cyclase system was tested by administration of synthetic derivatives of Mg-protoporphyrin IX monomethyl ester, in which the side-chain at the 6 position of the macrocycle was modified. Both β-hydroxy and β-keto derivatives (Fig. 9) were effective substrates for Mg-divinylpheoporphyrin a_5 formation. However, the acrylate derivative was ineffective. Only one enantiomer of the β-hydroxy derivative was effective as substrate. When Mg-protoporphyrin monomethyl ester was the substrate, a product from the incubation mixture having the transient kinetic behavior of an intermediate was isolated and identified as the β-hydroxy derivative [248]. This intermediate was not detected when the β-keto derivative was used as a substrate. When the β-keto derivative was used as a substrate, formation of Mg-divinylpheoporphyrin a_5 still required the presence of O_2 and NADPH as well as both the soluble and membrane-bound portions of the reconstituted cyclase system. The monovinyl (2-vinyl, 4-ethyl) form of the β-keto derivative was four times more active than the divinyl, β-keto derivative in the cyclization reaction [249]. When Mg-protoporphyrin was the substrate, the monovinyl and divinyl forms were equally effective.

A system catalyzing a similar reaction in isolated etioplasts of wheat, has now been characterized by a continuous spectroscopic assay [250]. While most of the properties of this preparation were similar to those of the cucumber preparation described above, there were some significant differences and additions. The wheat system has an absolute requirement for organelle intactness, and was inhibited by lipophilic Fe chelators and anaerobiosis. The Zn-porphyrin could replace the Mg-porphyrin as substrate, but the Cu-, Ni- and free porphyrins could not. Greening of etiolated tissue for 10 h did not increase or decrease the activity in subsequently isolated plastids.

In summary, formation of the isocyclic ring from Mg-protoporphyrin monomethyl ester requires O_2, NADPH, and enzyme(s) from the chloroplast stroma and membranes. Ring formation proceeds through 6-methyl-β-hydroxy- and 6-methyl-β-keto- propionate derivatives (Fig. 9), consistent with earlier proposals [236, 238, 239]. However, since the 6-methyl-acrylate derivative was ineffective as a substrate, the initial formation of the β-hydroxy derivative is probably not via hydration of a double bond as would be expected for a classical β-oxidation sequence. Instead, introduction of the O_2 may come about by a mixed function oxidase reaction, as suggested by Castelfranco and co-workers [244, 246]. The latter proposal is consistent with the O_2 and NADPH requirements for isocyclic ring formation, and with the recent finding that ^{18}O is incorporated into the product when ring formation occurs in the presence of $^{18}O_2$ [251].

2.4.1.4. Vinyl group reduction. Although plants may contain some chlorophyll end products having vinyl groups in both the 2 and 4 positions of the macrocycle [252], the majority of chlorophyll end products have the 4 side-chain reduced to an ethyl group. It appears likely that this reduction can occur at any number of points along the biosynthetic pathway, depending upon the nature of the plant and its growth history [234, 252, 253]. Despite the relative importance of the vinyl reduction step to the branched pathway proposal (see below), there has not been very much work reported on the biochemistry of this step.

Ellsworth and Hsing [254] used unpurified homogenates of etiolated wheat to catalyze incorporation of label from ^{3}H-NADH into the vinyl group. The original homogenization buffer had a relatively high salt concentration (0.2 M phosphate buffer) that might be expected to strip peripheral proteins off the membranes. Mg-protoporphyrin monomethyl ester was an effective porphyrin substrate, but protoporphyrin and Mg-divinylpheoporphyrin a_5 were not. NADPH was ineffective as a hydrogen donor. Incorporation of label into the 4 position rather than the 2 position was verified by analysis of the chromic acid oxidation products. Further analysis of this preparation indicated a pH optimum around 7.7 and a K_m for Mg-protoporphyrin monomethyl ester of 25 μM [255]. The K_m for NADH could not be determined. In preliminary reports, Richards and co-workers [226, 256] described the behavior of the 4-vinyl reductase on affinity columns. Most of the enzyme activity from a high-speed supernatant of etiolated wheat seedlings was retained on a column of Zn-protoporphyrin monomethyl ester, in the presence or absence of NAD. The enzyme was eluted by changing the pH from 7.7 to 9.0. A 70-fold purification was achieved compared to the activity of the applied extract.

The substrate specificity for the reduced pyridine nucleotide was also examined by enzymatic preparation of both 4-R and 4-S stereoisomers of 4-[^{3}H]-NADH and 4-[^{3}H]-NADPH. Although insufficient detail was presented to allow comparison of their relative effectiveness, both Mg-protoporphyrin monomethyl ester and Mg-divinylpheoporphyrin a_5 were substrates for the reductase [226, 256]. With either porphyrin substrate, [4R-^{3}H]-NADPH was best able to transfer label to the reaction product. The specificity experiments were performed with whole or broken etioplast preparations, rather than unpurified extracts or purified enzyme, and both of the reported substrate specificities conflict with the earlier results of Ellsworth and Hsing [254].

2.4.1.5. Protochlorophyllide reduction. Plants, requiring light for greening, accumulate small amounts of protochlorophyllide when grown in the dark. Upon exposure to white light, the double bond between carbons 7 and 8 of ring D is reduced to form chlorophyllide (compare Figs. 2 and 8). Most algae and gymnosperms do not require light for greening, and the conversion of protochlorophyllide to chlorophyllide is independent of light. In this conversion, two new chiral centers are created and the molecule retains its aromaticity.

2.4.1.5.1. Light-requiring reduction catalyzed by protochlorophyllide oxidoreductase. The light-dependent transformation can be followed by observing the spectral changes associated with the protein-bound pigment. In situ, the protochlorophyllide of etiolated plants has an absorbance maximum near 650 nm, and a shoulder at 638 nm. Following a 10-s illumination, the absorbance maximum shifts to 678 nm and there is a corresponding decrease in the 650 nm absorbing form [257]. During the next 30 s in the dark, the wavelength shifts to 684 nm, and finally to 672 nm after about 20 min [258]. This last transformation is commonly referred to as the Shibata shift. An action spectrum indicated that the photoreceptor for these transformations is protochlorophyllide [259]. When ALA was administered to the etiolated plants before illumination, another protochlorophyllide form, having a maximum at 630 nm, accumulated. This form was phototransformable only if the plants were given a series of flashes with intervening dark periods [260]. The physical state of protochlorophyllide in plant tissue has been reviewed [261], as have some of the theories relating the photoreduction to developmental processes [262]. The present discussion will focus on the biochemical explanation of these changes as well as some of the more rapid photochemical transformations.

Isolated etioplasts that are effective in phototransformation of protochlorophyllide have been studied [263] and used to identify the enzyme catalyzing the transformation as well as the required cofactors [264–267]. The most effective preparations are from membranes that are recovered by centrifugation from lysed etioplasts. Supplementation of the membranes with a NADPH-regenerating system resulted in increased photoreduction activity [264]. Specific incorporation of ^{3}H-NADPH into chlorophyllide proved that NADPH was the source of reducing equivalents [268]. Subsequently, it was shown that the hydride on the S face of NADPH was incorporated into the chlorophyllide [226]. Both the mono- and divinyl forms of protochlorophyllide were effective [232, 235], but protochlorophyll was not an effective substrate in vitro [269]. The formal name for the enzyme is NADPH:protochlorophyllide oxidoreductase (EC 1.6.99.1).

The polypeptide catalyzing the reaction was identified by labeling active membrane preparations with the sulfhydryl reagent, N-phenylmaleimide, while protected by substrates. Excess reagents were removed by gel filtration and the labeling was repeated in the light in the presence of ^{3}H-N-phenylmaleimide [266]. Light is required to photoreduce the bound protochlorophyllide and liberate the products, allowing the released products and the N-phenylmaleimide to compete for the site of the active thiol. The extent of photoconversion of the bound substrate was directly proportional to the extent of inhibition of activity upon subsequent re-incubation of the modified membranes with fresh substrates [267]. The modified membrane proteins were separated by SDS-PAGE and incorporation of label was detected by fluorography. Label incorporation into a 35,000–37,000 molecular weight band (perhaps a doublet) was proportional to the degree of photoconversion during the radioactive labeling period.

An active form of the enzyme could be solubilized from membranes with low concentrations of the non-ionic detergent Triton X-100 [270, 271]. The purified enzyme from etiolated oat seedlings is composed of 346 amino acids, contains one cysteine and no tryptophan residues, and has a molecular weight of 37, 800 [272]. Active enzyme could also be solubilized from membranes with the zwitterionic detergent, CHAPS [273]. In this case, the specific activity of the solubilized enzyme was greater than in the membranes. An affinity column with Mg-or Zn-divinylpheoporphyrin a_5

as the ligand could be used to enhance the purification of the CHAPS-solubilized enzyme. Although contaminating polypeptides were removed by this procedure, much of the activity was lost. The enzyme could be dissociated from the column by free ligand, but not by removal of NADP or by changing the pH. It can be concluded from these results that the enzyme does not require the presence of pyridine nucleotide for binding of the metalloporphyrin substrate. Enzyme activity was inhibited 50% by 20 μM of the flavin analog, quinacrine, and the activity in various preparations was proportional to the amount of extractable FAD [274]. However, more definitive information is required before it can be concluded that the reductase is a flavoprotein.

The cell-free preparations have been used to analyze some of the spectral changes observed in situ. The photoactive complex from freshly isolated etioplasts has protochlorophyllide and NADPH tightly bound to it, and it absorbs at 628 and 650 nm. This form has been called protochlorophyllide holochrome and is now known to be the substrate-loaded reductase. When the preparation is exposed to light, the protochlorophyllide is immediately photoconverted to chlorophyllide and the absorbance maximum changed to 678 nm. This form shifts to 672 nm upon further incubation in the dark. However, a 677 nm form can be stabilized in the dark if the preparation is supplemented with NADP. If the photoconversion is carried out in the presence of excess NADPH, or if a NADPH-regenerating system is added to a system containing the 677 nm form, the absorbance maximum shifts to 682–684 nm. Protochlorophyllide alone in aqueous detergent suspension absorbs maximally at 628–630 nm, and is considered to be the nonphototransformable or unbound form of the pigment observed in situ after ALA administration. Conversion to the 672 nm form is enhanced by addition of free protochlorophyllide to the 682–684 nm form, or by addition of N-phenylmaleimide to either the 677 or 682–684 nm forms.

These results have been interpreted in terms of a ternary complex of the enzyme, chromophore, and reduced or oxidized form of the bound pyridine nucleotide [275]. The model is illustrated in Fig. 10. The 628–630 nm form is unbound protochlorophyllide. The 638/650 nm form is the ternary complex between protochlorophyllide, enzyme, and NADPH. The 638/650 nm form is phototransformed by a brief flash of light to a 677–678 nm form which is the ternary complex of chlorophyllide, enzyme, and NADP. The NADP on the 677–678 nm form can be exchanged for fresh NADPH to form the 682–684 nm form. Finally, the chlorophyllide on the 684 nm ternary complex is slowly released (the Shibata shift) to become the 672 nm form, and allow the binary complex of enzyme and NADPH to accept a new molecule of protochlorophyllide and repeat the cycle.

The dependence upon the NADP/NADPH ratio for the 678 nm to 682–684 nm shift was also shown independently by El Hamouri and Sironval [276]. In intact leaves, the 682–684 nm form (enzyme-chlorophyllide-NADPH) can be driven back to the 678 nm form (enzyme-chlorophyllide-NADP) by a brief intense flash of red light [277]. The light-driven back reaction is much slower at colder temperatures. If

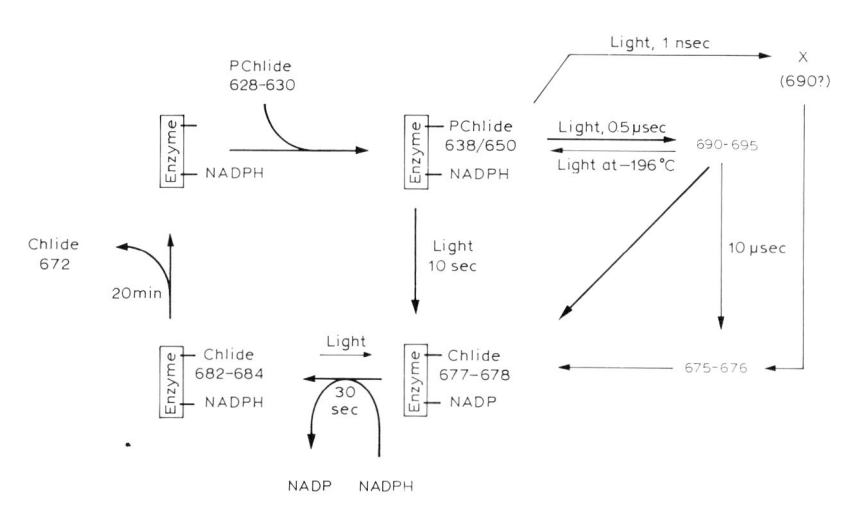

Fig. 10. The light-dependent protochlorophyllide reduction cycle. The enzyme is shown as a rectangle with binding sites for (proto)chlorophyllide and NADP(H). The pigment forms are designated by their light absorption maxima. The normal cycle, which occurs at room temperature in intact leaves or in reconstituted enzyme preparations, is shown on the left with heavy arrows. Intermediates in the light-dependent part of the cycle are shown in smaller letters and connected with lighter arrows. These intermediates are observed at low temperatures and/or with rapidly responding optical equipment. Abbreviations: PChlide, protochlorophyllide; Chlide, chlorophyllide.

the 672 nm form is similarly illuminated, there was a substantial O_2 uptake by the etioplasts accompanied by photodestruction of the pigment. When the 682–684 nm form was illuminated, both the O_2 uptake and photodestruction were significantly diminished [278]. The authors concluded that the in situ formation of the 682–684 nm complex is a mechanism which functions to protect the newly formed pigment from photodestruction.

Several other spectral shifts have been observed in situ or in vitro with rapid kinetic techniques and/or very low temperatures. An intermediate absorbing at 690–695 nm and formed from the 638/650 nm form was detected within 0.5 μs of a dye laser flash in crude extracts of bean leaves [279]. This intermediate was not fluorescent and partially decayed to a 675–676 nm form with a half-life of 7–10 μs. The half-life of the decay increased at temperatures below −10 °C. A similar flash-induced and non-fluorescent intermediate forms at −105 °C, and is stabilized at −196 °C [280]. If the temperature is raised to −20 °C, the intermediate decayed to the 678 nm form. The 690–695 nm intermediate could be converted back to the photoactive protochlorophyllide by a subsequent laser flash of 695 nm light given at −196 °C. Finally, when etioplast membranes were excited with a 35 ps pulse of light, the fluorescence at 657 nm due to the 638/650 nm-absorbing forms decayed within 1 ns [281]. That the flash-induced fluorescence decay of 1 ns represented an initial step in photoconversion was supported by the following observations: Preillumination for 1 min eliminated the flash-induced fluorescence decay, presumably by photoconverting all the photoactive protochlorophyllide. Addition of NADPH after the pre-illumination restored the flash-induced fluorescence decay, presumably by allowing the reformation of phototransformable from endogenous nonphototransformable protochlorophyllide. The fluorescence decay was also temperature dependent, increasing from 1 ns at 20 °C to 3.5 ns at −196 °C. It is not clear whether the fluorescence decay represents the formation of the 690–695 nm-absorbing intermediate. However, when repetitive pulses were given at −196 °C, fluorescence decay was still observed, but no product accumulated, suggesting that flash-induced formation of an intermediate is reversible at that temperature. This observation is consistent with the flash-induced reversibility of the 690–695 nm form at −196 °C.

2.4.1.5.2. Dark protochlorophyllide reduction. Some species, including gymnosperms and algae, do not require light for chlorophyll formation, although protochlorophyllide is an intermediate in the pathway [282, 283]. These organisms must have a light-independent mode of protochlorophyllide reduction. Pine embryos were reported to synthesize chlorophyll in the dark only if they are in contact with the megagametophyte tissue [282]. Recently, it was found that cytokinins can be substituted for the megagametophyte tissue [284].

Some progress in analyzing dark chlorophyll formation in extracts of cyanobacteria has been reported. *Anabaena* cells were lysed by osmotic adjustment, supplemented with protochlorophyllide and a NADPH regenerating system, and incubated in complete darkness. A linear relationship between protochlorophyllide consumed and chlorophyllide formed was observed, and the rate was temperature dependent [285]. Peschek et al. [286, 287] reported light-independent NADPH-protochlorophyllide oxidoreductase activity in purified plasma membrane preparations from *Anacystis nidulans*. The reaction was insensitive to illumination and was reversible in the presence of a high NADP/NADPH ratio. The reaction demonstrated a requirement for Ca^{2+} and was inhibited by the Ca chelator EGTA.

It should be noted that these organisms, as well as all others where the question has been examined, also have a light-dependent mode of protochlorophyllide reduction [288, 289].

If plants and algae that are capable of forming chlorophyll in the dark have two mechanisms for protochlorophyllide reduction, it might be asked whether the angiosperms also have a light-independent as well as the light-dependent pathway. The question becomes even more compelling when the phenomenon of the light-induced breakdown of the light-dependent reductase is considered (see section titled Regulation of Protochlorophyllide Photoreduction). A number of investigators have observed increases in chlorophyll accumulation in the dark when greening plants were transferred from the light to the dark. In some plants such as *Zostera* [290] and *Tradescantia* [291] dark accumulation can be quite substantial. Smaller dark increases have also been observed in green barley [292].

Attempts to correlate these increases with the dark synthesis and incorporation of newly formed tetrapyrroles into chlorophyll(ide) have had conflicting results. In one experiment, barley was germinated in the dark for 5 days, the intact plants were illuminated for various periods of time, and then the shoots were excised and incubated for 2 h in the dark with ^{14}C-ALA [293]. At this point, the shoots were either ground in liquid N_2 and the pigments extracted or the shoots were exposed to light for 10 min before extraction. Incorporation of label into protochlorophyllide, chlorophyllide, and chlorophyll were determined after separation of the pigments by reverse-phase HPLC. It was concluded that of the total incorporation into the three pigments, no more than 1.2% was incorporated into chlorophyllide in the dark, under conditions where pre-illumination ranged from 0 to 72 h. Similar experiments were performed by Packer and Adamson [294], except that in this case, the shoots were not excised from the seed and the dark incubation with label was for 18 h. After a more rigorous purification and degradation of the chlorophyll to the tetrapyrrole-derived portions, substantial dark incorporation into chlorophylls *a* and *b* was measured. When incorporation in intact seedlings was compared to that of excised shoots, no incorporation of label was detected in the chlorophylls from the excised shoots. On the contrary, the amount of chlorophyll present in the leaves decreased by more than 50% during the dark incubation. When incorporation of label into chlorophylls and protochlorophyllide was compared, the label in the chlorophyll was only 5% of the label in protochlorophyllide. Finally, both ^{14}C-ALA and ^{14}C-glutamate were incorporated into chlorophyll in the dark in intact barley leaves and isolated developing chloroplasts [295]. Again, the dark label incorporation into chlorophyll was minor compared to incorporation into protochlorophyllide. While it seems safe to conclude that some dark reduction of protochlorophyllide may occur in green barley, the extent is small compared to the light-dependent reduction.

2.4.1.6. Esterification of the ring-D propionate. All naturally occurring chlorophylls (with the exception of the *c*-type chlorophylls discussed below) are found with the ring-D propionate esterified to a long-chain alcohol. With rare exceptions, the alcohol is a polyisoprene. Chlorophylls *a* and *b* in plants and green algae are normally esterified only with the C_{20} alcohol 2-*E*,7-*R*,11-*R*-phyto, in contrast to the photosynthetic bacterial chlorophylls, which can be found esterified with phytol, geranylgeraniol, tetrahydrogeranylgeraniol, farnesol, and stearol.

Chlorophyllase, a ubiquitous plastid enzyme, can catalyze hydrolysis of the phytyl ester, transesterification (e.g., replacement of phytol with methanol), and esterification under some conditions in vitro [296]. At one time chlorophyllase was considered to have a biosynthetic role in phytylation. However, this enzyme is now generally believed to function only hydrolytically in vivo.

Chlorophyll synthetase from maize shoots or oat etioplast membranes catalyzes the esterification of chlorophyllide *a* with geranylgeranyl-pyrophosphate, but not with free geranylgeraniol [297, 298]. Chlorophyllide is also esterified with geranylgeraniol and its monophosphate if ATP is added to the incubations. In addition to geranylgeranyl-pyrophosphate, the pyrophosphates of phytol and farnesol can also serve as substrates, but those of geraniol or pentadecanol cannot [298]. Acceptable pigment substrates were chlorophyllides *a* and *b*, but not protochlorophyllide or bacteriochlorophyllide [299].

In the esterification of bacteriochlorophyllide *a* in *R. spheroides* and *Rhodospirillum rubrum*, both the bridge and the nonbridge oxygen atoms of the ester were found to be derived from the propionic acid of the chlorophyllide [300, 301].

In the early stages of greening of etiolated tissues, chlorophyll *a* appears to be initially esterified with all-*trans* geranylgeraniol, and the conversion of the geranylgeranyl group to phytol occurs after esterification [302]. Geranylgeraniol has double bonds at positions 2, 6, 10, and 14, whereas phytol retains only the one at position 2 [303]. The double bond in geranylgeraniol at position 6 is the first to be hydrogenated (Fig. 11), followed by that at position 10, and the double bond at position 14 is the last to be hydrogenated [303]. Illuminated etioplast fragments are able to esterify endogenous chlorophyllide a with exogenous geranylgeranyl-pyrophosphate, and then carry out the hydrogenations [304]. The hydrogenation reactions require the pyridine nucleotide NADPH. Certain herbicides interfere with the hydrogenation of geranylgeraniol, and cause the accumulation of chlorophyll molecules containing geranylgeraniol and dihydrogeranylgeraniol instead of phytol [305]. Anaerobiosis ap-

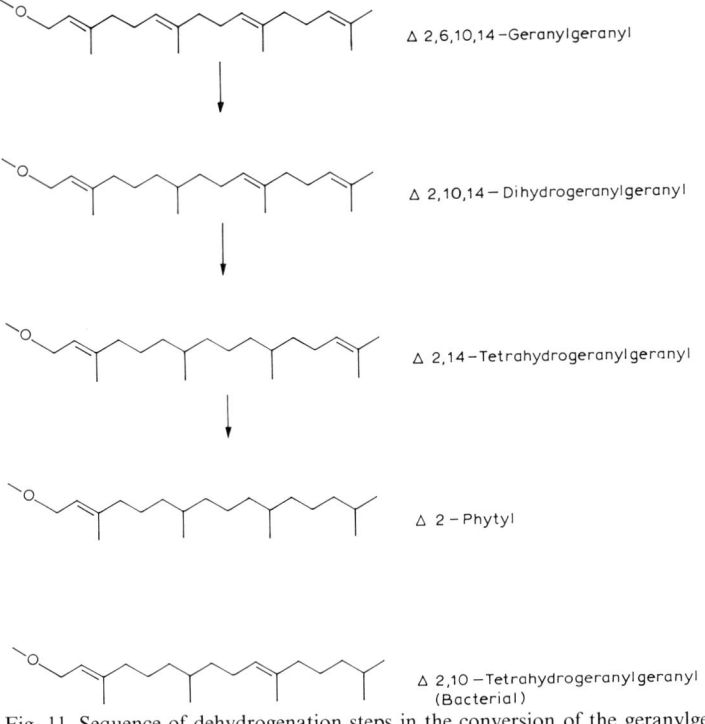

Fig. 11. Sequence of dehydrogenation steps in the conversion of the geranylgeranyl group to phytyl in plants. Also illustrated is the tetrahydrogeranylgeranyl group found in some bacteriochlorophylls.

pears to interfere with the hydrogenation of geranylgeranyl-chlorophyllide, but not with protochlorophyllide photoreduction or esterification with geranylgeraniol [306]. There exists a mutant strain of *Scenedesmus* that is deficient in vitamin E because it cannot form the phytol moiety of the vitamin [307]. Cells of this strain also are unable to complete chlorophyll synthesis, and instead accumulate near normal amounts of chlorophylls *a* and *b* that contain geranylgeraniol instead of phytol [308].

Minor amounts of bacteriochlorophyll *a* esterified with geranylgeraniol, dihydrogeranylgeraniol, and tetrahydrogeranylgeraniol were detected in purple photosynthetic bacteria [309]. These results suggest that, like greening etiolated plant tissues, those photosynthetic bacteria which form phytylated bacteriochlorophylls begin with the pigment esterified with geranylgeraniol.

Contrary to the case in greening etiolated tissues and photosynthetic bacteria, in green plants chlorophyllide appears to be esterified directly with phytol [310]. This direct phytylation may be able to occur in green plants because the plastids of these tissues contain a larger pool of phytyl-pyrophosphate than those of etiolated tissues.

2.4.2. Chlorophyll b formation

Higher plants and green algae contain, in addition to chlorophyll *a*, generally smaller amounts of chlorophyll *b*, which differs from chlorophyll a only by the presence of a formyl group in place of the methyl on ring B of the tetrapyrrole ring (Fig. 12). Chlorophyll *b* typically comprises close to 25% of the total chlorophyll present in green plant tissues.

2.4.2.1. In vivo studies. Early work suggested that the immediate precursor of chlorophyll *b* is chlorophyll *a*. Godnev et al. [311] reported that label from exogenous ^{14}C-chlorophyll *a* was transferred to chlorophyll *b* in greening onion leaves. The reverse transfer of label from chlorophyll *b* to chlorophyll *a* was not observed. In these

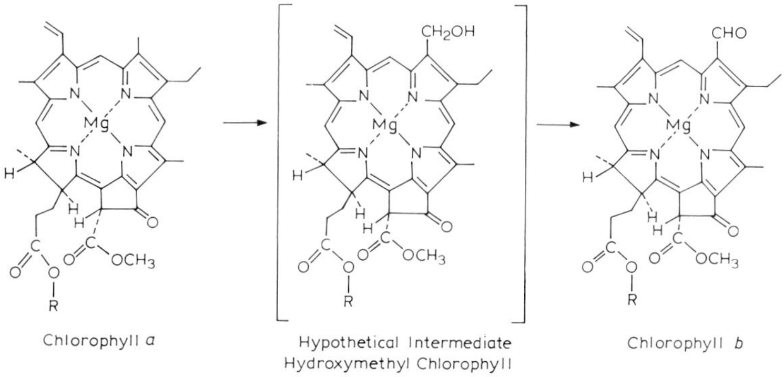

Fig. 12. Structures of chlorophylls *a* and *b*, and a hypothetical intermediate hydroxymethyl chlorophyll. It is uncertain at which stage in the phytylation process the conversion of the B-ring methyl to formyl occurs.

experiments, the pigments were dissolved in vegetable oil for introduction into the tissue.

Akoyunoglou et al. [312] generated ^{14}C-labeled chlorophyll *a* in situ by incubating excised etiolated barley leaves with ^{14}C-ALA and exposing the leaves to a series of light flashes separated by 15 min dark periods. Under the flashing light regime, very little chlorophyll *b* was formed, although considerable chlorophyll *a* appeared during 24 cycles extending to a total of 6 h [313]. The leaves were next washed free of exogenous ^{14}C-ALA, then exposed to 100 more flash cycles to use up any ^{14}C-ALA remaining within the leaves, and then the leaves were transferred to continuous light. During the continuous light phase the total radioactivity of the chlorophyll *a* pool decreased while the total radioactivity of newly formed chlorophyll *b* increased. The specific radioactivity of both chlorophylls decreased during the continuous light phase, that of chlorophyll *b* decreasing faster than chlorophyll *a*. The investigators concluded from these results that chlorophyll *b* is formed from a sub-population of newly formed chlorophyll *a* molecules.

Shlyk and Prudnikova [314] allowed greening barley leaves to assimilate ^{14}CO$_2$ and form radioactive chlorophyll. The leaves were then homogenized and incubated in the dark. Chlorophyll *b* became labeled during the dark incubation. Added unlabeled chlorophyll *a* lessened the degree of labeling of chlorophyll *b*. Light had no effect on the transfer of label [315]. The action spectrum for chlorophyll *b* formation in etiolated wheat leaves was found to be identical to the action spectrum for chlorophyll *a* formation, and both were similar to the absorption spectrum of protochlorophyllide [316]. Thus, although light was not specifically required for chlorophyll b formation, it was thought to stimulate the conversion by supplying a pool of newly synthesized chlorophyll *a* molecules.

The prevailing hypothesis that chlorophyll *b* is made from chlorophyll *a* has been challenged by a number of investigators. Oelze-Karow et al. [317] and Oelze-Karow and Mohr [318] observed that chlorophyllide *a* and chlorophyll *b* accumulation are both more sensitive to inhibition by far-red light than is chlorophyll *a* accumulation, and it was postulated that the precursor of chlorophyll *b* might be chlorophyllide *a* rather than chlorophyll *a*, with conversion of the 3-methyl moiety to a formyl group occurring before the phytylation step. This scheme is consistent with the observation that in broken etioplast preparations from dark grown oat seedlings, chlorophyllide *b* undergoes enzymatic esterification with geranylgeranyl pyrophosphate as readily as chlorophyllide *a* [299].

Duggan and Rebeiz [319] detected chlorophyllide *b* by fluorescence spectroscopy in extracts from greening and photoperiodically grown cucumber cotyledons, and confirmed the identification by synthesizing the oxime and the methyl ester, which had the expected spectrofluorometric and chromatographic properties. ^{14}C-chlorophyll *b* synthesized in vivo from ^{14}C-ALA and purified by thin layer chromatography was used to check how much of the observed chlorophyllide *b* could have arisen by chlorophyllase activity during pigment extraction. This approach indicated that no more than 13% of the chlorophyllide *b* detected spectrofluorometrically could have been an artifact of in vitro hydrolysis. Chlorophyllide b was also detected in *S. obliquus* cells [320].

Although these findings suggest that chlorophyllide *b* may be a biosynthetic intermediate on the route to chlorophyll *b*, an alternative explanation for the occurrence of chlorophyllide *b* in greening tissues is that chlorophylls and chlorophyll-protein complexes are subject to turnover before they become established at stable sites within the thylakoid membranes. It is known that chlorophylls do turn over. Chlorophyll *b* turns over to a greater extent than chlorophyll *a* in seedlings greening under intermittent illumination [321] or after transfer to the dark following a period of continuous light [322]. Both the protein and pigment components of newly formed light-harvesting chlorophyll-protein complexes are subject to degradation in the dark [322] or when chlorophyll *b* synthesis is blocked by mutation [323]. During the early stages of greening, chlorophyll *a* may not be made at a rate sufficient to saturate the requirements of the simultaneously forming light-harvesting complexes, and incomplete complexes might then be degraded, along with any attached molecules of chlorophyll *b*. The first step of chlorophyll *b* degradation is likely to be dephytylation to chlorophyllide *b*. In vitro chlorophyllase activity may not be an accurate measure of the potential for in vivo chlorophyll dephytylation, which could be accelerated by factors such as pigment association with unstable proteins.

Results from Rüduger's laboratory suggest that the immediate precursor of the formyl-containing pigment may be neither chlorophyllide *a* nor chlorophyll *a*, but rather the geranylgeranyl ester of chlorophyllide *a* [324]. Growing suspension cultures of tobacco cells can incorporate exogenous labeled monophosphate and pyrophosphate esters of geranylgeraniol, and also the pyrophosphate ester of phytol, into chlorophylls. With either of the geranylgeranyl compounds, label was incorporated at equal specific activity into chlorophylls *a* and *b*. However, phytol pyrophosphate was incorporated significantly better into chlorophyll *a* as compared to chlorophyll *b*. The investigators concluded from their results that the cells cannot convert phytylated chlorophyll *a* into chlorophyll *b*, but that chlorophyllide *a* esterified with geranylgeraniol can undergo the conversion. This implies that the natural biosynthetic sequence may be esterification with geranylgeraniol, then formyl group formation, and finally reduction of the geranylgeranyl group to phytol. Since the phytylation process lags behind, and occurs at a slower rate, than protochlorophyllide reduction, the proposed chlorophyll b biosynthetic sequence also explains the observed need for 'young' chlorophyll *a* molecules, since the young molecules would be newly esterified with geranylgeraniol, but not yet reduced to the phytyl level.

Shioi and Sasa [325] had earlier reported the existence in greening cucumber cotyledons of chlorophyll *b* species having all possible degrees of polyisoprene group hydrogenation between geranylgeranyl and phytyl, suggesting that the *a*-to-*b* transformation may occur before phytylation, or at least before the process is completed. However, these workers also detected protochlorophyllide pools having the same range of polyisoprene group hydrogenation [326], making interpretation of their results difficult.

In summary, it is presently not certain whether oxidative conversion of the ring-B methyl group occurs before or after phytylation, or if parts of the phytylation process take place both before and after the a-to-b conversion. If phytylation comes first, then chlorophyll *a* would be a biosynthetic intermediate. If the opposite is true, the reaction sequence would be chlorophyllide *a* to chlorophyllide *b* to chlorophyll *b*. If the last possibility occurs, the true conversion substrate would be geranylgeranyl chlorophyllide *a* or a chlorophyllide *a* with a partially reduced isoprenoid.

2.4.2.2. In vitro studies. Ellsworth et al. [327] reported that transfer of ^{14}C from exogenous chlorophyll *a* to chlorophyll *b* in soybean leaf homogenates occurred in all parts of the molecule, i.e., the phytyl chain, the methyl ester group on the cyclopentenone ring, and the tetrapyrrole portion. This result indicates that a true conversion had taken place, rather than transfer of methyl or phytyl groups from ^{14}C-chlorophyll *a* to newly formed chlorophyll b. These workers also reported that NADP+ stimulated the reaction. This latter observation was confirmed by Shlyk and coworkers et al. [328], who additionally showed that conversion of exogenous chlorophyll *a* to chlorophyll *b* can occur in leaves of barley, rye, and maize even without prior light exposure. In contrast to these results, Rebeiz and Castelfranco [329] reported that upon incubation with ^{14}C-ALA, homogenates of cucumber cotyledons incorporated ^{14}C into both chlorophyll *a* and chlorophyll *b* only if the cotyledons had been pre-illuminated for 4.5 h, while pre-illumination for 2.5 h resulted in label incorporation into chlorophyll *a* only.

C. reinhardtii strain *y-1* requires light for chlorophyll accumulation but can be grown heterotrophically in the dark. Bednarik and Hoober [330, 331] have made the interesting observation that when degreened cells of strain *y-1* are incubated in the dark at elevated temperatures (38 °C) with *o-* or *m*-phenanthroline a pigment rapidly accumulates and is excreted into the medium. The pigment was characterized as chlorophyllide *b*. Green cells treated the same way excreted a slightly different pigment that was characterized as divinylchlorophyllide *b*. The conversion required the presence of O_2. Cells incubated in the dark at elevated temperature without phenanthroline excreted protochlorophyllide. A membrane fraction was isolated from degreened cells which, when supplemented with phenanthroline, converted exogenous protochlorophyllide to chlorophyllide *b*. These results are consistent with the possibility that protochlorophyll(ide), rather than chlorophyll(ide) *a*, is the precursor to chlorophyll *b*, and that the reason why the presence of chlorophyll *a* is normally required for the formation of chlorophyll *b* is that chlorophyll *a* plays some regulatory role. It was suggested by the authors that phenanthroline stimulates chlorophyllide *b* formation by mimicking chlorophyll *a* at some regulatory site in the chlorophyllide *b* forming system. It will be of great interest to attempt to exploit this apparent breakthrough in the in vitro biosynthesis of chlorophyll *b*, toward understanding the mechanism of chlorophyll *b* formation and its regulation.

Attempts to replicate the above results with *S. obliquus* cells were unsuccessful [320]. However, a preparation from *Scenedesmus* was recently obtained that catalyzed conversion of chlorophyllide *a* and chlorophyll *a* to chlorophyllide *b* and chlorophyll *b*, respectively [332]. The activity was reported to be membrane bound and light independent.

2.4.3. Possible existence of multiple forms of chlorophylls a and b
A mutant strain of maize exists in which the ring-B ethyl groups are replaced by vinyls on both chlorophylls *a* and *b* [333, 334]. The divinyl structure of the variant chlorophyll *b* was confirmed [335]. The mutant (Olive Necrotic 8147) is necrotic, its leaves contain 70% less chlorophyll than wild-type leaves and its chlorophyll a-to-b ratio is three times higher, but it can carry out photosynthesis [336]. It is probable that the mutant lacks the enzyme that catalyzes the vinyl reduction step in the biosynthesis of chlorophyll *a*, and chlorophyll *b* is formed from the variant chlorophyll(ide) *a*. A pigment having the properties of divinyl chlorophyll *a* was also found in a free-living unicellular marine prochlorophyte [337]. This organism also contains apparently normal chlorophyll *b*, in approximately equal abundance as the novel chlorophyll *a*.

Several other novel chlorophyll types have been reported in greening barley leaves and cucumber cotyledons [233–235, 338, 339, 252]. These pigments have different low-temperature fluorescence excitation and emission maxima, and are partially separable by HPLC. The investigators have proposed that the chemical differences among the different species of chlorophylls include different states of oxidation at positions 2, 4 and 10, and different substitutions for phytol at position 7 [339, 252]. As many as 16 chemically and spectroscopically unique species of chlorophyll *a*, and perhaps the same number of chlorophyll *b* species have now been reported by Rebeiz et al.[339, 252]. In addition, these investigators have postulated that the ratio of end products and sequence of biosynthetic reactions is dependent upon the individual plant species and its growth history [340, 341]. It will be of interest to learn the precise structure of each chlorophyll type, and to possibly correlate these with different functional pools of chlorophyll and different chlorophyll-protein complexes. After the structures of the novel chlorophylls have been determined, it will still remain to be established whether the pigments are functionally significant in photosynthesis.

A study by Shioi and Sasa [325] identified four HPLC-separable chlorophyll species in extracts of spinach leaves as the intermediates between geranylgeranyl chlorophyllide *a* and chlorophyll *a*. No other anomalous pigments were detected, even though many chlorophyll breakdown products and derivatives were characterized. Four chlorophyll *a* species having different HPLC mobilities were also detected by Eskins and Harris [342]. The separated chlorophylls might be molecules having varying degrees of polyisoprene chain hydrogenation, i.e., the intermediates between geranylgeranyl chlorophyllide *a* and chlorophyll *a*, as described by Schoch et al. [302, 306] and Rüdiger et al. [297, 298].

2.4.4. Chlorophylls c

Chlorophylls c are accessory light-harvesting pigments found in many groups of eukaryotic algae that do not contain chlorophyll b, including diatoms, dinoflagellates, brown macrophytic algae, and cryptophytes (the latter also contain phycobiliprotein accessory pigments). There is also one report of a species that contains both chlorophyll b and chlorophyll c [343]. The chlorophyll c macrocycle is structurally not a chlorin, but a porphyrin, since there are no reduced pyrrole rings. Two c-type chlorophyll structures have been described, c_1 and c_2 (Fig. 13). A distinguishing feature of both of these pigments is that in place of the phytylated propionic acid group on ring D, there is a free, unesterified acrylic acid, which has a trans dehydrogenated structure. Chlorophyll c_1 has an ethyl group on ring B, like protochlorophyllide, and on chlorophyll c_2 the ethyl is replaced by vinyl, as on the precursor of protochlorophyllide, Mg-divinylpheoporphyrin a_5 [344–346]. Algae containing either chlorophyll c_1 or c_2 alone, or both, have been described [347]. In addition, several algal species have been found to contain another pigment, of undetermined structure, named chlorophyll c_3 [348]. This pigment has been found to occur together with chlorophylls a and c_2, and chlorophyll c_1 may be present as well. Yet another chlorophyll c-like pigment has been described in other algae [349]. This pigment has been reported to occur together with chlorophylls a and c_1, and chlorophyll c_2 may be present as well. The pigment is of undetermined structure, has an absorption spectrum similar to that of

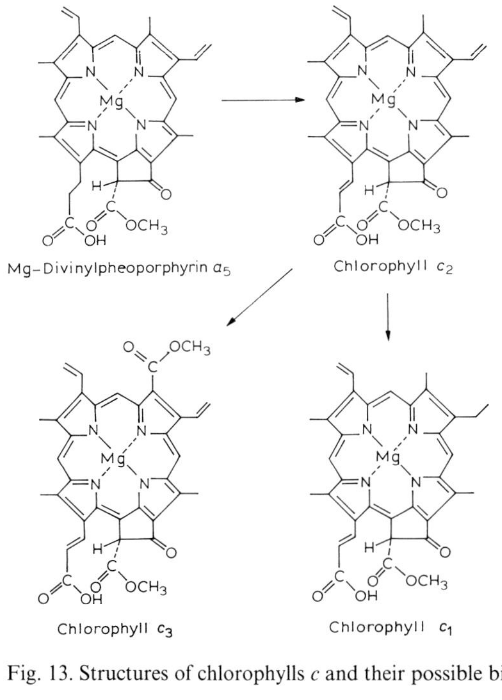

Fig. 13. Structures of chlorophylls c and their possible biosynthetic relationships.

chlorophyll c_2, and is chromatographically separable from the other three c chlorophylls. Finally, Mg-divinylpheoporphyrin a_5 itself accumulates to appreciable concentrations in some algae, where it occurs together with chlorophylls a and b, and appears to play a light harvesting role [350].

The close structural resemblance of chlorophylls c_2 and c_1 to Mg-divinylpheoporphyrin a_5 and protochlorophyllide, respectively, suggests that the pigments derive directly from these precursors (Fig. 13), rather than from chlorophyll a.

2.4.5. Reaction center chlorophylls

The most abundant of the tetrapyrroles found in oxygenic organisms is chlorophyll a, or one of its common derivatives. Functionally, most of this chlorophyll is used for light harvesting and transfer of the absorbed energy to specialized reaction centers where photochemistry occurs. Although they make up only a small percentage of the total, chlorophylls also occur in the reaction centers. Attempts have been made to explain the unique properties of the reaction centers in terms of modifications to the chlorophyll prosthetic groups, and/or by interactions of chlorophyll a (or bacteriochlorophyll a) with the specialized proteins which constitute the reaction center. This dilemma has been approached in two ways. Measured physical properties of the reaction center have been correlated with the measured physical properties of synthesized model compounds. In addition, the reaction center pigments have been extracted and examined for the presence of small, but consistent, amounts of modified pigment.

Various chlorophyll a derivatives have been suggested as the primary electron donor pigment of photosystem I (P_{700}). One proposed pigment is chlorophyll a', the C_{10}-epimer of chlorophyll a [351]. This pigment was reported to exist in approximately equimolar amounts as P_{700} in various plant and algal preparations [352]. Watanabe et al. [352] found that the ratio increased with harsher isolation treatments, but under mild conditions the ratio was always 1. Redox- and spectral-induced P_{700}-like spectral changes were observed when chlorophyll a' was added to a 65,000 molecular weight photosystem I apoprotein [352]. However, because epimerization about C_{10} is known to occur spontaneously in chlorophyll solutions, further work is required to establish whether chlorophyll a' is the natural P_{700} pigment.

2.4.6. Reaction center pheophytins

In the reaction centers of some groups of photosynthetic bacteria (but not others), Mg-free bacteriopheophytins serve as the primary electron acceptor [353, 354]. Similarly, photosystem II (but not photosystem I) reaction centers of oxygenic plants and algae contain contain pheophytin a as the primary acceptor. Although the only pheophytin found in photosystem II reaction centers is pheophytin a, bacterial reaction center bacteriopheophytin structures generally follow those of the corresponding reaction center bacteriochlorophylls, and can be of the a, b, or g type.

It has generally been assumed that the pheophytins are derived from the corresponding chlorophylls by loss of Mg accompanying incorporation into the reaction center proteins. However, it should be noted that the reaction center bacteriopheophytin *a* of *Rhodospirillum rubrum* is esterified with phytol, while the bacteriochlorophyll *a* molecules in the same reaction centers are esterified with geranylgeraniol [355]. This finding suggests that the pheophytin is formed by removal of Mg from newly esterified pigment molecules, before hydrogenation of the polyisoprene to phytol has begun.

3. Regulation

3.1. Environmental and physiological modulation of pigment content

3.1.1. Physiology of greening in plants and algae

When seeds of flowering plants are germinated in the dark, the seedlings are etiolated, i.e., they are tall and yellow in color. Hemes are present in these tissues, and although chlorophyll is absent, small amounts of protochlorophyll(ide) are present. The plastids, termed etioplasts, do not have the characteristic stromal and granal thylakoid system, nor do they have photosynthetic capacity. Instead, the etioplasts have a prolamellar body which appears 'paracrystalline' in electron micrographs, and has as its main protein constituent protochlorophyllide reductase [356, 357]. Upon exposure to light, the protochlorophyll(ide) is immediately photoconverted to chlorophyll(ide), the prolamellar bodies begin to break down and are replaced by thylakoids, and, after a lag period, a phase of rapid chlorophyll accumulation begins. The length of the lag period appears dependent upon the age of the etiolated seedling; the longer the plant was in the dark, the longer the lag period [358]. Chlorophyll accumulation is complete after 24–36 h. If the plants are returned to the dark during greening, chlorophyll accumulation stops, small amounts of protochlorophyllide build up again, and prolamellar bodies reform [359, 360]. Thus continuous chlorophyll accumulation requires light.

Similar studies, in which levulinic acid was included to prevent ALA utilization, demonstrated that ALA accumulation parallels chlorophyll accumulation in the absence of inhibitors [117, 361]. If ALA is administered in the dark to the dark-grown seedlings, protochlorophyllide and several porphyrin and Mg-porphyrin intermediates accumulate [362]. Even though ALA can be transformed to protochlorophyll(ide) in plants that have never been exposed to light, formation of ALA, and hence, chlorophyll, requires continuous light. Dark grown seedlings exposed to light for a short period and then returned to the dark for several hours, no longer exhibit the lag period in ALA or chlorophyll accumulation when the plants are subsequently brought back into the light [359].

The light effect upon chlorophyll accumulation can be divided into two components. There is an obvious and immediate effect upon the photoreduction of protochlorophyll(ide) to chlorophyll(ide), which may have indirect regulatory consequences (see below). In addition, it has been proposed that the enzymes responsible for ALA formation are induced by the phytochrome response. Red light can be substituted for white light and the effect of red light can, sometimes, be reversed by far red light [363, 364]. As with other phytochrome responses, Ca^{2+} has been implicated in the signal transduction chain for light-induced chlorophyll accumulation in etiolated cucumber cotyledons [365].

It has been shown that, although extractable ALA forming activity is present in dark-grown tissue, the extractable activity increases several fold during the first few hours of greening [57, 79, 82, 366]. Presumably, at least one of the enzymic components increases in response to light, because the RNA component does not increase in response to greening in these systems [58, 367]. Light quality was not investigated in the above reports.

The influence of light on the capacity of plants to form ALA could be measured during the later stages of greening by using isolated chloroplasts [368, 369]. The ability of isolated cucumber chloroplasts to synthesize ALA from exogenous substrates was dependent on the light regime immediately preceding organelle isolation. Chloroplasts isolated from cotyledons of plants that were germinated in the dark and greened for 24 h synthesized ALA to an extent greater than that required for protochlorophyllide regeneration [368]. If the seedlings were returned to darkness prior to chloroplast isolation, the capacity for ALA synthesis by the subsequently isolated chloroplasts diminished three fold. The decrease to the basal ALA-synthesizing capacity was time dependent and nearly complete by one h in the dark. The ALA-forming capacity recovered in a time-dependent manner when the dark-treated seedlings were re-illuminated. The in organello activity was independent of endogenous protochlorophyllide level, even in chloroplasts that were isolated from dark-treated seedlings. The light-dependent recovery of ALA-forming capacity was shown to be a phytochrome-mediated response, by the criteria of far-red light reversibility and the effectiveness of red, but not blue light in modulating the recovery of activity [369]. The light-induced recovery of ALA-forming capacity was blocked by prior administration of cycloheximide. In these studies it was apparent that the basal level of ALA- forming activity was not influenced by the light treatment.

The mRNA for the apoprotein of the nuclear-encoded chlorophyll a/b light-harvesting complex is also under phytochrome control [370, 371]. The presence of the chlorophyll-binding proteins is required for stabilization of the newly synthesized chlorophyll in the membrane. The light requirement for the phytochrome-mediated elimination of the lag phase in chlorophyll accumulation and the induction of transcription of the light-harvesting protein gene have been investigated in considerable detail [372, 373]. Both the kinetics and the fluence response for chlorophyll accumulation were different from those influencing the appearance of mRNA coding for the light- harvesting protein. These results suggested to the authors that availability of apoprotein is not the limiting factor during the early stages of greening.

Treatment of cotyledons in the dark with cytokinins can mimic the effect of light pretreatment with respect to overcoming the lag phase in chlorophyll and ALA accumulation [374, 375]. Cytokinin treatment increased the level of the mRNA for the light-harvesting complex apoprotein as well [376]. Pretreatment of cotyledons with cytokinins and/or gibberellic acid in the dark also stimulated the capacity of subsequently isolated plastids to convert exogenous ALA to protochlorophyllide [377, 378]. Both cytokinin and phytochrome pretreatment are time dependent with respect to whether the pretreatments will shorten or lengthen the lag period. Pretreatment within 24 h before continuous illumination eliminates or shortens the lag period, while pretreatment 72 h before continuous light lengthens the lag period [379]. The mechanism of these effects has not been established.

Etiolated tissues have been used for the investigation of greening and chloroplast development, because exposing the dark-grown tissue to light provides a means of initiating and synchronizing the myriad steps required for the developmental process. Under normal conditions, plants grow from seed in a diurnal light/dark cycle. The plastids begin as proplastids, which are capable of several developmental fates, including chromoplasts, amyloplasts, mature chloroplasts, etc. In plant tissues that normally develop above the ground, such as grass leaves, etioplasts are formed only under highly artificial conditions. It is reasonable to expect that regulatory aspects of tetrapyrrole biosynthesis observed in the transformation of etioplasts to chloroplasts may not be applicable to 'normal' chloroplast development. On the other hand, in tissues such as cucumber cotyledons, which carry out a considerable part of their development underground before emerging into the light, etioplasts may be a normal developmental stage of plastid development.

Unlike the angiosperms, most gymnosperms and algae do not require light for chlorophyll formation. However, several algal strains or mutants have been isolated that do require light for greening, and these have been used as convenient model systems for greening in the angiosperms. These include the C-10 mutant of *Chlorella vulgaris* [57], the *y-1* mutant of *C. reinhardtii* [330], and the C-2A' mutant of *S. obliquus* [380]. In addition, the phytoflagellate, *E. gracilis,* requires light for chlorophyll formation and chloroplast development [381]. These strains that require light for plastid development have been useful experimental models for studying light effects on plastid development. It is not known whether the nonlight-requiring wild-type algal strains contain two separate protochlorophyllide reducing systems, one light-requiring and the other light-independent, or if, instead, they contain an additional component that confers the ability of a universal light-requiring system to function in the dark.

3.1.2. Effects of Fe chelators and anaerobiosis on greening in intact tissue and chloroplasts

When etiolated bean leaves were treated in the dark with bidentate aromatic Fe chelators, such as α, α'-dipyridyl or *o*-phenanthroline, a doubling of protochlorophyllide accumulation was observed over a 21 h period [382]. During this treatment Mg-protoporphyrin methyl ester increased from an undetectable level to a level nearly three times the protochlorophyllide level in untreated tissue. Smaller amounts of other porphyrin intermediates were also detected. Accumulation of Mg-porphyrins was linear during the treatment period, and similar responses were observed with maize coleoptiles, cucumber cotyledons, and pea leaves. If the tissue was also treated with levulinic acid, chelator treatment caused a doubling of ALA accumulation. The ability of the chelators to enhance Mg-porphyrin and ALA accumulation was greater in greening tissue than in etiolated or fully greened tissue [383, 384]. In addition to raising the level of Mg-porphyrins accumulated, chelator treatment diminished by 38% the amount of extractable heme present in the tissue [383]. Finally, treatment of the leaves with salts of Fe^{2+}, Zn^{2+}, or Co^{2+} overcame the effects of chelator treatment. When the metal salts were added, Mg-protoporphyrin methyl ester disappeared in a time-dependent manner. In the light, the disappearance was accompanied by an increase in protochlorophyllide, but in the dark, the decline in Mg-porphyrin was apparently caused by enzyme-mediated degradation. The loss was slower in the cold and was inhibited by protein synthesis inhibitors [384]. It was postulated that the chelators have a two-fold effect on the pathway. First, chelation of Fe stimulates the synthesis of ALA, probably by removing an Fe-containing inhibitor such as heme. Second, chelation of Fe inhibits the conversion of Mg-protoporphyrin methyl ester to protochlorophyllide, perhaps at a specific step. These effects can be reversed by the above-mentioned transition metals, perhaps because they displace bound Fe.

Anaerobiosis can be expected to have a wide variety of effects on plant growth and development. Some of these effects can be localized to the tetrapyrrole biosynthetic pathway. In the early part of the pathway, O_2 is required for coproporphyrinogen and protoporphyrinogen oxidases. In the above experiments with α, α'-dipyridyl, O_2 was required for the conversion of Mg-protoporphyrin methyl ester to protochlorophyllide in the light when the metal salts were added [384]. In addition, anaerobiosis or Fe deficiency caused a decrease in chlorophyll, concomitant with an accumulation of small amounts of Mg-protoporphyrin. The same conditions inhibited the conversion of exogenously supplied ALA to protochlorophyllide [242].

Treatment of isolated chloroplasts with α, a'-dipyridyl resulted in an increase (20–30%) in the conversion of glutamate to ALA [243, 385]. Anaerobiosis was also shown to have an inhibitory effect on the process [386]. In this case, ALA synthesis in isolated chloroplasts could be supported in the dark with exogenous ATP and a NADPH regenerating system, or in the light with no exogenous cofactors. The light-driven synthesis was sensitive to DCMU, indicating a requirement for photosynthetic electron transport. Anaerobiosis caused a 26% inhibition of the dark reaction and a 90%

inhibition of the light-driven reaction. When ATP and reducing power were added to the anaerobic incubation in the light, ALA synthesis was restored to 70% of the light control value. Addition of 5 mM oxaloacetic acetic acid could replace the O_2 requirement in the light driven reaction. It was proposed that O_2 or oxaloacetic acid is required as an electron acceptor in the regeneration of ATP, by an unexplained mechanism. As noted before, O_2 was absolutely required for the cyclase reaction in intact chloroplasts [243, 244, 250].

3.1.3. Chlorophyll and heme turnover
The subject of tetrapyrrole degradation in plants has recently been reviewed [387]. Aspects of degradation which are related to senescence will not be discussed in the present review. When dark-grown barley is exposed to the light for 12 h, chlorophyll accumulation increases to an amount 100 times greater than the original amount of protochlorophyllide present in the tissue before illumination [43]. During this period there is only a very minor increase in the amount of extractable, non-covalently bound protoheme [388]. If ^{14}C-ALA is administered to the plants during a 5-h illumination period, the specific radioactivities of purified protoheme and pheophorbide (chlorophyll derivative) are almost identical [43]. Since new protoheme is synthesized but does not accumulate, it must turn over at a rate comparable to its rate of synthesis. Supplying ALA to etiolated shoots increased the amount of extractable protochlorophyllide, but not that of heme [389]. When ^{55}Fe^{2+} was administered to the shoots along with the ALA, the specific radioactivity of the extracted heme was 30% higher than that from the incubation without added ALA. Similarly, in rapidly dividing cultures of *Euglena*, the relative specific radioactivities of newly formed chlorophyll and heme indicated that heme must be subject to metabolic turnover [36]. If etiolated barley shoots are administered high, non-physiological concentrations of heme through the cut base, the plant material takes up a portion of the heme, and the subsequent disappearance of the heme from the tissue can be monitored [390]. Under these conditions, the exogenous heme was degraded faster in the dark than in the light, and the half-life of heme in the dark was estimated to be between 8 and 9 h. Although the conditions of the experiment are artificial, and the half-life calculations may not reflect physiological conditions, the experiment clearly shows the capacity of the tissue to eliminate excess heme.

The question of chlorophyll turnover is somewhat more complex. Turnover of newly formed chlorophyll in greening leaves of barley and wheat has been reported to be quite substantial [391, 392]. The method was based on the observation that after pre-illumination of levulinic acid-treated shoots (which are blocked from forming new chlorophyll) for 2 to 62 h, a further illumination (3600–4000 lux ≈ 50–55 μE m^{-2} s^{-1}, assuming similar light qualities) resulted in a decrease in the amount of chlorophyll present. In wheat, the half-life of chlorophyll during the first 6 h of greening was calculated to be 6.5–8.2 h. The half-life increased to 33–36 h after 2 days of greening [392]. These chlorophyll losses were not observed in the dark. Examination of chlorophyll levels in untreated greening barley revealed an increase from 37 nmol/leaf after 24 h in the light to 65 nmol/leaf after 48 h in the light [391]. Yet the calculated degradation rate constant was equivalent to or greater than the biosynthetic rate constant at 24 h, and the degradation rate was twice the biosynthetic rate at 48 h, suggesting that the total chlorophyll should decrease between 24 and 48 h. It therefore seems likely that levulinic acid treatment indirectly promoted chlorophyll breakdown. During the early phase of greening, when both apoprotein and pigments are being synthesized, there may be a mechanism to prevent the accumulation of incomplete complexes as might occur when chlorophyll synthesis is artificially blocked.

When kidney bean cotyledons were returned to the dark after 8.5 h of greening, chlorophyll *a* remained constant while chlorophyll *b* decreased by 80% in 50 h [393]. If the cotyledons were greened for 50 h before returning to darkness, chlorophyll *b* decreased only about 20%, while chlorophyll *a* again remained constant. During the dark period, the amount of light-harvesting complex decreased while the number of photosystem II reaction centers increased. It was suggested that the chlorophyll *a* from the light-harvesting-complexes was recycled and the chlorophyll *b* degraded. Treatment of etiolated barley segments for 5 min with ^{14}C-ALA of high specific radioactivity allowed the formation of protochlorophyllide in the dark and chlorophyll in the light to be monitored [394]. The specific radioactivity and total radioactivity of the isolated protochlorophyllide did not change over a period of 3 h in the dark, indicating no turnover. When the segments were brought into light of low intensity (13 μE m^{-2} s^{-1}), the total radioactivity remained constant until the end of the experiment at 4 h. The specific radioactivity began to decline after 1.5 h as new chlorophyll was made. At a higher light intensity (36 μE m^{-2} s^{-1}), there was a time-dependent decline to 40% of the original total radioactivity of the chlorophyllide. The decline leveled off in 45 min and the total radioactivity remained constant during the next 4 h of greening. The specific radioactivity declined after the onset of new chlorophyll biosynthesis. Thus the stability of newly formed chlorophyll(ide) is dependent upon time and light intensity. The results of the pulse-labeling experiments indicate that chlorophyll does not turn over rapidly during greening in barley shoots.

In summary, we know that flowering plants require light for the massive chlorophyll build-up that is necessary for the formation of mature chloroplasts. Without light, the only intermediate that accumulates is a small amount of protochlorophyll(ide) which is immediately photoconverted to chlorophyll(ide) when the plants are exposed to light. Even in the dark all enzymes are present that are necessary for formation of chlorophyll(ide) and hemes from ALA. Despite the massive ALA formation during greening, heme levels do not build up, and there is strong evidence that heme is turned over rapidly. Although there is conflicting evidence, it appears that the turnover of newly formed chlorophyll is minor unless the developmental process is interrupted by darkness or inhibitors. Before the enzymatic properties that may give rise to these developmental features are discussed, the demand for nonplastid hemes and the subcellular compartmentation of the enzymes will be considered.

3.1.4. Nonplastid heme synthesis and turnover

Mitochondrial development in plants has been studied mostly in non-green tissues, such as germinating seeds, root tips, and cauliflower buds. As is the case in chloroplasts, the mitochondria require hemes for their cytochromes. In mitochondria isolated from germinating peanut cotyledons, the absorbancies due to cytochromes increased several fold between 2 and 10 days after germination [395]. Similarly, the mitochondrial cytochrome content of germinating mung beans also increased several fold between 2 and 8 days of germination [396]. However, microsomal cytochrome b_5 rose to a peak at 4 days and fell drastically thereafter, while the level of microsomal cytochrome *P*-450 declined in the light but recovered in the dark [396, 397].

When sliced Jerusalem artichokes (tubers without chlorophyll) are aged in the dark, cytochromes b_5 and *P*-450 are induced [398]. Induction of these cytochromes, as well as other oxygenases, is enhanced by inclusion of $MnCl_2$, herbicides, or various xenobiotic materials [399, 400]. The total induced levels can be ten times greater than the basal level present in freshly cut tissues. The Mn-enhanced induction of these microsomal hemoproteins was effectively blocked by gabaculine, the inhibitor of ALA formation from glutamate [1]. Inclusion of 100 μM gabaculine in the aging medium also prevented the wound-induced build-up of cytochrome *P*-450 and guaiacol peroxidase, and caused a ten-fold decrease in the extractable heme content. The decrease in heme content in response to gabaculine inhibition of ALA formation occurred in the light or dark and indicates a substantial capacity for heme turnover. Although hemes appeared to turn over with gabaculine treatment, the respiratory capacity (as measured by O_2 consumption) was unaffected. Three possible reasons may be advanced. First, consumption of O_2 may occur by a mechanism which does not require cytochromes (alternate oxidase). Second, the hemes for the respiratory cytochromes may be formed through a pathway that is insensitive to gabaculine. Third, respiratory cytochromes may not be subject to the same turnover rates as other cellular hemes. The last reason is consistent with the evidence of Hendry et al. [396] which suggests that mitochondrial cytochromes are more stable than microsomal cytochromes. Labeling of the heme *a* moiety of cytochrome oxidase by glutamate in etiolated maize coleoptiles indicated that respiratory hemes are formed primarily or entirely *via* the five-carbon pathway [2], although the possible existence of a minor gabaculine-insensitive pathway cannot be ruled out.

It is clear from the foregoing discussion that a general model for the regulation of tetrapyrrole biosynthesis in plants must take into account the substantial pools of both mitochondrial cytochromes and cytosolic hemoproteins in addition to the chloroplast cytochromes and chlorophyll-protein complexes. Cytosolic hemoproteins can be induced and turned over independently of chloroplast development. In tubers and germinating seeds, mitochondrial cytochromes appear to be stable, although this question has not been directly addressed for mature tissue or tissue undergoing the transition from heterotrophic to autotrophic growth conditions. Finally, a general regulatory model must address the subcellular localization of the biosynthetic and regulatory processes.

3.2. Regulation of biosynthetic steps

3.2.1. Subcellular compartmentation of tetrapyrrole biosynthesis

Isolated chloroplasts are capable of converting exogenously supplied glutamate to chlorophyll(ide) [101, 295, 401]. By inference the chloroplasts must contain all the enzymes necessary for those transformations. Isolated etioplasts and mature chloroplasts also have ferrochelatase activity [158, 159, 161]. Thus, chloroplasts are theoretically capable of providing either hemes or heme precursors for all cellular tetrapyrroles. It is not clear to what extent, if any, mitochondria provide heme precursors or hemes for nonplastid hemoproteins. As indicated above, mitochondrial hemes in plants are formed via the five-carbon path of ALA biosynthesis. The same is true for cytosolic hemes, as was indicated by the gabaculine inhibition studies (above) and by direct labeling studies of the heme moiety of cationic peroxidase in peanut suspension cultures [51]. Thus it is not possible to determine the origin of the heme for specific cytosolic hemoproteins by differential labeling with precursors for either the five- carbon pathway or for ALA synthase. This approach is possible in *Euglena*, which has both pathways [36], but for phylogenetic reasons the results would not be applicable to higher plants or other algae.

In contrast to the situation for chloroplasts, it is not known if plant mitochondria are capable of independent tetrapyrrole biosynthesis. Three enzymes, coproporphyrinogen oxidase [146], protoporphyrinogen oxidase [151, 153] and ferrochelatase [158, 159] have been detected in both mitochondria and chloroplasts of plants. In studies of the latter two enzymes, the organelles were purified by sucrose density gradient centrifugation, and the putative mitochondrial activity coincided in the gradients with cytochrome oxidase activity [151, 398]. However, none of the above studies established conclusively that the mitochondrial membranes were totally free of chloroplast or etioplast contamination. In the most detailed of the studies (in terms of subcellular localization), a washed mitochondrial pellet from etiolated barley was subjected to equilibrium density centrifugation [158]. Ferrochelatase exhibited two bands of activity. The low density band (1.18–1.19 g/ml) also contained symmetrical bands of two mitochondrial markers, cytochrome oxidase and succinic dehydrogennase. However, when washed mitochondria were prepared from 4-h greened barley and subjected to the same centrifugation, chlorophyll was observed '... as a single broad peak between the densities of 1.18 and 1.24 g/ml ...' [158], indicating that the mitochondrial peak may have had some chloroplast membrane contamination. The fact that the ferrochelatase in the etioplast-enriched membranes had a different pH optimum than the one in the mitochondrial-enriched membranes supports the notion that each organelle has its own ferrochelatase.

Distribution of two soluble tetrapyrrole biosynthetic enzymes, ALA dehydratase and porphobilinogen deaminase, was investigated in green and etiolated leaves of peas, and in spadices of *Arum* [112]. The latter tissue is not green and undergoes a major increase in the activity of several Krebs cycle enzymes at the stage when they are harvested. The distribution of the dehydratase and deaminase in all three tissues most closely followed the markers for two soluble chloroplast enzymes, ADP-glucose pyrophosphorylase and alkaline pyrophosphatase. Even in etiolated peas and *Arum* spadices, the tetrapyrrole biosynthetic enzymes did not co-sediment with the mitochondrial marker (NAD-dependent malic enzyme) on density gradients. In addition, the tetrapyrrole biosynthetic enzymes did not behave as cytosolic enzymes. Although some activity always sedimented with a pellet, the percentage in the pellet corresponded to the percentage of soluble chloroplast marker in the pellet. Both ALA dehydratase and porphobilinogen deaminase are cytosolic enzymes in animal cells [402]. The refractory cell walls of most plant tissues preclude the isolation of intact organelles without breaking a substantial fraction of the organelles. Thus it would be impossible to differentiate between large amounts of activity released from broken organelles (in this case the plastids) and a small amount of activity that may normally be localized in the cytosol. Even if a cytosolic fraction could be prepared that is free of enzymes originating from broken organelles (e.g., by gentle lysis of protoplasts), the activity measured in the cytosol may be due to enzyme synthesized on cytoplasmic ribosomes and in the process of being transported to the organelle. The activity of this fraction, measured in vitro, may not have a physiological role.

In an attempt to determine the subcellular origin of the heme moiety of cationic peroxidase in cultured peanut cells, the cells were labeled with ^{14}C-ALA and the radioactivity of the hemes in various subcellular fractions was compared [403]. The hemes extracted from the mitochondrial fraction had 15 times more total radioactivity than those from the amyloplast fraction. However, no indication was given of the specific radioactivity of the hemes, nor was there an estimation of the purity of the subcellular fractions based on marker enzymes. While the results suggest a mitochondrial origin of the heme moiety of peroxidase, this conclusion is tentative. If the preceding results are accepted, it should also be determined whether the mitochondria supply heme for the peroxidase under conditions where the chloroplasts are actively synthesizing new tetrapyrroles.

It was recently found that intact plastids isolated from greening cucumber cotyledons excrete heme [404]. Although it has not been demonstrated that this heme excretion is of physiological significance, the observation supports the possibility that plastids may be capable of exporting heme to other cellular regions *in vivo*.

3.2.2. Regulation of metabolic activity
A model for the regulation of the flux of precursors through the tetrapyrrole pathway in the chloroplast must account for many of the aspects of greening and turnover which occur after the lag phase. Most of these changes occur on a more rapid time scale than changes that would result from enzyme turnover. Briefly, these aspects

include: (*a*) Porphyrin and Mg-porphyrin intermediates, except for protochlorophyllide, do no normally accumulate in the dark or the light; (*b*) Administration of ALA to isolated plastids or intact tissue causes the accumulation of several porphyrin and Mg-porphyrin intermediates, including protochlorophyllide. In the light, some of the accumulated protochlorophyllide is converted to chlorophyllide. The amount of extractable heme does not increase when exogenous precursors are administered; (*c*) During greening of etiolated barley, the ratio of extractable Mg-porphyrins to Fe-porphyrins increases from 1.7 to almost 70 after 24 h in the light [43, 388]. During this change there is only a marginal increase in heme content, so that the increase in chlorophyll content accounts for most of the change in the ratio; (*d*) Newly synthesized protochlorophyllide and chlorophyll are relatively stable, while hemes undergo substantial turnover; (*e*) Transfer of greening material to the dark results in an almost complete cessation of chlorophyll accumulation, and only a small amount of protochlorophyllide accumulates; (*f*) During light/dark transitions, ALA accumulation in levulinic acid- or dioxoheptanoic acid-treated tissue parallels chlorophyll accumulation in untreated tissue. Many of these observations can be explained in terms of effector-mediated control of individual enzymes and/or conversion of the effectors to non-inhibitory compounds. The following discussion of effector-mediated control of enzyme activity refers only to plants which require light for greening, and which have proceeded beyond the initial lag phase in the greening curve.

3.2.2.1. Effector-mediated regulation of ALA-forming activity. A model for regulation has been proposed by Castelfranco and Jones [43] and, in an expanded form, by Chereskin and Castelfranco [243] and Castelfranco and Beale [405]. The model accounts for regulation of chlorophyll synthesis during greening after a dark-to-light transition. In this model, the primary point of regulation is the formation of ALA from glutamate. Heme is a potent inhibitor of this step in intact plastids [243] and most of the soluble enzyme systems which have been characterized [55, 57, 59, 61, 406]. There is a substantial body of evidence that heme directly inhibits ALA formation in vivo as well.

Exogenous heme administered to intact barley shoots [390] and to cultures of *Chlamydomonas* [407] inhibited the formation of chlorophyll in vivo, although in both cases the concentration of exogenous heme was extremely high. It is likely that the increased ALA accumulation noted in intact tissue treated with aromatic Fe chelators [382] was due to interference with heme formation, causing a decrease in the concentration of this feedback inhibitor. Similar effects were observed in intact plastids, where the Fe chelator eliminated the inhibition caused by added protoporphyrin [243]. It has been shown that chloroplasts are capable of heme biosynthesis [161, 398], and that hemes undergo significant turnover in vivo [43, 390]. If heme breakdown is an aerobic [174, 396] and ongoing process, then anaerobic conditions might be expected to prevent normal heme breakdown and thereby inhibit ALA biosynthesis. This effect has been observed in isolated plastids. Light-driven ALA formation in isolated plastids requires O_2. Although most of the O_2 requirement could be overcome by addition of ATP and NADPH, there was still a 30% inhibition of ALA formation under anaerobic conditions in the light or the dark [386]. It is possible that anaerobic conditions prevented O2-dependent heme breakdown, allowing the build up of inhibitory concentrations of heme.

Genetic evidence also supports the role of heme in regulating ALA formation in vivo. A light-brown mutant of *Chlamydomonas* that is deficient in chlorophyll synthesis accumulated small amounts of protoporphyrin [408]. Introduction of a second mutation resulted in a dark-brown phenotype in which the organism accumulated 15 times more protoporphyrin than the first mutant. The specific activity of the ALA-forming enzymes in the unpurified extract of the first mutant was about 16% of the wild type, while the double mutant had 64% of the activity of the wild type. Because both extracts were equally sensitive to heme, the authors proposed that the phenotype of the original mutation was caused by a block at a step that utilizes protoporphyrin. In the original mutant, accumulation of a small amount of protoporphyrin results in repression of the ALA-forming enzyme system. A defective regulatory gene in the double mutant allows higher expression of the enzymes. The authors further suggested that heme is the co-repressor in this system, although no direct evidence was presented to support this view. Introduction of a different second mutation, non-allelic with the first, also resulted in a dark-brown phenotype with elevated protoporphyrin accumulation. However, in this case, the enzyme activity was much less sensitive to heme. Thus in *Chlamydomonas*, heme appears to regulate both the activity and expression of ALA synthesizing enzymes.

Heme concentrations which inhibit in vitro activity by 50% range from 0.05 μM in partially purified *Chlamydomonas* extracts [408] to 1.2 μM in *Chlorella* extracts [57]. In contrast to most of the organisms examined, very high concentrations (25 μM) of heme were required for 50% inhibition in *Euglena* [54]. It does not appear likely that heme plays a significant role in regulating ALA formation from glutamate in this organism.

Except for the special case of intact plastids, the heme concentrations required for 50% inhibition appear to be higher in the more highly purified preparations. This observation suggests that an additional component is required for maximum heme sensitivity, and that this component is lost upon enzyme purification. The sensitivity of *Chlorella* extracts to heme inhibition was increased eight- to ten-fold by the presence of 1 mM glutathione [409]. The sensitizing effect was observed with both reduced and oxidized glutathione, but was not obtained with other sulfhydryl-containing compounds.

Although Mg-protoporphyrin was inhibitory in intact plastids [243], relatively high concentrations were required for 50% inhibition in the solubilized systems [57, 408]. Other potential feedback inhibitors (protoporphyrin, protochlorophyllide, and chlorophyllide) were not effective, except at very high concentrations. In experiments designed to test for a possible in vivo regulatory role of protochlorophyllide, the dark accumulation of ALA in levulinic acid-treated barley leaves was found to be inversely proportional to the level of protochlorophyllide bound in the ternary complex with the reductase enzyme [410, 411]. However, ALA accumulation could still take place in leaves that were artificially manipulated to have relatively high levels of free reductase and free protochlorophyllide. To explain these results, the authors proposed an indirect relationship between protochlorophyllide reduction and ALA synthesis via a common supply of NADPH, whereby photoreduction of the ternary protochlorophyllide complex would release bound NADP, which could then be reduced and used for ALA biosynthesis from glutamate. It should be noted that these experiments did not take into account variations in the level of free heme. In addition, experiments comparing the pool sizes of NADP(H) and protochlorophyllide reductase were not reported.

In contrast to the absence of observable effects of protochlorophyllide in the above cases, protochlorophyllide was reported to inhibit glutamyl-tRNA synthetase in extracts of *S. obliquus* [412]. Half-maximal inhibition was achieved at approximately 1 μM protochlorophyllide, and over 80% inhibition was reached at protochlorophyllide concentrations above 10 μM. In view of the absence of effects of protochlorophyllide on ALA synthesis in the in vivo cases described above, the relative ineffectiveness of protochlorophyllide as an inhibitor of *in vitro* ALA synthesis [54, 57], and the generally nonrate-limiting activity of glutamyl-tRNA synthetase in the systems examined thus far, it is premature to conclude that the inhibition of *Scenedesmus* glutamyl-tRNA synthetase by protochlorophyllide is of physiological significance.

It was reported that of the three enzyme activities required for ALA formation in vitro, heme exerts its effect only on the enzyme responsible for reduction of the glutamyl-tRNA adduct [76]. However, more recent work with separated fractions from *Chlamydomonas* revealed that both the glutamyl-tRNA synthetase and the dehydrogenase activities were inhibited by heme, with 50% inhibition occurring at 2 and 10 μM, for the synthetase and dehydrogenase steps, respectively [413]. Finally, it was reported that preincubation of partially purified extracts with ATP stimulated ALA formation upon subsequent addition of the remaining substrates and cofactors [57]. Although the behavior was consistent with a protein kinase phosphoprotein phosphatase modulation of activity, no direct evidence was obtained to support that hypothesis.

3.2.2.2. Expression and turnover of ALA-forming enzymes. Direct measurement of synthesis and turnover of the proteins catalyzing ALA synthesis has not yet been accomplished. The most reliable method for determining the level of enzyme protein is by immunochemical techniques which require purified protein to elicit the antibodies. Although antibodies to the glutamyl-tRNA synthetase from barley chloroplasts and *Chlamydomonas* [65, 414] have now been prepared, they have not yet been used for turnover studies. However, there have been several studies based on the effects of inhibitors of protein synthesis on ALA accumulation in vivo, or the levels of extractable ALA synthesis activity during greening, which have been interpreted in terms of expression and turnover of enzyme protein. It should be emphasized that with either of these approaches, changes in ALA accumulated in vivo or changes in extractable activity in vitro can be the result of processes other than the turnover of new enzyme (e.g., changes in substrate supply or effector concentration in vivo, or covalent modification in vitro).

Levulinic acid-dependent accumulation of ALA in *Chlorella* was inhibited by treatment with the protein synthesis inhibitor, cycloheximide [415]. Inhibition was complete after 30 min, suggesting that continuous synthesis of enzyme is required in this organism. The phytochrome-mediated elimination of the lag phase in chlorophyll synthesis is also thought to involve new enzyme synthesis. Thus, the normally rapid accumulation of ALA that occurs upon illumination in levulinic acid-treated, dark- germinated maize or beans is prevented by application of cycloheximide 2 h prior to light treatment [361]. Application of cycloheximide to intact plants at the same time as illumination, or later, did not significantly affect ALA accumulation for another 4 or 6 h. The interpretation was that cycloheximide treatment prior to illumination prevented the synthesis of ALA biosynthetic enzymes during the normal lag phase. Cycloheximide treatment during the lag phase allowed enzyme synthesis to continue while the inhibitor was penetrating and reaching the sites of protein synthesis. The plants could then accumulate ALA until the newly synthesized enzyme turned over. A half-life of 80 min for the activity was estimated by returning greening leaves to the dark and relating the length of the dark-period to the level of ALA accumulation when the plants were subsequently brought back into the light [359].

The plant and algal systems which require light for greening, and from which active cell-free ALA synthesizing systems have been obtained, had a basal amount of activity even in dark-grown cells or plants. The level of activity in relatively unpurified extracts increased three to four fold when the plants or algae were exposed to light for 2 to 4 h prior to extraction. This phenomenon was observed in barley [79], cucumber plastids [366], maize [82], and *Chlorella* [57]. It is not clear which of the enzyme components increase(s) upon light exposure, but in *Chlorella* [58] and barley [367], the tRNA component does not increase during greening.

Regulation of ALA synthesis in *Euglena* is a particularly perplexing problem. It is the only organism known in which both routes to ALA exist and operate simultaneously, albeit in separate compartments. In the dark, the level of ALA synthase was relatively high and decreased drastically when cells were brought into the light [35]. Light-grown cells had low ALA synthase activity which increased four fold after 4 h in the dark. Activity of the five-carbon route was absent in dark-grown cells and high in light-grown cells [54]. The interplay between the two biosynthetic routes was demonstrated by treatment of light-grown cells with gabaculine, the inhibitor of the glutamate pathway. The gabaculine treatment caused a two-fold increase in extractable ALA synthase activity in the light [416]. Pretreatment with gabaculine also prevented the decrease in extractable ALA synthase activity which normally occurs upon transfer of dark-grown cells to the light. Thus it is clear that the cells have some mechanism to maintain a supply of porphyrin precursors and that this mechanism is not solely a light-dependent phenomenon. To complicate the matter further, both ALA synthase [35] and the five-carbon route to ALA [54] are insensitive to heme inhibition in vitro.

As yet no physiological effectors have been characterized which modulate the level of activity of the enzymes that catalyze the reactions between ALA and protoporphyrin. However, enzymatic activities of ALA dehydratase, porphobilinogen deaminase, and uroporphyrinogen III synthase were present in etiolated peas, and increased three to four fold during 60 h of greening [417]. In the case of ALA dehydratase and porphobilinogen deaminase, there was a corresponding increase in enzyme protein, as detected by immunochemical techniques.

3.2.2.3. Regulation of ALA dehydratase activity. Beyond the steps catalyzing formation of ALA, the only enzyme before the branch point that has been shown to have regulatory properties is ALA dehydratase. The enzyme from *R. spheroides* [418], but not *R. capsulatus* [419], is inhibited by heme.

ALA dehydratase activity was present in etiolated peas, and increased three to four fold during 60 h of greening [417]. There was a corresponding increase in enzyme protein, as detected by immunochemical techniques. Several other laboratories have reported increased ALA dehydratase levels in etiolated tissues after exposure to light [420–423]. However, the dark levels of the enzyme are generally sufficient to catalyze PBG formation at the highest rates measured in vivo. Thus a physiological regulatory role for ALA dehydratase appears unlikely.

3.2.2.4. Branch point regulation. Insertion of the central metal ion into protoporphyrin is the step that controls the flux of porphyrins to either hemes or chlorophylls. As such it is a branch point in the metabolic pathway, and it would be expected that the activities of the two enzymes are highly regulated. However, activities of ferrochelatase and Mg-chelatase have never been compared in the same tissue under similar conditions. Nor has either enzyme been systematically investigated with respect to changes in activity during greening.

The membrane-bound ferrochelatase requires only its substrates and no other cofactors. Both its reaction product, heme, and Mg-protoporphyrin are inhibitory in the micromolar concentration range [158]. In contrast, Mg-chelatase requires ATP as a co-substrate. Mg-chelatase activity in intact plastids was effectively inhibited by exogenous protochlorophyllide and chlorophyllide, with 50% inhibition occurring at 1.0 and 3.0 μM, respectively [424]. A role for protochlorophyllide acting as a feedback inhibitor of Mg-chelatase is also supported by mutant analysis in *Chlamydomonas* [425]. In addition, Mg-chelatase was inhibited by AMP; 50% inhibition at 3.5 mM AMP in the presence of 10 mM ATP, suggesting that the energy charge of the cell may have some effect in regulating chlorophyll biosynthesis [212]. Thus it is possible to account for the shut-down of chlorophyll biosynthesis and continued heme production for nonplastid hemoproteins, by postulating a build-up of protochlorophyllide (see below) and/or ATP depletion. However, it is still difficult to explain the high ratio of chlorophyll to heme accumulation which occurs during greening.

The argument that the lower K_m for protoporphyrin of Mg-chelatase compared to ferrochelatase (0.025 μM versus 0.2 μM) in plastid membranes can account for the preferential synthesis of Mg-porphyrins during greening is deficient. The catalytic rate and molar amount of each enzyme present must also be considered. Moreover, it is still not certain that Mg-porphyrins are synthesized at a higher rate than hemes during greening. It is possible that the preferential accumulation of Mg-porphyrins is due solely to the turnover of hemes.

No information is available on the regulation of the steps between Mg-protoporphyrin and protochlorophyllide. However, it is known that S-adenosyl-methionine, which is required for the methylation step, is not made in the plastid, and must originate in the cytosol [426].

3.2.2.5. Regulation of protochlorophyllide photoreduction. In plants that require light for chlorophyll biosynthesis, the photoreduction of protochlorophyllide is the signal that initiates the biosynthetic process. This is the only step that has a direct light requirement. NADPH:protochlorophyllide oxidoreductase is present in etiolated tissue and is localized in the etioplast membranes. The enzyme behaves in a peculiar manner in response to light. Upon exposure to light, the enzyme activity [427, 428], the amount of enzyme protein [428, 429], and the amount of poly-A mRNA from which the protein is translated [430, 431] all decrease dramatically and continuously. The proteolysis of the reductase was also observed when isolated etioplasts or etioplast membranes obtained from dark-grown barley were exposed to continuous light [422, 423]. The protease is membrane bound and apparently specific for the reductase, as other plastid proteins are not degraded in a similar light-dependent manner. In the experiments with etioplasts and etioplast membranes, the light-dependent proteolysis was prevented by inclusion of exogenous substrates, and promoted (in the dark) by inclusion of an inhibitor which prevented substrate binding [434]. The authors suggested that proteolysis can occur equally well in the dark or light, whereas in the dark, proteolysis is prevented by the binding of substrate to the free enzyme. The reductase could be localized by immunoblotting of specific fractions and immunogold labeling of leaf slices [429, 435–437]. As expected, the reductase was mostly localized to the etioplast membranes. However, a significant portion was also observed in the cytoplasmic membranes. The reductase in developing plastid membranes was more susceptible to light-induced breakdown than the cytoplasmic enzyme. After a 72-h illumination almost 90% of the immunoreactive reductase was no longer detectable, and the 10% remaining was mostly, but not entirely, localized in the cytoplasm [429]. If the plants were placed on a 12-h light/12-h dark cycle for 72 h, the level of reductase in the plastids was 20% of that in dark grown tissue. Thus even in fully greened young tissue the amount of reductase remaining would be sufficient to account for the amount of chlorophyll synthesis at this stage.

A double role can be postulated for the large amount of protochlorophyllide reductase present in etiolated tissue and its subsequent proteolysis in the light. Relatively large amounts of the enzyme in the etioplast are required to prevent the accumulation of free protochlorophyllide, since the free pigment would be phototoxic when the seedlings emerge into the light. The rapid photoreduction of the large amount of bound protochlorophyllide can supply pigment for the immediate assembly of reaction centers to get photosynthesis underway. The bulk of the enzyme can then serve as a plastid-localized source of free amino acids as it undergoes proteolysis in the absence of substrate. The remaining activity will be sufficient to catalyze steady-state chlorophyll synthesis during greening.

A model for the regulatory steps which account for the rapid response of chlorophyll and ALA synthesis to light while accommodating the need for heme in both photosynthetic and nonphotosynthetic processes is presented in Fig. 14. This model does not include phytochrome- and/or phytohormone-mediated responses on de novo synthesis of enzymes. In the model, ALA synthesis is controlled by substrate supply and feedback inhibition by heme at possibly two enzymatic steps, the formation and subsequent reduction of glutamyl-tRNA. There appears to be no regulatory restriction on protoporphyrin formation from ALA. Ferrochelatase is probably subject to product inhibition by heme and possibly by Mg-protoporphyrin as well. In the dark, modulation of ALA synthesis is controlled by the level of a heme pool which

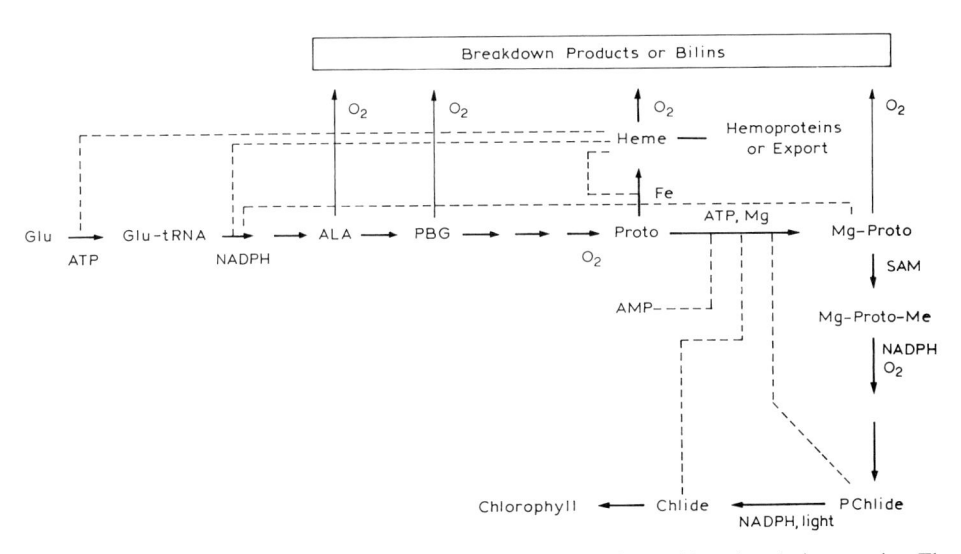

Fig. 14. Model for the regulation of chlorophyll and heme synthesis in the chloroplast during greening. The predominant flux of intermediates is indicated by heavy arrows, and minor fluxes by lighter arrows. Each feedback inhibition is indicated by a broken line connecting the inhibitor to the arrow representing the step that is inhibited. Abbreviations: Chlide, chlorophyllide; Proto, protoporphyrin; Mg-Proto, Mg-protoporphyrin; Mg-Proto-Me, Mg-protoporphyrin monomethyl ester; PChlide, protochlorophyllide; SAM, S-adenosyl-methionine.

constantly turns over. Thus an increased demand for cytosolic heme could probably be supplied by the plastid even in the dark, simply by depletion of the pool. The mechanism by which heme might be transported out of the plastid is unknown. Protochlorophyllide also accumulates in the dark. Most of it will be bound to the reductase, but there is both in vitro and genetic evidence to suggest that some of it serves as a feedback inhibitor to inhibit Mg-chelatase. Inhibition of Mg- chelatase by protochlorophyllide forces the flux of protoporphyrin through the Fe branch, thus contributing to a constant supply of heme. In contrast to heme, protochlorophyllide probably does not turn over in the dark. In the light, the bound protochlorophyllide is immediately photoreduced, thus freeing up sites on the reductase for free protochlorophyllide, and the lowering of free protochlorophyllide concentration in turn relieves the inhibition of Mg-chelatase. Activation of Mg-chelatase diminishes the flux of protoporphyrin through the Fe branch, causing a depletion of the heme pool, thereby relieving the inhibition of ALA synthesis.

In this model, ALA formation is modulated by light indirectly through the effect of protochlorophyllide on Mg-chelatase and the heme pool. The model also allows for modulation of ALA synthesis in a light-independent manner. It predicts that there will be a competition for heme between apoproteins and heme-degrading systems. It is not known if the system responsible for heme degradation is localized in the plastid or cytosol or both. Nor is it known if it functions in a manner similar to the heme oxygenase of red and blue-green algae, where it is a biosynthetic enzyme system.

3.3. Regulation of phycobilin content by light and nutritional status

3.3.1. Light effects
Wild-type cells of *Cyanidium* form neither chlorophyll nor phycocyanin in the dark, although the cells are capable of vigorous dark growth in glucose-supplemented medium. When cells are transferred to the light, both bilin and apoprotein moieties of phycocyanin are synthesized de novo in stoichiometric amounts, rather than being assembled from pre-existing components [438].

Administration of exogenous ALA in the dark causes the cells to synthesize and excrete free phycocyanobilin [190], indicating that there is no light-requiring step in bilin biosynthesis from ALA. Nevertheless, the normal inhibition of phycobilin formation by darkness does indicate some sort of photoregulation of the process. Mutant strains which normally are unable to form phycobiliproteins in the light or dark accumulate the phycobilin moiety, but not the apoprotein, if supplied with exogenous ALA. On the other hand, if levulinic acid is administered to wild-type cells in vivo to block pigment formation, apoprotein is still synthesized to some extent [439]. These results imply that apoprotein formation is not linked directly to bilin formation. However, neither bilin nor apoprotein normally accumulates in cells undergoing wide changes in phycobiliprotein level. Therefore, some sort of indirect co-regulation must occur to keep the components in approximate stoichiometric ratio.

In cells that are accumulating both hemes and phycobilins, the first biosynthetic step unique to the phycobilin branch would be the heme oxygenase reaction. This reaction therefore is a logical candidate for a physiological control point, regulating the cellular level of phycobilins.

Nichols and Bogorad [440] examined the action spectrum for phycobilin formation in a nonphotosynthetic, chlorophyll-less mutant of *Cyanidium*. Light of 420 nm wavelength was most effective in promoting bilin formation. A second peak in the action spectrum occurred at about 580 nm. The authors proposed that the photoreceptor may be a heme compound. They also reported that dark phycocyanin synthesis could be induced by unidentified factors present in some culture medium components. Another mutant strain of *Cyanidium* accumulates chlorophyll *a* but not phycocyanin. The action spectrum for chlorophyll synthesis corresponds to the absorption spectrum of protochlorophyllide [441]. Wild-type cells, which contain both chlorophyll and phycocyanin, have a single action spectrum for synthesis of both pigments, and this resembles the sum of the action spectra exhibited by the two mutant strains.

Many algae have been reported to change both their absolute phycobilin content and the relative abundances of phycocyanin and phycoerythrin in response to variations in the light regime. Marine intertidal red macroalgae contain increasing amounts of phycobilin pigments when grown at decreasing average light levels which occur at increasing water depths [442]. In bluegreen algae, at least three mechanisms are known to operate in bringing about changes in phycobilin pigment composition. In some species, the ratio of phycoerythrin to phycocyanin changes during chromatic adaptation [443]. Other species also vary the subunit number and composition in phycocyanin [444]. Still other organisms have been reported to incorporate different bilins into apoprotein subunits under different growth conditions [445].

Among cyanophytes that exhibit complementary chromatic adaptation, cells growing in red light contain little or no phycoerythrin, but this becomes the major phycobiliprotein during growth in green light. A search for pigments that might mediate the response to variations in spectral light distribution has resulted in the detection of several photochromic species, both in vivo and in vitro [446]. Isolated photochromic pigments appear to be similar or identical to some phycobiliprotein subunits. They undergo photoreversible transformations, with peak positions that are close to the peaks of the action spectra for the chromatic adaptation responses, i.e., 520 to 550 nm for the green light effects and 640 to 660 nm for the red light effects [447]. The photochromic species were named phycochromes by Björn and Björn [448]. There is no agreement yet on which of the isolated photochromic pigments (at least four have been described) correspond(s) to the physiologically active light quality sensing pigment(s) in vivo. One report indicates that the photoreversible pigments of *Tolypothrix tenuis* are chromoproteins that are derived from phycobiliproteins upon photobleaching [449]. Because of their close structural and functional resemblance to phytochrome, the phycochromes are attractive candidates for its evolutionary predecessors.

Since phycobiliproteins differ from each other in the protein moieties as well as in their phycobilin complement, differential synthesis or breakdown of proteins must be a requirement for complementary chromatic adaptation. Gendel et al. [450] have shown that the induction of phycoerythrin synthesis by green light in the bluegreen alga *Fremyella* diplosiphon is blocked by rifamycin, indicating a requirement for gene transcription in the response. Transfer from green light to darkness slowed, but did not stop, phycoerythrin synthesis, indicating that differential phycobiliprotein synthesis is induced by green light, but that continuous irradiation is not required for the response to occur. Cells that were transferred from green to red light stopped synthesizing phycoerythrin within 45 min. The same time course for cessation of phycoerythrin synthesis was observed when rifamycin was administered to cells growing in green light, suggesting that the chromatic adaptation response is mediated at the transcriptional level [450]. Green light was shown to specifically induce the transcription of a phycoerythrin operon containing the genes coding for both the *a* and *b* subunits of phycoerythrin [451].

Egelhoff and Grossman [452] reported that in eukaryotic algae, the phycobilin apoproteins are synthesized on plastid ribosomes. This conclusion was based on differential sensitivity to inhibitors of protein synthesis. The same conclusion is supported by the results of Belford et al. [453], who found that cell-free synthesis of *Cyanidium* phycocyanin subunits occurred when a reticulocyte lysate translation system was supplied with the non-poly(A) fraction of mRNAs isolated from the algae.

Other evidence suggests that in eukaryotic cells the rate-limiting steps of phycobilin synthesis are catalyzed by enzymes synthesized on cytoplasmic ribosomes. For example, it is known from exogenous feeding experiments that ALA availability is rate-limiting for phycobilin synthesis and excretion. ALA formation is rapidly inhibited in the algae by the administration of cycloheximide, which blocks cytoplasmic protein synthesis.

Yu et al. [445] reported that the red alga *Callithamnion roseum*, when grown under different light intensities, acquires phycoerythrin with different ratios of phycoerythrobilin to phycourobilin, while the protein subunit composition remains constant. The mechanism responsible for the substitution of bilins is not yet known.

Beguin et al. [454] described a mutant strain of *F. diplosiphon* that does not accumulate phycoerythrin under any light regime. In red light, the cells are phenotypically indistinguishable from wild-type. In green light, some of the phycocyanin subunits contain a bilin of unknown structure, that has a distinctive violin-like visible absorption spectrum. These interesting observations have been interpreted as supporting the hypothesis that bilins other than phycocyanobilin are derived from protein-bound phycocyanobilin, either by enzymatic isomerization or nonenzymatically as a result of protein folding [454]. However, a clearer understanding of these results may have to await determination of the structure of the novel bilin and the nature of the defect in the mutant cells.

Successful cloning of a DNA fragment from the bluegreen alga *Synechococcus* sp. PCC 7002 (*Agmenellum quadruplicatum*) containing the genes for both the *a* and *b* subunits of phycocyanin has been reported [455]. The cloned genes will make possible the study of the regulation of expression of the genes for the apoproteins [206]. Also, the expression of these genes in *E. coli* may make it possible to obtain substrate quantities of apoprotein for in vitro studies of the chromophore ligation reactions [456].

When the cloned cyanobacterial phycocyanin genes were reintroduced into *Synechococcus* at higher multiplicity, overexpression of the genes resulted. The cells accumulated correspondingly greater than normal amounts of chromophore-containing phycocyanin [457]. Apparently, the cells were able to synthesize sufficient phycobilins to accommodate the increased demand for pigment. This result suggests that bilin formation does not limit the rate of phycocyanin accumulation, but that, instead, the bilin biosynthetic rate may respond to the availability of apoprotein.

A different mode of coordination of apoprotein and chromophore synthesis appears to occur in *Cyanidium*. In this eukaryotic species, the cellular levels of the chloroplast-encoded apoproteins appear to be regulated at the transcriptional level by heme. Wild-type cells do not normally accumulate phycobiliproteins in the dark, and the mRNAs coding for the apoproteins are also absent in the dark. However, if the cells are administered the phycobilin precursors, ALA, protoporphyrin, or heme, in the dark, they accumulate phycocyanin and allophycocyanin apoproteins and mRNAs [458]. Other genes which are normally expressed only in the light were not induced in the dark by phycobilin precursors. Biliverdin, which is also a phycobilin precursor, did not stimulate phycobiliprotein apoprotein or mRNA accumulation in the dark. The authors concluded that heme is a specific regulatory factor involved in the transcriptional regulation of phycobiliprotein gene expression.

At present, the only study at the enzyme level to reveal regulation of a specific step of phycobilin synthesis is that of Brown et al. [162], which reported a large increase in the activity of ferrochelatase paralleling the rise in phycocyanin accumulation when dark-grown wild-type *Cyanidium* cells are transferred from the dark to the light.

3.3.2. Responses to nitrogen status
Phycobiliproteins comprise a significant proportion of total cellular proteins of cyanobacteria when the cells are grown in medium containing non-limiting amounts of nitrogen. When the cells are shifted to growth on limiting nitrogen, they specifically degrade much of their phycobiliproteins [459]. Foulds and Carr [460] have described a proteolytic enzyme activity in extracts of *Anabaena cylindrica* that is specific for phycocyanin. Enzyme activity was associated with both the $100,000 \times g$ pellet and supernatant fractions. Activity was several-fold higher in extracts of cells harvested

during heterocyst development in response to growth at low nitrogen than in extracts of cells grown on non-limiting nitrogen. It was suggested that the in vitro proteolytic activity is physiologically related to in vivo phycocyanin degradation, and the enzyme responsible for the activity was named phycocyaninase. The fate of the bilin chromophores was not reported.

Acknowledgments

We thank the many investigators who made results available to us before publication.

Note added in proof

The field of plant and algal photosynthetic pigment biosynthesis is in a period of accelerated advance, and many important new results have appeared after the literature review for this article was completed. Some of these new results are briefly described below. The enzymes required for ALA biosynthesis from glutamate have been purified and partially characterized from *Chlamydomonas reinhardtii* [461–464] and *Synechocystis* sp. PCC 6803 [465,466]. Glutamyl-tRNA reductase (dehydrogenase from *Synechocystis* was purified to homogeneity and the N-terminal amino acid sequence was determined [466]. No flavin or other light-absorbing cofactor was detected. Glutamate-1-semialdehyde aminotransferase has been purified from *Synechococcus* sp. PCC 6301 [464], and the gene for this enzyme has been cloned and sequenced from barley [468] and *Synechococcus* [469]. A gabaculine-resistant mutant of *Synechococcus* was determined to contain an altered aminotransferase gene that encodes a protein containing a short deletion at the N-terminal region and an internal amino acid substitution [470]. It is now generally agreed that the aminotransferase from all sources require a pyridoxal cofactor. Porphobilinogen deaminse has been purified to homogeneity from pea chloroplasts and the N-terminal amino acid sequence has been determined [471]. cDNA encoding this enzyme has been cloned and sequenced from *Euglena gracilis* [472], and the enzyme has been localized within the plastids of *Euglena* by immunogold [473]. Overall activity catalyzing protoporphyrin IX formation from ALA sedimented with the membrane fraction of cucumber cortyledon etioplasts that were lysed in hypotonic buffer containing high Mg concentration [474]. Diphenylether herbicides have been shown to powerfully inhibit protoporphyrinogen oxidase in plants plastids [475, 476]. The protoporphyrinogen that is synthesized in the presence of the inhibitors appears to diffuse out of the plastids and accumulate in the cell membrane, where it is oxidized to protoporphyrin IX. Reactions of the protoporphyrin IX with light and O_2 lead to membrane damage and cell death. Biosynthesis of phycocyanobilin from biliverdin in *Cyanidium caldarium* extracts was found to proceed via two sequential, ferredoxin-mediated two-electron reduction steps followed by isomerizations [477]. The first two-electron reduction yields 15,16-dihydrobiliverdin IXα. This product is then reduced to 3(Z)-phycoerythrobilin. The 3(Z)-phycoerythrobilin is isomerized to 3(Z)-phycoerythrobilin, and the ethylidine group then undergoes a glutathione-requiring cis-trans isomerization to yield the final product, 3(E)-phycocyanobilin. Intact cucumber cotyledon plastids catalyzed the conversion of biliverdin IXα to phytochromobilin [478]. The reaction was stimulated by NADPH and ATP, and the product was able to assemble with apophytochrome to yield photoreversible phytochrome that was spectrally indistinguishable from natural phytochrome. The substrate requirements (porphyrin and ATP) of the Mg-chelatase in intact cucumber cotyledon chloroplasts have been investigated [479]. It has been hypothesized that Mg chelation is a key regulatory step of chlorophyll synthesis, and that this step is activated by light and inhibited by protochlorophyllide or other pigments. In recent studies of Mg-chelatase in intact isolated cucumber plastids, activity increased with increasing time of light exposure of the cotyledons before isolation of the plastids. However, inhibition of activity was not caused by added protochlorphyllide, chlorophyllide, Mg-protoporphyrin IX, or heme. Broken pea plastids yielded in vitro Mg-chelatase which required ATP, Mg^{2+}, and porphyrin for activity [480]. The activity has been separated into two heat-labile fractions, one soluble and the other membrane associated. Chlorophyll-free chromoplasts isolated from daffodil petals were capable of converting ALA to Mg-protoporphyrin IX monomethyl ester in the presence of S-adenosylmethionine [481]. Isocyclic ring-containing products were not detected.

The soluble and membrane-associated components of the isocyclic ring-forming system from cucumber cotyledon chloroplasts were separated and partially characterized [482]. The chromatographic behavior of the soluble component on affinity columns suggested that this fraction binds the Mg-porphyrin substrate but not NADPH. The membrane-associated fraction appears to bind a required metal ion. Although protochlorophyllide reductase is located in the prolamellar bodies of etioplasts, the enzyme in mature spinach chloroplasts was found to be located on the cytosolic side of the outer envelope membrane [483]. cDNA coding for protochlorophyllide reductase has been cloned and sequenced from oat [484], barley [485], and *Arabidopsis* [486]. The full-length barley cDAN has been expressed in *E. coli* to yield enzymatically-active enzyme. In contrast to the reports for other plants, there was no decline in the level of protochlorophyllide reductase mRNA in *Arabidopsis* after exposure to light [487]. Intact chloroplasts isolated from green cucumber cotyledons were reported to synthesize both chlorphyll *a* and chlorophyll *b* [488]. Requirements include light, ALA, and S-adenosylmethionine. Previous studies have localized the genes encoding all components required for chlorophyll synthesis within the nuclear genome of plants and algae, with the exception of the encoding plastid tRNA^{Glu}. Recently, a chloroplast gene has been implicated in light-independent protochlorophyllide reduction in *Chlamydomonas* [489]. Most previous studies on light regulation of ALA biosynthetic enzyme levels have been done with tissues that contain etioplasts when grown in the dark. In *Euglena* cells, which contain proplastids rather than etioplasts when grown in the dark, all three enzymes of the five-carbon ALA biosynthetic pathway, as well as specific tRNA^{Glu} required for this process, increased when cells were transferred from the dark to the light [490]. Either red or blue light induced these components to levels near those induced by white light. The component whose level was profoundly affected by light was glutamate-1-semialdehyde aminotransferase. Nongreening mutant cells that were previously shown not to form ALA from glutamate are completely devoid of the required tRNA^{Glu}, whether grown in the light or dark [491]. However, the mutant cells contain all three of the enzymes of the five-carbon pathway. Moreover, the levels of all three enzymes in the mutant cells were increased by growth in the light. Blue light was required for enzyme induction in the mutant cells, and red light was ineffective.

A model for functional heterogeneity of the plastid ALA pool has been proposed [492]. The model is based on the differential sensitivity of ALA and protochlorophyllide accumulation to inhibitors of ALA formation.

References

1 Werck-Reichhart, D., Jones, O.T.G. and Durst, F. (1988) Biochem. J. 249, 473–480.
2 Schneegurt, M.A. and Beale, S.I. (1986) Plant Physiol. 81, 965–971.
3 Zumft, W.G. (1972) Biochim. Biophys. Acta 276, 363–375.
4 Murphy, M.J., Siegel, L.M., Tove, S.R. and Kamen, H. (1974) Proc. Natl. Acad. Sci. USA 71, 612–616.
5 Hucklesby, D.P., James, D.M., Banwell, M.J. and Hewitt, E.J. (1976) Phytochemistry 15, 599–603.
6 Dunlap, J.C., Hastings, J.W. and Shimomura, O. (1981) FEBS Lett. 135, 273–276.
7 Kikuchi, G., Kumar, A., Talmage, P. and Shemin, D. (1958) J. Biol. Chem. 233, 1214–1219.
8 Sawyer, E. and Smith, R.A. (1958) Bacteriol. Proc. 1958, 111.
9 Nandi, D.L. and Shemin, D. (1977) J. Biol. Chem. 252, 2278–2280.
10 Viale, A.A., Wider, E.A. and Batlle, A.M. del C. (1987) Comp. Biochem. Physiol. 87B, 607–613.
11 Zaman, Z., Jordan, P.M. and Akhtar, M. (1973) Biochem. J. 135, 257–263.
12 Abboud, M.M., Jordan, P.M. and Akhtar, M. (1974) J. Chem. Soc. Chem. Commun. 1974, 643–644.
13 Laghai, A. and Jordan, P.M. (1976) Biochem. Soc. Trans. 4, 52–53.

14 Laghai, A. and Jordan, P.M. (1977) Biochem. Soc. Trans. 5, 299–300.
15 Miller, J.W. and Teng, D. (1967) Proceedings of the 7th International Congress of Biochemistry Tokyo, p. 1059.
16 Barreiro, O.L.C. de (1975) Phytochemistry 14, 2165–2168.
17 Batlle, A.M. del C., Llambias, E.B.C., Wider de Xifra, E. and Tigier, H.A. (1975) Int. J. Biochem. 6, 591–606.
18 Wider de Xifra, E.A., Batlle, A.M. del C. and Tigier, H. (1971) Biochim. Biophys. Acta 235, 511–517.
19 Wider de Xifra, E.A., Stella, A.M. and Batlle, A.M. del C. (1978) Plant Sci. Lett. 11, 93–98.
20 Ramaswamy, N.K. and Nair, P.M. (1973) Biochim. Biophys. Acta 293, 269–277.
21 Ramaswamy, N.K. and Nair, P.M. (1974) Plant Sci. Lett. 2, 249–256.
22 Ramaswamy, N.K. and Nair, P.M. (1976) Indian J. Biochem. Biophys. 13, 394–397.
23 Meller, E. and Gassman, M.L. (1982) Plant Sci. Lett. 26, 23–29.
24 Klein, O. and Senger, H. (1978) Photochem. Photobiol. 27, 203–208.
25 Klein, O. and Senger, H. (1978) Plant Physiol. 62, 10–13.
26 Meller, E. and Harel, E. (1978) in: Chloroplast Development (G. Akoyunoglou and J.-H. Argyroudi-Akoyunoglou, Eds.) pp. 51–57, Elsevier, Amsterdam.
27 Porra, R.J. and Grimme, L.H. (1974) Arch. Biochem. Biophys. 164, 312–321.
28 Harel, E. (1978) in: Chloroplast Development (G. Akoyunoglou and J.-H. Argyroudi-Akoyunoglou, Eds.) pp. 33–44, Elsevier, Amsterdam.
29 Humbeck, K. and Senger, H. (1981) in: Photosynthesis (G. Akoyunoglou, Ed.) Vol. 5, pp. 161–170, Balaban, Philadelphia.
30 Kah, A., Dörnemann, D., Ruhl, D. and Senger, H. (1981) in: Photosynthesis (G. Akoyunoglou, Ed.) Vol. 5, pp. 137–144, Balaban, Philadelphia.
31 Weinstein, J.D. and Beale, S.I. (1984) Plant Physiol. 74, 146–151.
32 Beale, S.I., Foley, T. and Dzelzkalns, V. (1981) Proc. Nat. Acad. Sci. USA 78, 1666–1669.
33 Dzelzkalns, V., Foley, T. and Beale, S.I. (1982) Arch. Biochem. Biophys. 216, 196–203.
34 Beale, S.I. and Foley, T. (1982) Plant Physiol. 69, 1331–1333.
35 Foley, T., Dzelzkalns, V. and Beale, S.I. (1982) Plant Physiol. 70, 219–226.
36 Weinstein, J.D. and Beale, S.I. (1983) J. Biol. Chem. 258, 6799–6807.
37 Beale, S.I. and Castelfranco, P.A. (1974) Plant Physiol. 53, 297–303.
38 Meller, E., Belkin, S. and Harel, E. (1975) Phytochemistry 14, 2399–2402.
39 Beale, S.I., Gough, S.P. and Granick, S. (1975) Proc. Nat. Acad. Sci. USA 72, 2719–2723.
40 Avissar, Y.J. (1980) Biochim. Biophys. Acta 613, 220–228.
41 Kipe-Nolt, J.A. and Stevens, S.E., Jr. (1980) Plant Physiol. 65, 126–128.
42 Jurgenson, J.E., Beale, S.I. and Troxler, R.F. (1976) Biochem. Biophys. Res. Commun. 69, 149–157.
43 Castelfranco, P.A. and Jones, O.T.G. (1975) Plant Physiol. 55, 485–490.
44 Laycock, M.V. and Wright, J.C.C. (1981) Phytochemistry 20, 1265–1268.
45 McKie, J., Lucas, C. and Smith, A. (1981) Phytochemistry 20, 1547–1549.
46 Oh-hama, T., Seto, H., Otake, N. and Miyachi, S. (1982) Biochem. Biophys. Res. Commun. 105, 647–652.
47 Porra, R.J., Klein, O. and Wright, P.E. (1983) Eur. J. Biochem. 130, 509–516.
48 Oh-hama, T., Seto, H. and Miyachi, S. (1986) Arch. Biochem. Biophys. 246, 192–198.
49 Oh-hama, T., Seto, H. and Miyachi, S. (1986) Eur. J. Biochem. 159, 189–194.
50 Smith, K.M. and Huster, M.S. (1987) J. Chem. Soc. Chem. Commun. 1987, 14–16.
51 Chibbar, R.N. and van Huystee, R.B. (1983) Phytochemistry 22, 1721–1723.
52 Beale, S.I. (1984) in: Chloroplast Biogenesis (N.R. Baker and J. Barber, Eds.) pp. 133–205, Elsevier, Amsterdam.
53 Gough, S.P. and Kannangara, C.G. (1977) Carlsberg Res. Commun. 42, 459–464.
54 Mayer, S.M., Weinstein, J.D. and Beale, S.I. (1987) J. Biol. Chem. 262, 12541–12549.

55 Wang, W.-Y., Huang, D.-D., Stachon, D., Gough, S.P. and Kannangara, C. G. (1984) Plant Physiol. 74, 569–575.
56 Breu, V. and Dörnemann, D. (1988) Biochim. Biophys. Acta 967, 135–140.
57 Weinstein, J.D. and Beale, S.I. (1985) Arch. Biochem. Biophys. 237, 454–464.
58 Weinstein, J.D., Mayer, S.M. and Beale, S.I. (1986) in: Regulation of Chloroplast Differentiation (G. Akoyunoglou and H. Senger, Eds.) pp. 43–48, Alan R. Liss, New York.
59 Rieble, S. and Beale, S.I. (1988) J. Biol. Chem. 263, 8864–8871.
60 O'Neill, G.P., Peterson, D.M., Schön, A, Chen, M.-W. and Söll, D. (1988) J. Bacteriol. 170, 3810–3816.
61 Rieble, S., Ormerod, J.G. and Beale, S.I. (1988) Plant Physiol. 86(Suppl), 60.
62 Houen, G., Gough, S.P. and Kannangara, C.G. (1983) Carlsberg Res. Commun. 48, 567–572.
63 Hoober, J.K., Kahn, A., Ash, D., Gough, S. and Kannangara, C.G. (1988) Carlsberg Res. Commun. 53, 11–25.
64 Jordan, P.M., Sharma, R.P. and Warren, M.J. (1989) Tet. Lett. (submitted).
65 Bruyant, P. and Kannangara, C.G. (1987) Carlsberg. Res. Commun. 52, 99–109.
66 Weinstein, J.D., Mayer, S.M. and Beale, S.I. (1987) Plant Physiol. 84, 244–250.
67 Rieble, S. and Beale, S.I. (1989) Plant Physiol. 89(Suppl.), 51.
68 Kannangara, C.G., Gough, S.P., Oliver, R.P. and Rasmussen, S.K. (1984) Carlsberg Res. Commun. 49, 417–437.
69 Huang, D.-D., Wang, W.-Y., Gough, S.P. and Kannangara, C.G. (1984) Science 225, 1482–1484.
70 Weinstein, J.D. and Beale, S.I. (1985) Arch. Biochem. Biophys. 239, 87–93.
71 Schön, A., Krupp, G., Gough, S., Berry-Lowe, S., Kannangara, C.G. and Söll, D. (1986) Nature 322, 281–284.
72 Schön, A., Kannangara, C.G., Gough, S. and Söll, D. (1988) Nature 331, 187–190.
73 Lapointe, J., Duplain, L. and Proulx, M. (1986) J. Bacteriol. 165, 88–93.
74 Schneegurt, M.A. and Beale, S.I. (1988) Plant Physiol. 86, 497–504.
75 Schneegurt, M.A., Rieble, S. and Beale, S.I. (1988) Plant Physiol. 86(Suppl), 61.
76 Huang, D.-D. and Wang, W.-Y. (1986) J. Biol. Chem. 261, 13451–13455.
77 Avissar, Y.J. and Beale, S.I. (1988) Plant Physiol. 88, 879–886.
78 Kannangara, C.G. and Gough, S.P. (1978) Carlsberg Res. Commun. 43, 185–194.
79 Kannangara, C.G. and Gough, S.P. (1979) Carlsberg Res. Commun. 44, 11–20.
80 Kah, A. and Dörnemann, D. (1987) Z. Naturforsch. 42c, 209–214.
81 Meisch, H.-U. and Maus, R. (1983) Z. Naturforsch. 38c, 563–570.
82 Harel, E. and Ne'eman, E. (1983) Plant Physiol. 72, 1062–1067.
83 Avissar, Y.J. and Beale, S.I. (1989) Plant Physiol. 89, 852–859.
84 Kannangara, C.G. and Schouboe, A. (1985) Carlsberg Res. Commun. 50, 179–191.
85 Wang, W.-Y., Gough, S.P. and Kannangara, C.G. (1981) Carlsberg Res. Commun. 46, 243–257.
86 Kannangara, C.G., Gough, S.P. and Girnth, C. (1981) in: Photosynthesis (G. Akoyunoglou, Ed.) Vol. 5, pp. 117–127, Balaban, Philadelphia.
87 Kannangara, C.G., Gough, S.P., Bruyant, P., Hoober, J.K., Kahn, A. and von Wettstein, D. (1988) Trends Biochem. Sci. 13, 139–143.
88 Kahn, A. and Kannangara, C.G. (1987) Carlsberg Res. Commun. 52, 73–81.
89 Bull, A.D., Pakes, J.F., Hoult, R.C., Rogers, L.J. and Smith, A.J. (1989) Biochem. Soc. Trans. 17, 911–912.
90 Grimm, B., Bull, A., Welinder, K.G., Gough, S.P. and Kannangara, C. G. (1989) Carlsberg Res. Commun. 54, 67–79.
91 Mau, Y.-H.L. and Wang, W.-Y. (1988) Plant Physiol. 86, 793–797.
92 Avissar, Y.J., Ormerod, J.G. and Beale, S.I. (1989) Arch. Microbiol. 151, 513–519.
93 Sato, K., Ishida, K., Mutsushika, O. and Shimizu, S. (1985) Agric. Biol. Chem. 49, 3415–3421.
94 Shioi, Y. and Doi, M. (1988) Arch. Biochem. Biophys. 266, 470–477.

95 Oh-hama, T., Stolowich, N.J. and Scott, A.I. (1988) FEBS Lett. 228, 89–93.
96 Friedmann, H.C. and Thauer, R.K. (1986) FEBS Lett. 207, 84–88.
97 Friedmann, H.C., Thauer, R.K., Gough, S.P. and Kannangara, C.G. (1987) Carlsberg Res. Commun. 52, 363–371.
98 Li, J.-M., Brathwaite, O., Cosloy, S.D. and Russell, C.S. (1989) J. Bacteriol. 171, 2547–2552.
99 Avissar, Y.J. and Beale, S.I. (1989) J. Bacteriol. 171, 2919–2924.
100 O'Neill, G.P., Chen, M.-W. and Söll, D. (1989) FEMS Microbiol. Lett. 60, 255–260.
101 Fuesler, T.P. Castelfranco, P.A. and Wong, Y.-S. (1984) Plant Physiol. 74, 928–933.
102 Huang, L. and Castelfranco, P.A. (1986) Plant Physiol. 82, 285–288.
103 Tamai, H., Shioi, Y. and Sasa, T. (1979) Plant Cell Physiol. 20, 435–444.
104 Shibata, H. and Ochiai, H. (1977) Plant Cell Physiol. 18, 421–429.
105 Schneider, H.A.W. (1970) Z. Pflanzenphysiol. 62, 328–342.
106 Schneider, H.A.W. (1970) Z. Pflanzenphysiol. 62, 133–145.
107 Liedgens, W., Grützmann, R. and Schneider, H.A.W. (1980) Z. Naturforsch. 35c, 958–962.
108 van Heyningen, S. and Shemin, D. (1971) Biochemistry 10, 4676–4682.
109 Nandi, D.L. and Waygood, E.R. (1967) Can. J. Biochem. 45, 327–336.
110 Shetty, A.S. and Miller, G.W. (1969) Biochem. J. 114, 331–337.
111 Nasri, F., Huault, C. and Belangé, A.P. (1988) Phytochemistry 27, 1289–1295.
112 Smith, A.G. (1988) Biochem. J. 249, 423–428.
113 Tigier, H.A., Batlle, A.M. del C. and Locascio, G.A. (1970) Enzymologia 38, 43–56.
114 Jordan, P.M. and Seehra, J.S. (1980) J. Chem. Soc. Chem. Commun. 1980, 240–242.
115 Abboud, M.M. and Akhtar, M. (1976) J. Chem. Soc. Chem. Commun. 1976, 1007–1008.
116 Beale, S.I. (1970) Plant Physiol. 45, 504–506.
117 Beale, S.I. and Castelfranco, P.A. (1974) Plant Physiol. 53, 291–296.
118 Williams, D.C., Morgan, G.S., McDonald, E. and Battersby, A.R. (1981) Biochem. J. 193, 301–310.
119 Juknat, A.A., Dörnemann, D. and Senger, H. (1988) Z. Naturforsch. 43c, 351–356.
120 Juknat, A.A., Dörnemann, D. and Senger, H. (1988) Z. Naturforsch. 43c, 357–362.
121 Higuchi, M. and Bogorad, L. (1975) Ann. N.Y. Acad. Sci. 244, 401–418.
122 Shioi, Y., Nagamine, M., Kuroki, M. and Sasa, T. (1980) Biochim. Biophys. Acta 616, 300–309.
123 Jordan, P.M. and Shemin, D. (1973) J. Biol. Chem. 248, 1019–1024.
124 Kotler, M.L., Fumagalli, S.A., Juknat, A.A. and Batlle, A.M. del C. (1987) Comp. Biochem. Physiol. 87B, 601–606.
125 Rossetti, M.V., Juknat, A.A. and Batlle, A.M. del C. (1989) Z. Naturforsch. 44c, 578–580.
126 Sharif, A.L., Smith, A.G. and Abell, C. (1989) Eur. J. Biochem. 184, 353–359.
127 Jordan, P.M. and Seehra, J.S. (1979) FEBS Lett. 104, 364,366.
128 Battersby, A.R., Fookes, C.J.R., Matcham, G.W.J. and McDonald, E. (1979) J. Chem. Soc. Chem. Commun. 1979, 539–541.
129 Jordan, P.M. and Berry, A. (1981) Biochem. J. 195, 177–181.
130 Battersby, A.R., Fookes, C.J.R., Matcham, G.W.J., McDonald, E. and Hollenstein, R. (1983) J. Chem. Soc. Perkin Trans. I 1983, 3031–3040.
131 Battersby, A.R., Fookes, C.J.R., Gustafson-Potter, K.E., Matcham, G.W.J. and McDonald, E. (1979) J. Chem. Soc. Chem. Commun. 1979, 1155–1158.
132 Scott, A.I., Burton, G., Jordan, P.M., Matsumoto, H., Fagerness, P.E. and Pryde, L.M. (1980) J. Chem. Soc. Chem. Commun. 1980, 384–387.
133 Hart, G.J., Miller, A.D., Leeper, F.J. and Battersby, A.R. (1987) J. Chem. Soc. Chem. Commun. 1987, 1762–1765.
134 Jordan, P.M. and Warren, M.J. (1987) FEBS Lett. 225, 87–92.
135 Warren, M.J. and Jordan, P.M. (1988) Biochem. Soc. Trans. 16, 963–965.
136 Castelfranco, P.A., Thayer, S.S., Wilkinson, J.Q. and Bonner, B.A. (1988) Arch. Biochem. Biophys. 266, 219–226.

137 Bogorad, L. (1958) J. Biol. Chem. 233, 501–509.
138 Hart, G.J. and Battersby, A.R. (1985) Biochem. J. 232, 151–160.
139 Kohashi, M., Clement, R.P., Tse, J. and Piper, W.N. (1984) Biochem. J. 220, 755–765.
140 Frydman, R.B. and Feinstein, G. (1974) Biochim. Biophys. Acta 350, 358–373.
141 Battersby, A.R., Fookes, C.J.R., Matcham, G.W.J., McDonald, E. and Gustafson-Potter, K.E. (1979) J. Chem. Soc. Chem. Commun. 1979, 316–319.
142 Rosé, S., Frydman, R.B., de los Santos, C., Sburlati, A., Valasinas, A. and Frydman, B. (1988) Biochemistry (USA) 27, 4871–4879.
143 Chen, T.C. and Miller, G.W. (1974) Plant Cell Physiol. 15, 993–1005.
144 Weinstein, J.D. and Castelfranco, P.A. (1977) Arch. Biochem. Biophys. 178, 671–673.
145 Castelfranco, P.A., Weinstein, J.D., Schwarcz, S., Pardo, A.D. and Wezelman, B.E. (1979) Arch. Biochem. Biophys. 192, 592–598.
146 Hsu, W.P. and Miller, G.W. (1970) Biochem. J. 117, 215–220.
147 Cavaleiro, J.A.S., Kenner, G.W. and Smith, K.M. (1974) J. Chem. Soc. Perkin Trans. I 1974, 1188–1194.
148 Tait, G.H. (1972) Biochem. J. 128, 1159–1169.
149 Poulson, R. and Polglase, W.J. (1974) J. Biol. Chem. 249, 6367–6371.
150 Jacobs, J.M. and Jacobs, N.J. (1981) Arch. Biochem. Biophys. 211, 305–311.
151 Jacobs, J.M., Jacobs, N.J. and De Maggio, A.E. (1982) Arch.. Biochem. Biophys. 218, 233–239.
152 Jacobs, J.M. and Jacobs, N.J. (1984) Arch. Biochem. Biophys. 229, 312–319.
153 Jacobs, J.M. and Jacobs, N.J. (1987) Biochem. J. 244, 219–224.
154 Jacobs, J.M. and Jacobs, N.J. (1984) Biochem. Biophys. Res. Commun. 123, 1157–1164.
155 Dailey, H.A. (1982) J. Biol. Chem. 257, 14714–14718.
156 Dailey, H.A., Fleming, J.E. and Harbin, B.M. (1986) J. Bacteriol. 165, 1–5.
157 Goldin, B.R. and Little, H.N. (1969) Biochim. Biophys. Acta 171, 321–332.
158 Little, H.N. and Jones, O.T.G. (1976) Biochem. J. 156, 309–314.
159 Porra, R.J. and Lascelles, J. (1968) Biochem. J. 108, 343–348.
160 Jones, O.T.G. (1967) Biochem. Biophys. Res. Commun. 28, 671–674.
161 Jones, O.T.G. (1968) Biochem. J. 107, 113–119.
162 Brown, S.B., Holroyd, J.A., Vernon, D.I. and Jones, O.T.G. (1984) Biochem. J. 220, 861–863.
163 Brown, S.B., Holroyd, J.A., Vernon, D.I., Troxler, R.F. and Smith, K.M. (1982) Biochem. J. 208, 487–491.
164 Beale, S.I. and Chen, N.C. (1983) Plant Physiol. 71, 263–268.
165 Brown, S.B., Holroyd, J.A., Troxler, R.F. and Offner, G.D. (1981) Biochem. J. 191, 137–147.
166 Schuster, A., Köst, H.-P., Rüdiger, W., Holroyd, J.A. and Brown, S.B. (1983) Plant Cell Rep. 2, 85–87.
167 Köst, H.-P. and Benedikt, E. (1982) Z. Naturforsch. 37c, 1057–1063.
168 Beale, S.I. and Cornejo, J. (1983) Arch. Biochem. Biophys. 227, 279–286.
169 Holroyd, J.A., Vernon, D.I. and Brown, S.B. (1985) Biochem. Soc. Trans. 13, 209–210.
170 Troxler, R.F. (1972) Biochemistry 11, 4235–4242.
171 Brown, S.B., Holroyd, J.A. and Troxler, R.F. (1980) Biochem. J. 190, 445–449.
172 Troxler, R.F., Brown, A.S. and Brown, S.B. (1979) J. Biol. Chem. 254, 3411–3418.
173 Beale, S.I. and Cornejo, J. (1984) Plant Physiol. 76, 7–15.
174 Cornejo, J. and Beale, S.I. (1988) J. Biol. Chem. 263, 11915–11921.
175 Drummond, G.S. and Kappas, A. (1981) Proc. Natl. Acad. Sci. USA 78, 6466–6470.
176 Beale, S.I. and Cornejo, J. (1984) Arch. Biochem. Biophys. 235, 371–384.
177 Cornejo, J. and Beale, S.I. (1990) Plant Physiol. 92(Suppl.), (in press).
178 Knox, W.E. (1960) in: The Enzymes (P.D. Boyer, H. Lardy and K. Myrback, Eds.) Vol 2 Part A, pp. 253–294, Academic Press, New York.

179 Seltzer, S. (1972) in: The Enzymes (P.D. Boyer, Ed.) 3rd ed., Vol. 6, pp. 381–406, Academic Press, New York.
180 Fu, E., Friedman, L. and Siegelman, H.W. (1979) Biochem. J. 179, 1–6.
181 Weller, J.-P. and Gossauer, A. (1980) Chem. Ber. 113, 1603–1611.
182 Benedikt, E. and Köst, H.-P. (1985) Z. Naturforsch. 40c, 755–759.
183 Gossauer, A., Nydegger, F., Benedikt, E. and Köst, H.-P. (1989) Helv. Chim. Acta 72, 518–529.
184 Siegelman, H.W., Chapman, D.J. and Cole, W.J. (1968) in: Porphyrins and Related Compounds (T.W. Goodwin, Ed.) pp. 107–120, Academic Press, New York.
185 Glazer, A.N. and Hixson, C.S. (1977) J. Biol. Chem. 252, 32–42.
186 O'Carra, P., Murphy, R.F. and Killilea, S.D. (1980) Biochem. J. 187, 303–309.
187 Bishop, J.E., Rapoport, H., Klotz, A.V., Chan, C.F., Glazer, A.N., Füglistaller, P. and Zuber, H. (1987) J. Am. Chem. Soc. 109, 875–881.
188 Killilea, S.D. and O'Carra, P. (1985) Biochem. J. 226, 723–731.
189 Nagy, J.O., Bishop, J.E., Klotz, A.V., Glazer, A.N. and Rapoport, H. (1985) J. Biol. Chem. 260, 4864–4868.
190 Troxler, R.F. and Bogorad, L. (1966) Plant Physiol. 41, 491–499.
191 Troxler, R.F. and Lester, R. (1967) Biochemistry 6, 3840–3846.
192 Paul, K.G. (1951) Acta Chem. Scand. 5, 389–405.
193 Brockmann, H. Jr. and Knobloch, G. (1973) Chem. Ber. 106, 803–811.
194 Gossauer, A. (1983) Tetrahedron 39, 1933–1941.
195 Klein, G. and Rüdiger, W. (1978) Liebigs Ann. Chem. 1978, 267–279.
196 Schoenleber, R.W., Leung, S.-L., Lundell, D.J., Glazer, A.N. and Rapoport, H. (1983) J. Am. Chem. Soc. 105, 4072–4076.
197 Bishop, J.E., Lagarias, J.C., Nagy, J.O., Schoenleber, R.W., Rapoport, H., Klotz, A.V. and Glazer, A.N. (1986) J. Biol. Chem. 261, 6790–6796.
198 Klotz, A.V., Glazer, A.N., Bishop, J.E., Nagy, J.O. and Rapoport, H. (1986) J. Biol. Chem. 261, 6797–6805.
199 Schoenleber, R.W., Lundell, D.J., Glazer, A.N. and Rapoport, H. (1984) J. Biol. Chem. 259, 5481–5484.
200 Klotz, A.V. and Glazer, A.N. (1985) J. Biol. Chem. 260, 4856–4863.
201 Schmidt, G., Siebzehnrübl, S., Fischer, R., Rüdiger, W., Scheer, H., Schirmer, T., Bode, W. and Huber, R. (1987) Z. Naturforsch. 42c, 845–848.
202 Wilbanks, S.M., Wedemayer, G.J. and Glazer, A.N. (1989) J. Biol. Chem. 264, 17860–17867.
203 Elich, T.D., McDonagh, A.F., Palma, L.A. and Lagarias, J.C. (1989) J. Biol. Chem. 264, 183–189.
204 Smith, H. and Kendrick, R.E. (1976) in: Chemistry and Biochemistry of Plant Pigments (T.W. Goodwin, Ed.) 2nd ed., Vol. 1, pp. 377–424, Academic Press, New York.
205 Dring, M.J. (1974) in: Algal Physiology and Biochemistry (W.D.P. Stewart, Ed.) pp. 814–837, University of California Press, Berkeley.
206 Elich, T.D. and Lagarias, J.C. (1987) Plant Physiol. 84, 304–310.
207 Gorchein, A. (1972) Biochem. J. 127, 97–106.
208 Gorchein, A. (1973) Biochem. J. 134, 833–845.
209 Ellsworth, R.K. and Lawrence, G.D. (1973) Photosynthetica 7, 73–86.
210 Smith, B.B. and Rebeiz, C.A. (1977) Arch. Biochem. Biophys. 180, 178–185.
211 Richter, M.L. and Rienits, K.G. (1980) FEBS Lett. 116, 211–216.
212 Pardo, A.D., Chereskin, B.M., Castelfranco, P.A., Franceschi, V.R. and Wezelman, B.E. (1980) Plant Physiol. 65, 956–960.
213 Fuesler, T.P., Wright, L.A. Jr. and Castelfranco, P.A. (1981) Plant Physiol. 67, 246–249.
214 Fuesler, T.P., Hanamoto, C.M. and Castelfranco, P.A. (1982) Plant Physiol. 69, 421–423.
215 Richter, M.L. and Rienits, K.G. (1982) Biochim. Biophys. Acta 717, 255–264.

216 Fuesler, T.P., Wong, Y.-S. and Castelfranco, P.A. (1984) Plant Physiol. 75, 662–664.
217 Radmer, R.J. and Bogorad, L. (1967) Plant Physiol. 42, 463–465.
218 Gibson, K.D., Neuberger, A. and Tait, G.H. (1963) Biochem. J. 88, 325–334.
219 Hinchigeri, S.B., Nelson, D.W. and Richards, W.R. (1984) Photosynthetica 18, 168–178.
220 Ellsworth, R.K., Dullaghan, J.P. and St. Pierre, M.E. (1974) Photosynthetica 8, 375–384.
221 Ellsworth, R.K. and St. Pierre, M.E. (1976) Photosynthetica 10, 291–301.
222 Hinchigeri, S. B, Chan, J.C.-S. and Richards, W.R. (1981) Photosynthetica 15, 351–359.
223 Ebbon, J.G. and Tait, G.H. (1969) Biochem. J. 111, 573–582.
224 Shieh, J., Miller, G.W. and Psenak, M. (1978) Plant Cell Physiol. 19, 1051–1059.
225 Richards, W.R., Fung, M., Wessler, A.N. and Hinchigeri, S.B. (1987) in: Progress in Photosynthesis Research (J. Biggins, Ed.) Vol. 4, pp. 475–482, Martinus Nijhoff, Boston.
226 Yee, W.C., Eglsaer, S.J. and Richards, W.R. (1989) Biochem. Biophys. Res. Commun. 162, 483–490.
227 Hinchigeri, S.B. and Richards, W.R. (1982) Photosynthetica 16, 554–560.
228 Jones, O.T.G. (1963) Biochem. J. 88, 335–343.
229 Jones, O.T.G. (1963) Biochem. J. 89, 182–189.
230 Richards, W.R. and Lascelles, J. (1969) Biochemistry 8, 3473–3482.
231 Pradel, J. and Clement-Metral, J.D. (1976) Biochim. Biophys. Acta 430, 253–264.
232 Griffiths, W.T. and Jones, O.T.G. (1975) FEBS Lett. 50, 355–358.
233 Belanger, F.C. and Rebeiz, C.A. (1979) Biochem. Biophys. Res. Commun. 88, 365–372.
234 Belanger, F.C. and Rebeiz, C.A. (1980) J. Biol. Chem. 255, 1266–1272.
235 Belanger, F.C. and Rebeiz, C.A. (1980) Plant Sci. Lett. 18, 343–350.
236 Granick, S. (1950) Harvey Lectures 44, 220–245.
237 Bogorad, L. (1966) in: The Chlorophylls (L.P. Vernon and G.R. Seeley, Eds.) pp. 481–510, Academic Press, New York.
238 Ellsworth, R.K. and Aronoff, S. (1968) Arch. Biochem. Biophys. 125, 269–277.
239 Ellsworth, R.K. and Aronoff, S. (1969) Arch. Biochem. Biophys. 130, 374–383.
240 Ellsworth, R.K. and Hervish, P.V. (1975) Photosynthetica 9, 125–139.
241 Mattheis, J.R. and Rebeiz, C.A. (1977) J. Biol. Chem. 252, 4022–4024.
242 Spiller, S.C., Castelfranco, A.M. and Castelfranco, P.A. (1982) Plant Physiol. 69, 107–111.
243 Chereskin, B.A. and Castelfranco, P.A. (1982) Plant Physiol. 69, 112–116.
244 Chereskin, B.A., Wong, Y.-S. and Castelfranco, P.A. (1982) Plant Physiol. 70, 987–993.
245 Chereskin, B.A., Castelfranco, P.A., Dallas, J.L. and Straub, K.M. (1983) Arch. Biochem. Biophys. 226, 10–18.
246 Wong, Y.-S. and Castelfranco, P.A. (1984) Plant Physiol. 75, 658–661.
247 Wong, Y.-S. and Castelfranco, P.A. (1985) Plant Physiol. 79, 730–733.
248 Wong, Y.-S., Castelfranco, P.A., Goff, D.A. and Smith, K.M. (1985) Plant Physiol. 79, 725–729.
249 Walker, C.J., Mansfield, K.E., Rezzano, I.N., Hanamoto, C.H., Smith, K.M. and Castelfranco, P.A. (1988) Biochem. J. 255, 685–692.
250 Nasrulhaq-Boyce, A., Griffiths, W.T. and Jones, O.T.G. (1987) Biochem. J. 243, 23–29.
251 Walker, C.J., Mansfield, K.E., Smith, K.M. and Castelfranco, P.A. (1989) Biochem. J. 257, 599–602.
252 Rebeiz, C.A., Wu, S.M., Kuhadja, M., Daniell, H. and Perkins, E.J. (1983) Mol. Cell Biochem. 57, 97–125.
253 Belanger, F.C. and Rebeiz, C.A. (1982) J. Biol. Chem. 257, 1360–1371.
254 Ellsworth, R.K. and Hsing, A.S. (1973) Biochim. Biophys. Acta 313, 119–129.
255 Ellsworth, R.K. and Hsing, A.S. (1974) Photosynthetica 8, 228–234.
256 Kwan, L.Y.-M., Darling, D.L. and Richards, W.R. (1986) in: Regulation of Chloroplast Differentiation (G. Akoyunoglou and H. Senger, Eds.) pp. 57–62, Alan R. Liss, Inc., New York.
257 Gassman, M.L., Granick, S. and Mauzerall, D. (1968) Biochem. Biophys. Res. Commun. 32, 295–309.

258 Shibata, K. (1957) J. Biochem. (Tokyo) 44, 147–173.
259 Koski, V.M., French, C.S. and Smith, J.H.C. (1951) Arch. Biochem. Biophys. 31, 1–17.
260 Gassman, M.L. (1973) Plant Physiol. 52, 295–300.
261 Virgin, H.I. (1981) Annu. Rev. Plant Physiol. 32, 451–463.
262 Kasemir, H. (1983) Photochem. Photobiol. 37, 701–708.
263 Horton, P. and Leech, R.M. (1972) FEBS Lett. 26, 277–280.
264 Griffiths, W.T. (1974) FEBS Lett. 46, 301–304.
265 Griffiths, W.T. (1978) Biochem. J. 174, 681–692.
266 Oliver, R.P. and Griffiths, W.T. (1980) Biochem. J. 191, 277–280.
267 Griffiths, W.T. and Oliver, R.P. (1984) in: Chloroplast Biogenesis (R.J. Ellis, Ed.) pp. 245–258, Cambridge University Press, Cambridge.
268 Griffiths, W.T. (1981) in: Photosynthesis (G. Akoyunoglou, Ed.) Vol. 5, pp. 65–71, Balaban, Philadelphia.
269 Griffiths, W.T. (1974) FEBS Lett. 49, 196–200.
270 Beer, N.S. and Griffiths, W.T. (1981) Biochem. J. 195, 83–92.
271 Apel, K., Santel, H.-J., Redlinger, T.E. and Falk, K. (1980) Eur. J. Biochem. 111, 251–258.
272 Roper, U., Prinz, H. and Lutz, C. (1987) Plant Sci. 52, 15–19.
273 Richards, W.R., Walker, C.J. and Griffiths, W.T. (1987) Photosynthetica 21, 462–471.
274 Walker, C.J. and Griffiths, W.T. (1988) FEBS Lett. 239, 259–262.
275 Oliver, R.P. and Griffiths, W.T. (1982) Plant Physiol. 70, 1019–1025.
276 El Hamouri, B. and Sironval, C. (1980) Photobiochem. Photobiophys. 1, 219–223.
277 Franck, F. and Inoue, Y. (1984) Photobiochem. Photobiophys. 8, 85–96.
278 Franck, F. and Schmid, G.H. (1985) Z. Naturforsch. 40c, 699–704.
279 Franck, F. and Mathis, P. (1980) Photochem. Photobiol. 32, 799–803.
280 Dujardin, E., Franck, F., Gysemberg, R. and Sironval, C. (1986) Photobiochem. Photobiophys. 12, 97–105.
281 van Bochove, A.C., Griffiths, W.T. and van Grondelle, R. (1984) Photochem. Photobiol. 39, 101–106.
282 Bogorad, L. (1950) Bot. Gaz. 111, 221–241.
283 Bogdanović, M. (1973) Physiol. Plant. 29, 17–18.
284 Jelié, G. and Bogdanović, M. (1989) Plant Sci. 61, 197–202.
285 Adamson, H., Walker, C., Bees, A. and Griffiths, T. (1987) in: Progress in Photosynthesis Research (J. Biggins, Ed.) Vol. 4, pp. 483–486, Martinus Nijhoff, Boston.
286 Peschek, G.A., Hinterstoisser, B., Wastyn, M., Kuntner, O., Pineau, B., Missbichler, A. and Lang, J. (1989) J. Biol. Chem. 264, 11827–11832.
287 Peschek, G.A., Hinterstoisser, B., Pineau, B. and Missbichler, A. (1989) Biochem. Biophys. Res. Commun. 162, 71–78.
288 Selstam, E., Widell, A. and Johansson, L.B.-Åring;. (1987) Physiol. Plant. 70, 209–214.
289 Senger, H. and Brinkmann, G. (1986) Physiol. Plant. 68, 119–124.
290 Adamson, H. and Packer, N. (1987) Photosynthetica 21, 472–481.
291 Adamson, H.Y., Hiller, R.G. and Vesk, M. (1980) Planta 150, 269–274.
292 Adamson, H., Griffiths, T., Packer, N. and Sutherland, M. (1985) Physiol. Plant. 64, 345–352.
293 Apel, K., Motzkus, M. and Dehesh, K. (1984) Planta 161, 550–554.
294 Packer, N. and Adamson, H. (1986) Physiol. Plant. 68, 220–230.
295 Tripathy, B.C. and Rebeiz, C.A. (1987) in: Progress in Photosynthesis Research (J. Biggins, Ed.) Vol. 4, pp. 439–443, Martinus Nijhoff, Boston.
296 Ellsworth, R.K. (1971) Photosynthetica 5, 226–232.
297 Rüdiger, W., Hedden, P., Köst, H.-P. and Chapman, D.J. (1977) Biochem. Biophys. Res. Commun. 74, 1268–1272.
298 Rüdiger, W., Benz, J. and Guthoff, C. (1980) Eur. J. Biochem. 109, 193–200.

299 Benz, J. and Rüdiger, W. (1981) Z. Naturforsch. 36c, 51–57.
300 Emery, V.C. and Akhtar, M. (1985) J. Chem. Soc. Chem. Commun. 1985, 600–601.
301 Ajaz, A.A., Corina, D.L. and Akhtar, M. (1985) Eur. J. Biochem. 150, 309–312.
302 Schoch, S., Lempert, U. and Rüdiger (1977) Z. Pflanzenphysiol. 83, 427–436.
303 Schoch, S. and Schäfer, W. (1978) Z. Naturforsch. 33c, 408–412.
304 Benz, J., Wolf, C. and Rüdiger, W. (1980) Plant Sci. Lett. 19, 225–230.
305 Rüdiger, W., Benz, J., Lempert, U. and Steffens, D. (1976) Z. Pflanzenphysiol. 80, 131–143.
306 Schoch, S., Hehlein, C. and Rüdiger, W. (1980) Plant Physiol. 66, 576–579.
307 Bishop, N.I. and Wong, J. (1974) Ber. Deutsch. Bot. Gaz. 87, 359–371.
308 Henry, A., Powls, R. and Pennock, J.F. (1986) Biochem. Soc. Trans. 14, 958–959.
309 Shioi, Y. and Sasa, T. (1984) J. Bacteriol. 158, 340–343.
310 Rüdiger, W. (1987) in: Progress in Photosynthesis Research (J. Biggins, Ed.) Vol. 4, pp. 461–467, Martinus Nijhoff, Boston.
311 Godnev, T.N., Rotfarb, R.M. and Shlyk, A.A. (1960) Dokl. Akad. Nauk. USSR 130, 663–666.
312 Akoyunoglou, G., Argyroudi-Akoyunoglou, J.H., Michel-Wolwertz, M.R. and Sironval, C. (1967) Chim. Chron. 32, 5–8.
313 Akoyunoglou, G., Argyroudi-Akoyunoglou, J.H., Michel-Wolwertz, M.R. and Sironval, C. (1966) Physiol. Plant. 19, 1101–1104.
314 Shlyk, A.A. and Prudnikova, I.V. (1967) Photosynthetica 1, 157–170.
315 Shlyk, A.A., Vlasenok, L.I., Akhramovich, N.I., Vrubel, S.V. and Akulovich, E.M. (1975) Dokl. Akad. Nauk. SSSR 221, 1234–1236.
316 Ogawa, T., Inoue, Y., Kitajima, M. and Shibata, K. (1973) Photochem. Photobiol. 18, 229–235.
317 Oelze-Karow, H., Kasemir, H. and Mohr, H. (1978) in: Chloroplast Development (G. Akoyunoglou, G. and J.-H. Argyroudi-Akoyunoglou, Eds.) pp. 787–792, Elsevier, Amsterdam.
318 Oelze-Karow, H. and Mohr, H. (1978) Photochem. Photobiol. 27, 189–193.
319 Duggan, J.X. and Rebeiz, C.A. (1982) Biochim. Biophys. Acta 714, 248–260.
320 Kotzabasis, K. and Senger, H. (1989) Bot. Acta 102, 173–177.
321 Thorne, S.W. and Boardman, N.K. (1971) Plant Physiol. 47, 252–261.
322 Bennett, J. (1981) Eur. J. Biochem. 118, 61–70.
323 Bellemare, G., Bartlett, S.G. and Chua, N.-H. (1982) J. Biol. Chem. 257, 7762–7767.
324 Benz, J., Lempert, U. and Rüdiger, W. (1984) Planta 162, 215–219.
325 Shioi, Y. and Sasa, T. (1983) Biochim. Biophys. Acta 756, 127–131.
326 Shioi, Y. and Sasa, T. (1983) Arch. Biochem. Biophys. 220, 286–292.
327 Ellsworth, R.K., Perkins, H.J., Detwiller, J.P. and Liu, K. (1970) Biochim. Biophys. Acta 223, 275–280.
328 Shlyk, A.A., Prudnikova, I.V. and Malashevich, A.V. (1971) Dokl. Akad. Nauk. SSSR 201, 1481–1484.
329 Rebeiz, C.A. and Castelfranco, P.A. (1971) Plant Physiol. 47, 33–37.
330 Bednarik, D.P. and Hoober, J.K. (1985) Science 230, 450–453.
331 Bednarik, D.P. and Hoober, J.K. (1986) in: Regulation of Chloroplast Differentiation (G. Akoyunoglou, and H. Senger, Eds.) pp. 105–114, Alan R. Liss, New York.
332 Kotzabasis, K., and Senger, H. (1989) Physiol. Plant. 76, 474–478.
333 Bazzaz, M.B., Bradley, C.V. and Brereton, R.G. (1982) Tet. Lett. 23, 1211–1214.
334 Brereton, R.G., Bazzaz, M.B., Santikarn, S. and Williams, D.H. (1983) Tet. Lett. 24, 5775–5778.
335 Wu, S.-M. and Rebeiz, C.A. (1985) J. Biol. Chem. 260, 3632–3634.
336 Bazzaz, M.B., Govindje and Paolillo, D.J., Jr. (1974) Z. Pflanzenphysiol. 72, 181–192.
337 Chisholm, S.W., Olson, R.J., Zettler, E.R., Goericke, R., Waterbury, J.B. and Welschmeyer, N.A. (1988) Nature 334, 340–343.
338 Belanger, F.C. and Rebeiz, C.A. (1980) Biochemistry 19, 4875–4883.
339 Rebeiz, C.A., Belanger, F.C., Freyssinet, G. and Saab, D.G. (1980) Biochim. Biophys. Acta 590,

234–247.
340 Tripathy, B.C. and Rebeiz, C.A. (1986) J. Biol. Chem. 261, 13556–13564.
341 Tripathy, B.C. and Rebeiz, C.A. (1988) Plant Physiol. 87, 89–94.
342 Eskins, K. and Harris, L. (1981) Photochem. Photobiol. 33, 131–133.
343 Wilhelm, C. (1987) Biochim. Biophys. Acta 892, 23–29.
344 Dougherty, R.C., Strain, H.H., Svec, W.A., Uphaus, R.A. and Katz, J.J. (1966) J. Am. Chem. Soc. 88, 5037–5038.
345 Dougherty, R.C., Strain, H.H., Svec, W.A., Uphaus, R.A. and Katz, J.J. (1970) J. Am. Chem. Soc. 92, 2826–2833.
346 Budzikiewicz, H. and Taraz, K. (1971) Tetrahedron 27, 1447–1460.
347 Anderson, R.A. and Mulkey, T.J. (1983) J. Phycol. 19, 289–294.
348 Jeffrey, S.W. and Wright, S.W. (1987) Biochim. Biophys. Acta 894, 180–188.
349 Fawley, M.W. (1989) Plant Physiol. 91, 727–732.
350 Brown, J.S. (1985) Biochim. Biophys. Acta 807, 143–146.
351 Watanabe, T., Nakazato, M., Mazaki, H., Hongu, A., Konno, M., Saitoh, S. and Honda, K. (1985) Biochim. Biophys. Acta 807, 110–117.
352 Watanabe, T., Kobayashi, M., Nakazato, M., Ikegami, I. and Hiyama, T. (1987) in: Progress in Photosynthesis Research (J. Biggins, Ed.) Vol. 1, pp. 303–306, Martinus Nijhoff, Boston.
353 Amesz, J. (1987) Photosynthetica 21, 225–235.
354 Glazer, A.N. and Melis, A. (1987) Annu. Rev. Plant Physiol. 38, 11–45.
355 Walter, E., Schreiber, J., Zass, E. and Eschenmoser, A. (1979) Helv. Chim. Acta 62, 899–920.
356 Ikeuchi, M. and Murakami, S. (1983) Plant Cell Physiol. 24, 71–80.
357 Dehesh, K. and Ryberg, M. (1985) Planta 164, 396–399.
358 Sisler, E.C. and Klein, W.H. (1963) Physiol. Plant. 16, 315–322.
359 Fluhr, R., Harel, E., Klein, S. and Meller, E. (1975) Plant Physiol. 56, 497–501.
360 Minkov, I.N., Ryberg, M. and Sundqvist, C. (1988) Physiol. Plant. 72, 725–732.
361 Klein, S., Harel, E., Ne'eman, E., Katz, E. and Meller, E. (1975) Plant Physiol. 56, 486–496.
362 Nadler, K. and Granick, S. (1970) Plant Physiol. 46, 240–246.
363 Klein, S., Katz, E. and Ne'eman, E. (1977) Plant Physiol. 60, 335–338.
364 Masoner, M. and Kasemir, H. (1975) Planta 126, 111–117.
365 Reiss, C. and Wayne, R. (1989) Plant Physiol. 89(Suppl.), 21.
366 Weinstein, J.D. (1979) Ph.D. Dissertation, University of California, Davis, 123 pp.
367 Berry-Lowe, S. (1987) Carlsberg Res. Commun. 52, 197–210.
368 Huang, L. and Castelfranco, P.A. (1989) Plant Physiol. 90, 996–1002.
369 Huang, L., Bonner, B.A. and Castelfranco, P.A. (1989) Plant Physiol. 90, 1003–1008.
370 Mösinger, E., Batschauer, A., Schäfer, E. and Apel, K. (1985) Eur. J. Biochem. 147, 137–142.
371 Silverthorne, J. and Tobin, E.M. (1984) Proc. Nat. Acad. Sci. USA 81, 1112–1116.
372 Briggs, W.R., Mosinger, E. and Schäfer, E. (1988) Plant Physiol. 86, 435–440.
373 Horwitz, B.A., Thompson, W.F. and Briggs, W.R. (1988) Plant Physiol. 86, 299–305.
374 Fletcher, R.A., Teo, C. and Ali, A. (1973) Can. J. Bot. 51, 937–939.
375 Arnold, V. and Fletcher, R.A. (1986) Physiol. Plant. 68, 169–174.
376 Teyssendier de la Serve, B., Axelos, M. and Peaud-Lenoel, C. (1985) Plant Mol. Biol. 5, 155–163.
377 Daniell, H. and Rebeiz, C.A. (1982) Biochem. Biophys. Res. Commun. 104, 837–843.
378 Daniell, H. and Rebeiz, C.A. (1986) in: Regulation of Chloroplast Differentiation (G. Akoyunoglou, and H. Senger, Eds.) pp. 63–70, Alan R. Liss, Inc., New York.
379 Cohen, L., Arzee, T. and Zilberstein, A. (1988) Physiol. Pant. 72, 57–64.
380 Oh-hama, T. and Hase, E. (1980) Plant Cell Physiol. 21, 1263–1272.
381 Schiff, J.A. (1974) in: Proceedings of the Third International Congress on Photosynthesis (M. Avron, Ed.) pp. 1691–1717, Elsevier, Amsterdam.
382 Duggan, J. and Gassman, M. (1974) Plant Physiol. 53, 206–215.

383 Gassman, M. and Duggan, J. (1974) in: Proceedings of the Third International Congress on Photosynthesis (M. Avron, Ed.) pp. 2105–2113, Elsevier, Amsterdam.
384 Vlcek, L.M. and Gassman, M.L. (1979) Plant Physiol. 64, 393–397.
385 Weinstein, J.D. and Castelfranco, P.A. (1978) Arch. Biochem. Biophys. 186, 376–382
386 Huang, L. and Castelfranco, P.A. (1988) Plant Sci. 54, 185–192.
387 Hendry, G.A.F., Houghton, J.D. and Brown, S.B. (1987) New Phytol. 107, 255–302.
388 Stillman, L.C. and Gassman, M.L. (1978) Plant Physiol. 62, 182–184.
389 Nasrulhaq-Boyce, A. and Jones, O.T.G. (1981) Phytochemistry 20, 1005–1009.
390 Hendry, G.A.F. and Stobart, A.K. (1978) Phytochemistry 17, 73–77.
391 Hendry, G.A.F. and Stobart, A.K. (1986) Phytochemistry 25, 2735–2737.
392 Stobart, A.K. and Hendry, G.A.F. (1984) Phytochemistry 23, 27–30.
393 Argyroudi-Akoyunoglou, J.H., Akoyunoglou, A., Kalosakas, K. and Akoyunoglou, G. (1982) Plant Physiol. 70, 1242–1248.
394 Manetas, Y. and Akoyunoglou, G. (1981) Photosynthetica 15, 534–539.
395 Breidenbach, R.W., Castelfranco, P.A. and Criddle, R.S. (1967) Plant Physiol. 42, 1035–1041.
396 Hendry, G.A.F., Houghton, J.D. and Jones, O.T.G. (1981) Biochem. J. 194, 743–751.
397 Hendry, G.A.F., Houghton, J.D. and Jones, O.T.G. (1981) Biochem. J. 196, 825–829.
398 Benveniste, I., Salaun, J.P. and Durst. (1977) Phytochemistry 16, 69–73.
399 Reichhart, D., Salaun, J.P., Benveniste, I. and Durst, F. (1980) Plant Physiol. 66, 600–604.
400 Fonne-Pfister, R., Simon, A., Salaun, J.-P. and Durst, F. (1988) Plant Sci. 55, 9–20.
401 Gomez-Silva, B., Timko, M.P. and Schiff, J.A. (1985) Planta 165, 12–22.
402 Granick, S. and Beale, S.I. (1978) Adv. Enzymol. 46, 33–203.
403 Chibbar, R.N. and van Huystee, R.B. (1986) Phytochemistry 25, 585–587.
404 Thomas, J. and Weinstein, J.D. (1990) Plant Physiol. 94, 1414–1423.
405 Castelfranco, P.A. and Beale, S.I. (1983) Annu. Rev. Plant Physiol. 34, 241–278.
406 Gough, S.P. and Kannangara, C.G. (1979) Carlsberg Res. Commun. 44, 403–416.
407 Hoober, J.K. and Stegeman, W.J. (1973) J. Cell Biol. 56, 1–12.
408 Huang, D.-D. and Wang, W.-Y. (1986) Mol. Gen. Genet. 205, 217–220.
409 Weinstein, J.D., Howell, R., Leverette, R. and Brignola, P. (1989) Plant Physiol. 89(Suppl.), 74.
410 Stobart, A.K. and Ameen-Bukhari, I. (1984) Biochem. J. 222, 419–426.
411 Stobart, A.K. and Ameen-Bukhari, I. (1986) Biochem. J. 236, 741–748.
412 Dörnemann, D., Kotzabasis, K., Richter, P., Breu, V. and Senger, H. (1989) Bot. Acta 102, 112–115.
413 Lin, J.M., Wegmann, B., Chang, T.N. and Wang, W.Y. (1989) Plant Physiol. 89(Suppl.), 50.
414 Chang, T.-E., Wang, W -Y. and Wegmann, B. (1988) Plant Physiol. 86S, 60.
415 Beale, S.I. (1971) Plant Physiol. 48, 316–319.
416 Corriveau, J.L. and Beale, S.I. (1986) Plant Sci. 45, 9–17.
417 Smith, A.G. (1986) in: Regulation of Chloroplast Differentiation (G. Akoyunoglou and H. Senger, Eds.) pp. 49–54, Alan R. Liss, New York.
418 Nandi, D.L., Baker-Cohen, K.F. and Shemin, D. (1968) J. Biol. Chem. 243, 1224–1230.
419 Nandi, D.L. and Shemin, D. (1973) Arch. Biochem. Biophys. 158, 305–311.
420 Steer, B.T., and Gibbs, M. (1969) Plant Physiol. 44, 781–783.
421 Schneider, H.A.W. (1971) Phytochemistry 10, 319–321.
422 Balangé, A.P. and Rollin, P. (1979) Physiol. Veg. 17, 153–166.
423 Hampp, R., and Ziegler, H. (1975) Planta 124, 255–260.
424 Fuesler, T.P. (1984) Ph.D. Dissertation, University of California, Davis, 123 pp.
425 Wang, W.-Y., Boynton, J.E. and Gillham, N.W. (1977) Mol. Gen. Genet. 152, 7–12.
426 Wallsgrove, R.M., Lea, P.J. and Miflin, B.J. (1983) Plant Physiol. 71, 780–784.
427 Mapleston, E.R. and Griffiths, W.T. (1980) Biochem. J. 189, 125–133.
428 Santel, H.-J. and Apel, K. (1981) Eur. J. Biochem. 120, 95–103.

429 Dehesh, K., Kreuz, K. and Apel, K. (1987) Physiol. Plant. 69, 173–181.
430 Apel, K. (1981) Eur. J. Biochem. 120, 89–93.
431 Batschauer, A. and Apel, K. (1984) Eur. J. Biochem. 143, 593–597.
432 Kay, S.A. and Griffiths, W.T. (1983) Plant Physiol. 72, 229–236.
433 Hauser, I., Dehesh, K. and Apel, K. (1984) Arch. Biochem. Biophys. 228, 577–586.
434 Walker, C.J. and Griffiths, W.T. (1986) in: Regulation of Chloroplast Differentiation (G. Akoyunoglou and H. Senger, Eds.) pp. 99–104, Alan R. Liss, New York.
435 Shaw, P., Henwood, J., Oliver, R. and Griffiths, T. (1985) Eur. J. Cell Biol. 39, 50–55.
436 Dehesh, K., Klaus, M., Hauser, I. and Apel, K. (1986) Planta 169, 162–171.
437 Dehesh, K., van Cleve, B., Ryberg, M. and Apel, K. (1986) Planta 169, 172–183.
438 Troxler, R.F. and Brown, A. (1970) Biochim. Biophys. Acta 215, 503–511.
439 Schuster, A., Köst, H.-P., Rüdiger, W. and Eder, J. (1983) Arch. Microbiol. 135, 30–35.
440 Nichols, K.E. and Bogorad, L. (1962) Bot. Gaz. 124, 85–93.
441 Schneider, H.A.W. and Bogorad, L. (1978) in: Chloroplast Development (G. Akoyunoglou, and J.-H. Argyroudi-Akoyunoglou Eds.) pp. 823–826, Elsevier, Amsterdam.
442 Ramus, J., Beale, S.I., Mauzerall, D. and Howard, K.L. (1976) Mar. Biol. 37, 223–229.
443 Bennett, A. and Bogorad, L. (1973) J. Cell Biol. 58, 419–435.
444 Bryant, D.A. and Cohen-Bazire, G. (1981) Eur. J. Biochem. 119, 415–424.
445 Yu, M.-H., Glazer, A.N., Spencer, K.G. and West, J.A. (1981) Plant Physiol. 68, 482–488.
446 Björn, L.O. and Björn, G.S. (1980) Photochem. Photobiol. 32, 849–852.
447 Vogelmann, T.C. and Scheibe, J. (1978) Planta 143, 233–239.
448 Björn, G.S. and Björn, L.O. (1976) Physiol. Plant. 36, 297–304.
449 Ohki, K. and Fujita, Y. (1979) Plant Cell Physiol. 20, 1341–1347.
450 Gendel, S., Ohad, I. and Bogorad, L. (1979) Plant Physiol. 64, 786–790.
451 Mazel, D., Guglielmi, G., Houmard, J., Sidler, W., Bryant, D.A. and Tandaeu de Marsac, N. (1986) Nuc. Acids Res. 14, 8279–8290.
452 Egelhoff, T. and Grossman, A. (1983) Proc. Natl. Acad. Sci. USA 80, 3339–3343.
453 Belford, H.S., Offner, G.D. and Troxler, R.F. (1983) J. Biol. Chem. 258, 4503–4510.
454 Beguin, S., Guglielmi, G., Rippka, R. and Cohen-Bazire, G. (1985) Biochemie 67, 109–117.
455 de Lorimier, R., Bryant, D.A., Porter, R.D., Liu, W.-Y., Jay, E. and Stevens, S.E., Jr. (1984) Proc. Natl. Acad. Sci. USA 81, 7946–7950.
456 Bryant, D.A., Dubbs, J.M., Fields, P.I., Porter, R.D. and de Lorimier, R. (1985) FEMS Microbiol. Lett. 29, 343–349.
457 de Lorimier, R., Wang, Y.-J. and Yeh, M.-L. (1988) Proc. 3rd. Annu. Penn. State Symp. Plant Physiol. pp. 332–336.
458 Troxler, R.F., Lin, S. and Offner, G.D. (1989) J. Biol. Chem. 264, 20596–20601.
459 Allen, M.M. and Smith, A.J. (1969) Arch. Mikrobiol. 69, 114–120.
460 Foulds, I.J. and Carr, N.G. (1977) FEMS Microbiol. Lett. 2, 117–119.
461 Chang, T.-E., Wegmann, B., and Wang, W.-Y. (1990) Plant Physiol. 93, 1641–1649.
462 Chen, M.W., Jahn, D., Schön, A, O'Neill, G.P., and Söll, D. (1990) J. Biol. Chem. 265, 4054–4057.
463 Chen, M.-W., Jahn, D., O'Neill, G.P., and Söll, D. (1990) J. Biol. Chem. 265, 4058–4063.
464 Jahn, D., Chen, M.-W., and Söll, D. (1991) J. Biol. Chem. 266, 161–167.
465 Rieble, S., and Beale, S.I. (1991) Arch. Biochem. Biophys. 289, 289–297.
466 Rieble, S., and Beale, S.I. (1991) J. Biol. Chem. 266, 9740–9745.
467 Bull, A.D., Breu, V., Kannangara, C.G., Rogers, L.J., and Smith, A.J. (1990) Arch. Microbiol. 154, 56–59.
468 Grimm, B. (1990) Proc. Natl. Acad. Sci. USA 87, 4169–4173.
469 Grimm, B., Bull, A., and Breu, V. (1991) Mol. Gen. Genet. 225, 1–10.
470 Grimm, B., Smith, A.J., Kannangara, C.G., and Smith, M. (1991) J. Biol. Chem. 266, 12495–12501.
471 Spano, A.J., and Timko, M.P. (1991) Biochim Biophys. Acta 1076, 29–36.

472 Sharif, A.L., Smith, A.G., and Abell, C. (1989) Eur. J. Biochem. 184, 353–359.
473 Shashidhara, L.S., and Smith, A.G. (1991) Proc. Natl. Acad. Sci. USA 88, 63–67.
474 Lee, H., Ball, J., and Rebeiz, C.A. (1991) Plant Physiol. 96, 910–915.
475 Jacobs, J.M., Jacobs, N.J., Borotz, S.E., and Guerinot, M.L. (1990) Arch. Biochem. Biophys. 280, 369–375.
476 Camadro, J.-M., Matringe, M., Scalla, R., and Labbe, P. (1991) Biochem. J. 277, 17–21.
477 Beale, S.I., and Cornejo, J. (1991) J. Biol. Chem. 266, (three papers in press).
478 Terry, M.J., and Lagarias, J.C. (1991) J. Biol. Chem. 266, (in press).
479 Walker, C.J., and Weinstein, J.D. (1991) Plant Physiol. 95, 1189–1196.
480 Walker, C.J., and Weinstein, J.D. (1991) Proc. Natl. Acad. Sci. USA 88, 5789–5793.
481 Lützow, M., and Kleining, H. (1990) Arch. Biochem. Biophys. 277, 94–100.
482 Walker, C.J., Castelfranco, P.A., and Whyte, B.J. (1991) Biochem. J. 276, 691–697.
483 Joyard, J., Block, M., Pineau, B., Albrieux, C., and Douce, R. (1990) J. Biol. Chem. 265, 21820–21827.
484 Darrah, P.M., Kay, S.A., Teakle, G.R., and Griffiths, W.T. (1990) Biochem. J. 265, 789–798.
485 Schulz, R., Steinmüller, K., Klaas, M., Forreiter, C., Rasmussen, S., Hiller, C., and Apel, K. (1989) Mol. Genet. 217, 355–361.
486 Benli, M., Schulz, R., and Apel, K. (1991) Plant Mol. Biol. 16, 615–625.
487 Forrieter, C., van Cleve, B., Schmidt, A., and Apel, K. (1991) Planta 183, 126–132.
488 Huang, L., and Hoffman, N.E. (1990) Plant Physiol. 94, 375–379.
489 Roitgrund, C., and Mets, L.J. (1990) Curr. Genet. 17, 147–153.
490 Mayer, S.M., and Beale, S.I. (1990) Plant Physiol. 94, 1365–1375.
491 Mayer, S.M., and Beale, S.I. (1991) Plant Physiol. 97, (in press).
492 Huang, L., and Castelfranco, P.A. (1990) Plant Physiol. 92, 375–379.

CHAPTER 6

The structure and biosynthesis of bacteriochlorophylls

KEVIN M. SMITH

Department of Chemistry, University of California, Davis, CA 95616, U.S.A.

1. Nomenclature

Two systems of tetrapyrrole nomenclature are currently in wide use. The Fischer system for chlorophyll (Chl) derivatives is shown in Structure (1), and features eight peripheral positions, numbered 1–8, two additional positions (9,10) associated with the iso-cyclic ring E, and four interpyrrolic (or *meso*) positions, designated α, β, γ, and δ. Other carbons are numbered with primes (') inward, or with alphabetical letters outward from the central chelating core. This system is used mainly in the USA, and has the merit that it enables contemporary and historical work to be integrated and allows continued use of a large number of classical and indispensable trivial names. The second nomenclature system, as shown in Structure 2 is recommended by the I.U.P.A.C. and is based on the corrin (1–20) system of nomenclature. Through personal preference of the author, the Fischer system (1) will be used in this review.

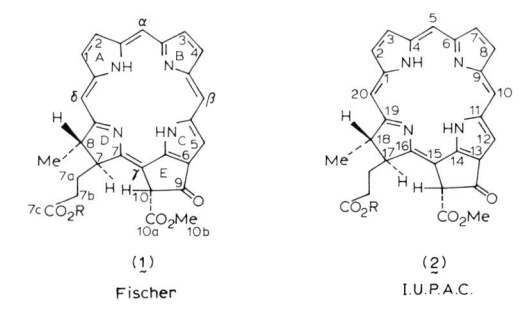

(1) Fischer

(2) I.U.P.A.C.

2. Occurrence and structures

2.1 Bacteriochlorophylls a, b

Bacteriochlorophylls *a* (see Structure 3) and *b* (Structure 4) are the best known Chls produced by photosynthetic bacteria. Specifically, these bacteriochlorophylls (BChl) are the photosynthetic pigments of purple bacteria, and their purple color results from the fact that the parent porphyrin macrocycle is 'reduced' not only in ring D [which gives rise to the green color characteristic of Chl *a* (Structure 5)], but also in ring B. Thus, one might regard the BChl *a* and *b* as 'tetrahydroporphyrins'. This is certainly true for the case of BChl *a*, but the reduction state of BChl b might more correctly be described as being at the same 'dihydro' level as for the normal plant Chls. Simple migration of the ring B ethylidene double bond into ring B proper would actually yield a macrocycle at the same reduction level as Chl a (5). However, BChl *b* also possesses a purple color because the ring B ethylidene moiety still constitutes a structural interference of the π-electron delocalization pathway. BChls *a* from different sources have been shown to possess variability in their 7-side chain esterifying alcohol; for example, the BChl *a* from *Rhodobacter sphaeroides* possesses a phytyl ester (i.e. BChl a_{phy}) while the same pigment from *Rhodospirillum rubrum* features geranylgeraniol (i.e. BChl a_{gg}). The best known purple BChl, BChl *a* (3), is found in Rhodospirillaceae. BChl *b* is the major pigment in *Rhodopseudomonas viridis* [1].

2.2 Bacteriochlorophylls c, d, e, f

The BChls *c*, *d*, and *e* are antenna pigments and were originally termed *Chlorobium* Chls, and are found in the Chlorobiaceae [2]. The BChls *c* are found in strains such as *Prosthecochloris aestuarii*, which usually grows best symbiotically with *Desulfuromonas acetoxidans*. The gliding filamentous bacterium *Chloroflexus aurantiacus* produces only one homologue (4-Et,5-Me) of the BChl *c*. A typical BChl *d* producing organism is *Chlorobium vibrioforme* forma *thiosulphatophilum*, while BChl *e* are the antenna pigments in, for example, *C. pheovibrioides* and *C. pheobacteroides*. It should be mentioned that the reaction center pigments in all of the Chlorobiaceae is BChl *a*.

Very considerable effort has been expended in establishing the structures of the BChls *c*, *d*, and *e*, which are found in green and brown sulfur bacteria. Early work by Holt and co-workers [3a,b,c,d] led to the establishment of the gross structures for the homologous mixtures of the BChls *c* and *d*. The absolute stereochemistry in ring D has been shown to be the same as for Chl *a* (i.e. 7-S,8-S) for the BChls *c* [4], *d* [5], and *e* [6]. Stereochemistry at the 2-(1-hydroxyethyl) for these pigments was subsequently established on the basis of HPLC, NMR, and X-ray crystallography for the BChl *c* [7] and BChls *d* [8]. The BChls *e* were discovered some years later [9a,b], and were characterized, with the exception of the chirality of the 2-(1-hydroxyethyl) this was subsequently established for the BChl *e* on the basis of NMR, HPLC, and chemical transformation into pheophorbide derivatives of the BChls *c* [10]. Tables 1, 2, and 3 show the structures established for the BChls *c*, *d*, and *e*, respectively. All of these BChls are devoid of the 10-methoxycarbonyl present in Chl *a* , and are said to be 'pyro' compounds. Notably, most of the homologues possess extra methionine derived [11] carbons on the 4, and 5 side chains, such that the 4-ethyl normally present in Chl *a* is extended to become a propyl, iso-butyl, or even neo-pentyl substituent, while the methyl at position 5 is extended, on occasion, to become an ethyl. The BChls *c* and *e* also possess δ *meso*-methyl substituents, and the relationship between the BChls *c* and *e* is similar to that between Chl *a* and Chl *b*. The esterifying alcohols on the 7-side chain are mostly farnesol in the BChls *c*, *d*, and *e*, (instead of phytol which is found in Chl *a*), though the BChls *c* [12] and *d* [13] have been shown to contain a variety of other alcohols, depending on the age of the culture. All of the Chlorobiaceae possess a 2-(1-hydroxyethyl) in place of the Chl *a* vinyl, and this has been suggested to be important in construction of the antenna array in green bacteria [14]. The BChl *c* isolated from *Chloroflexus aurantiacus* has been shown [15] to contain only one homologue (4-Et,5-Me) and to possess stearol as the esterifying alcohol on the 7-side chain. BChl *f* has been assigned a generic structure (see Structure 9), though this series of pigments has not yet been observed.

[Structure (6) shown with HO, Me, H at 2a position, Me groups, R4, R5, Mg center, CO2Farnesyl, and porphyrin ring system]

	R⁴	R⁵	Chirality at 2a posn.	Approx. amount present (%)*
a	i-Bu	Et	S	4.5
b	i-Bu	Et	R	<0.1
c	n-Pr	Et	S	5.3
d	n-Pr	Et	R	18.3
e	Et	Et	R	71.7
f	Et	Me	R	0.2

i-Bu = CH$_2$CH(CH$_3$)$_2$; n-Pr = CH$_2$CH$_2$CH$_3$; Et = CH$_2$CH$_3$; Me = CH$_3$
*Percentages can vary slightly [7].

TABLE 1
Structures [7] of the BChl c

2.3 Bacteriochlorophyll g

BChl g (9) was only recently isolated from *Heliobacterium chlorum* [16]. The structure originally assigned to BChl g had geranylgeraniol as the esterifying alcohol, but this was subsequently corrected [17] to farnesol, as indicated in (9). Thus, BChl g is quite similar to BChl b, except for its esterifying alcohol, and the fact that it possesses a vinyl at the 2-position in place of acetyl.

3. Biosynthesis

3.1 Bacteriochlorophylls a, b, g

Using carbon-14 labeled glycine and glutamate, 5-aminolevulinic acid (ALA) biosynthesis in *Rhodopseudomonas palustris* (under conditions causing ALA accumulation) was shown to proceed via the glycine and succinyl-CoA (Shemin; ALA synthase) route [18]. Using analysis of carbon-13 NMR spectra, Oh-Hama and col-

[Structure (7): chlorophyll-type macrocycle with HO, Me, 2a, Me, R⁴, Mg, R⁵, CO₂Farnesyl substituents]

	R⁴	R⁵	Chirality at 2a posn.
a	neo-Pn	Et	S
b	neo-Pn	Me	S
c	i-Bu	Et	S
d	i-Bu	Me	S
e	n-Pr	Et	R
f	n-Pr	Me	R
g	Et	Et	R
h	Et	Me	R

neo-Pn = $CH_2C(CH_3)_3$; i-Bu = $CH_2CH(CH_3)_2$;
n-Pr = $CH_2CH_2CH_3$; Et = CH_2CH_3; Me = CH_3

TABLE 2
Structures [8] of the BChl d

leagues have also shown that ALA used in biosynthesis of BChl *a* in *R. sphaeroides* is derived exclusively from glycine and succinyl-CoA and not via the glutamate (C-5) pathway; the methoxyl of the isocyclic ring methyl ester was derived from the C-2 carbon of glycine, presumably via methionine [19]. In contrast, the same group of workers demonstrated that ALA biosynthesis in *Chromatium vinosum* proceeds through the glutamate (C-5) pathway [20]. Full discussion of the glutamate pathway can be found in Chapters 1 and 5 of this Volume.

Dioxygen appears to be a requirement in the conversion of coproporphyrinogen III into protoporphyrinogen IX, in the subsequent aromatization to protoporphyrin IX in a number of Chl *a* synthesizing systems and as the source of the keto-oxygen atom in the isocyclic ring [21a,b]. In photosynthetic bacteria, which carry out anoxygenic photosynthesis under anaerobic circumstances it is unlikely that oxygen is available for these steps. Some oxygen dependent activity has been observed in the cytoplasmic fraction of *R. sphaeroides* and *Chromatium vinosum*, [22] but although the anaerobic transformation of coproporphyrinogen III into protoporphyrinogen IX in phototrophically grown *R. sphaeroides* has been observed [23], the requirements are complex and the electron acceptor has not been identified.

With the exception of the variability in ALA biosynthesis discussed above, there

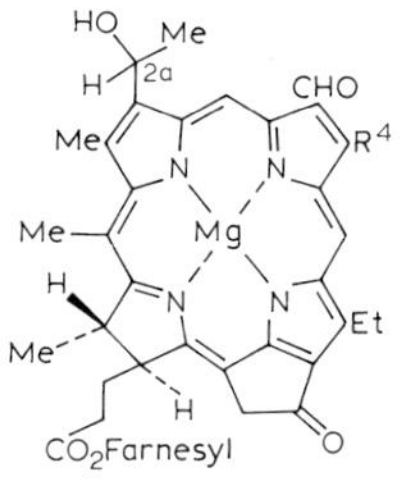

(8)

	R^4	Chirality at 2a posn.	Percentages* of (R) or (S) for each R^4 homologue (%)
a	i-Bu	S	98
b	i-Bu	R	2
c	n-Pr	S	60
d	n-Pr	R	40
e	Et	S	5
f	Et	R	95

i-Bu = $CH_2CH(CH_3)_2$; n-Pr = $CH_2CH_2CH_3$;
Et = CH_2CH_3; Me = CH_3
*Determined by transformation into the corresponding BChl c pheophorbides [10].

TABLE 3
Structures [9, 10] of the BChl e

has been little progress in BChl biosynthesis since early experiments using *Chlorella* and *R. sphaeroides* mutants. This work has been reviewed extensively [24]. Magnesium 2,4-divinylpheoporphyrin a_5 (11), several metal-free derivatives, pheophorbide *a*, 2-hydrated chlorophyllide *a* (12), the related bacteriochlorophyllide *a* (13) and bacteriochlorophyllide *a* (14) have all been isolated after excretion from the mutants [25]. Scheme 1 shows the currently accepted pathway from magnesium(II) protoporphyrin IX monomethyl ester (10), and this is almost entirely based on these mutant

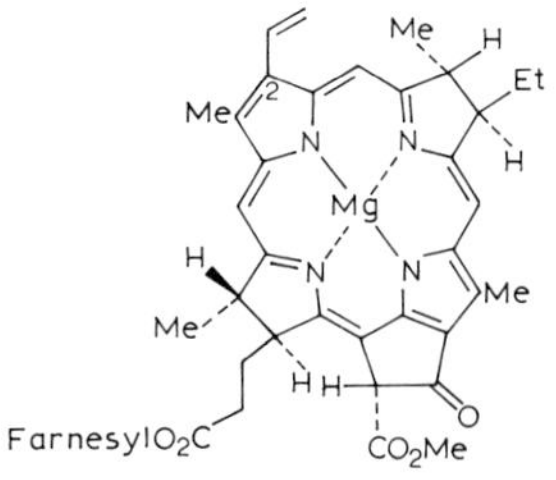

(9)

studies. A possible alternate pathway has also been proposed [26] on account of the fact that a *R. sphaeroides* mutant has been shown to excrete a protein-pigment containing 2-deacetyl-2-vinylbacteriopheophorbide *a* (15); this suggests that reduction of chlorin to tetrahydroporphyrin can take place before, or after, hydration of the 2-vinyl in chlorophyllide *a* to 2-(1-hydroxyethyl). It should be mentioned here that BChl *g* (9) possesses a 2-vinyl in place of the 2-acetyl found in BChl *a*, and thus would, by necessity, require 2-vinyl precursors similar to (15) for its biosynthesis. Much of what is shown in Scheme 1 has already been discussed in this Volume (Chapter 5) because it relates to the biosynthesis of Chl *a*. The reason for the apparent lack of activity in the past two decades, and particularly involving feeding of labeled substrates to enzyme systems, relates to the difficulty in synthesis of radio-

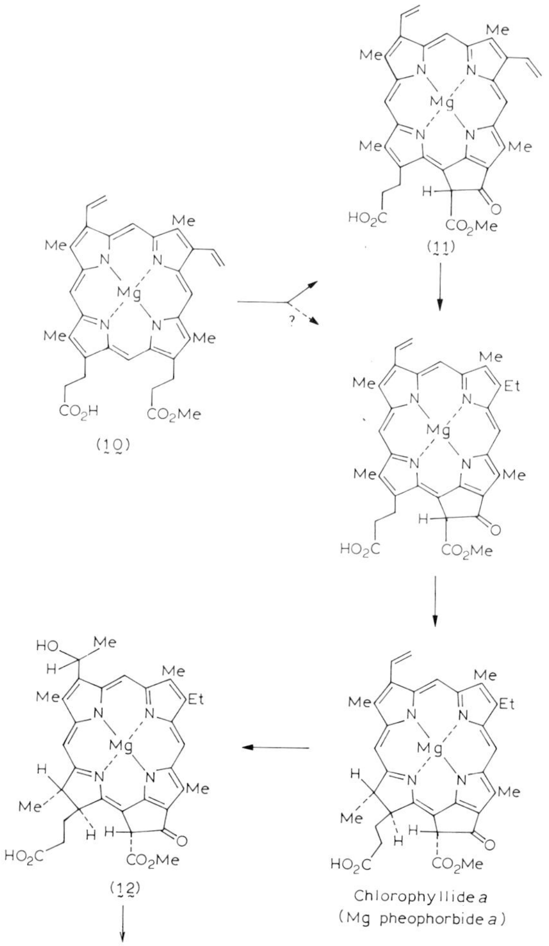

Scheme 1. Currently accepted biosynthetic pathway from magnesium(II) protoporphyrin IX monomethyl ester (10) to BChl *a* (3).

chemically labeled advanced metabolites, as well as problems in obtaining enzymically active systems to which the substrates could be fed. In addition, pigment synthesis has been shown to vary inversely with light intensity, and to be highly sensitive to the presence of oxygen [27].

Emery and Akhtar [28] have studied incorporation of (2R,3R)-ALA-2,3-T_2-2,3-$^{14}C_2$ into BChl *a* in *R. sphaeroides* and have shown that the 4-ethyl group is formed by addition of hydrogen to the si-face of the corresponding double bond, resulting in an overall, formal, *trans*-addition. Facile photoreduction reactions of vinyl-chlorins [29] and porphyrins [30] have demonstrated that formal vinyl reduction to ethyl can be accomplished by initial reduction of the *pyrrole subunit double bond* to give (16) (Scheme 2), followed by double bond migration to give the ethylidene derivative (17). Further facile bond migration results in formation of the ethyl-substituted unsaturated compound (18); the whole sequence of transformations proceeds in an energetically favourable manner (Scheme 2, Path A). These observations can also be applied as models for the biosynthetic formation of the ring B ethylidene function in BChl *b* (4); thus, Scheme 2 (Path A) can provide an effective alternative to the generally accepted, but thermodynamically uphill alternative (Scheme 2; Path B) involving initial direct reduction of the 4-vinyl group [to give (18)], followed by migration of the ring double bond out of conjugation to give (17). On the basis of observations on photoisomerism reactions of the bacteriopheophytin derived from the BChl g (9) of *Heliobacterium chlorum*, Michalski et al. [17]. have also suggested that all Chls are biosynthesized from a common intermediate bearing an ethylidene group, as is present in BChl b (4) and BChl g (9).

The biosynthetic steps involving addition of the 7-ester have been investigated in some detail. For example, phytol formation in *R. sphaeroides* has been shown, from feeding experiments, to be closely coupled to the biosynthesis of the chromophore [31]. It has also been determined that the linking (i.e. non-carbonyl) oxygen atoms of both the 7-phytyl in BChl a_{phy} and 7-geranylgeranyl (in BChl a_{gg}) esters are derived from the carboxylate oxygen originally present in the precursor ALA [32a,b, 33]; the same results were obtained for the methoxyl oxygen of the 10-CO_2Me group, leading to the conclusion that both of the ester bonds in BChl a are produced by carboxylate attack upon an activated isoprenyl (for the 7-side chain) or methyl (for the 10-side chain) moiety.

The biosynthesis of BChl b (4) presumably proceeds similarly to that of BChl a (3) as far as the 2-(1-hydroxyethyl)-chlorophyllide a [compound (12), Scheme 1]. Then, reduction of the ring B double bond (cf. Scheme 2) would afford the corresponding 2-(1-hydroxyethyl)-4-vinylbacteriochlorin which would suffer double bond migration (Scheme 2) to give the ethylidene derivative. Oxidation of the 2-(1-hydroxyethyl) to 2-acetyl, followed by esterification of the 7-side chain would then give BChl b (4).

BChl g (9) biosynthesis has not yet been studied. One might anticipate that its biosynthesis follows that of BChl a (3) as far as chlorophyllide a (Scheme 1), and that the branch point involves reduction of the chlorophyllide a ring B double bond to give a vinyl bacteriochlorin [cf. compound (16), Scheme 2]. Migration of the double bond would then give the ethylidene compound analogous to (17), and the remaining steps to (9) would parallel those in Scheme 1 to BChl a, or as discussed in the preceding paragraph for BChl b biosynthesis. Indeed, compound (15), isolated from a *R.*

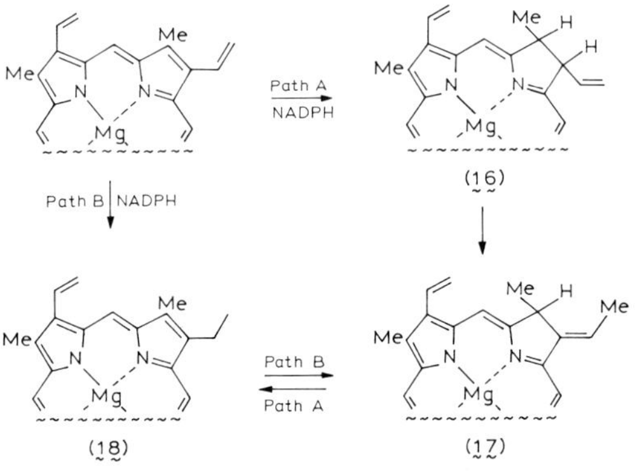

Scheme 2. Pathways for vinyl reduction in Chl and BChl biosynthesis, and possible route to formation of ring B ethylidene substituents in BChl b (4) and BChl g (9) biosynthesis.

sphaeroides mutant [26] might be regarded as a demetallated further reduction product of the ethylidene precursor of BChl *g*.

3.2 Bacteriochlorophylls c, d, e, f

Carbon-14 labeled glycine and glutamate were used to show that ALA is biosynthesized almost exclusively from glutamate in *C. limicola* under conditions favoring ALA accumulation [18]. On the other hand, and very surprisingly, neither glycine nor glutamate were the source of ALA in *Chloroflexus aurantiacus* [18]; However, very recent labeling evidence has shown that chloroflexus does use the glutamate pathway [42]. Two separate carbon-13 NMR studies of BChl *c* biosynthesis have also shown the source of the ALA-derived macrocyclic carbons to be from glutamate through the C-5 pathway. In the first study, Oh-Hama and colleagues [34] fed [1-^{13}C]-(L)-glutamate to *Prosthecochloris aestuarii* and showed that eight macrocyclic carbon atoms were enriched as expected. Feeding [2-^{13}C]-glycine under the same conditions resulted in no enrichment whatsoever. In the second study, [35] feeding of [1-^{13}C]-(L)-glutamate to a strain of *C.vibrioforme* forma *thiosulphatophilum* (normally a producer of BChl *d*, which has adapted to production of BChl *c* [36]) confirmed the results of Oh-Hama et al. that the Chlorobiaceae follow the C-5 pathway, but was in conflict with the glycine results. Smith and Huster [35] clearly showed that glycine is very effectively incorporated at the termini of the 4- and 5- substituents, as well as into the methionine-derived δ-*meso*- methyl. The glycine is presumably being initally transformed efficiently via methyl-tetrahydrofolate into methionine. The difference in the conclusions of these two studies are probably not associated with the different strains of bacteria used, but are more likely due to the fact that Oh-Hama and colleagues used short incubation times with glycine, whereas Smith and Huster incubated for several days. Onset of 4- and 5-ethyl biosynthetic homologation has been shown to be dependent upon external environmental factors, such as availability of light, and alkylation appears to be induced in old cultures by lack of light due to reduced transmission through highly populated medium [36].

Carbon-13 labeled methionine, and NMR detection, have been used to show that all of the 'extra' carbon atoms [structure (19)] in the BChl *c* are derived from methionine [11]. In the same study, small but significant incorporations of synthetic carbon-

14 labeled porphobilinogen (20) were also demonstrated. Not much is known with certainty about the latter stages in the biosynthesis of the BChls c, d, and e, so much of what follows should be regarded as conjecture rather than substantiated fact. Early chemical inhibition studies were carried out by Russian workers [37], using *C. thiosulphatophilum* and *Chloropseudomonas ethylicum* (now renamed as *Prosthecochloris aestuarii*); they concluded that the early stages in BChl c biosynthesis follow those of normal Chls, and that the process branches from this after formation of magnesium(II) protoporphyrin IX. Inhibition studies of Richards and Rapoport [38a,b] led to the conclusion that the pathway proceeded through magnesium(II) protoporphyrin IX monomethyl ester (10), and also through BChl a, or one of its immediate precursors. However, these results could be reinterpreted in terms of total inhibition of BChl c biosynthesis very early in the pathway such that the metabolites being excreted into the medium were simply intermediates in normal BChl a biosynthesis. In particular, alkylation of the terminal position of the unactivated ethyl side chain at the 4-position of BChl a is unlikely, and it seems much more feasible that either a propionate or a vinyl side chain (i.e. somewhere between uroporphyrinogen III and protoporphyrin IX) is first alkylated.

The propionate alkylation pathway is outlined in Scheme 3; thus, standard transformation of the propionate (21) into vinyl (22), followed by '4-vinyl' reduction (cf. Scheme 2) would give the 4-ethyl substituent (23) in a manner entirely analogous with production of the 4-ethyl in Chl a or BChl a biosynthesis. S-Adenosylmethionine-

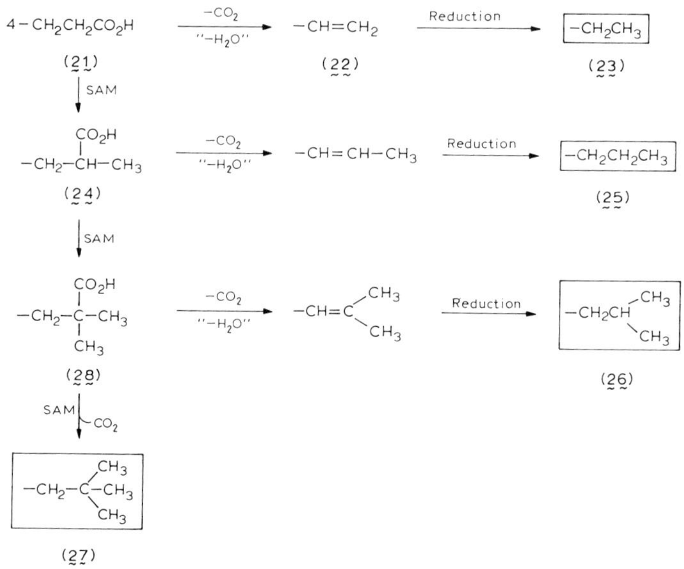

Scheme 3. Possible biosynthetic route for R^4 homologation in the BChl c, d, and e, (6–8) assuming a 4-propionic acid precursor.

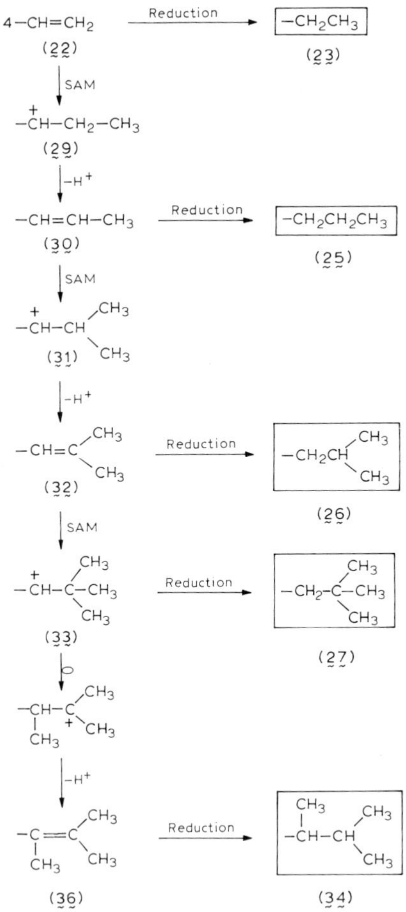

Scheme 4. Possible biosynthetic route for R[4] homologation in the BChl c, d, and e, (6–8) assuming a 4-vinyl precursor.

mediated methylation of the methylene adjacent of the propionate carbonyl, which would presumably be enzymatically activated, would lead to the methyl-substituted propionate side chain (24) which, as for the 4-ethyl homologue, could be vinylated and reduced to give the n-propyl homologue (25). Repetition of this same process of methylation, vinylation, and reduction would similarly give the 4-isobutyl derivative (26). Finally, for formation of the 4-neopentyl compound (27), it is necessary to postulate a rather cumbersome concomitant decarboxylation and methylation of (28). The possibly more likely vinyl alkylation pathway is shown in Scheme 4. Here, reduction of the 4-vinyl compound (22), as before, would furnish the 4-ethyl homologue (23). On the other hand, S-adenosylmethionine mediated methylation of the vinyl would yield the carbocation (29) which could deprotonate to the corresponding alkene (30) and then be reduced to (25). Further methylation of the alkene (30) would

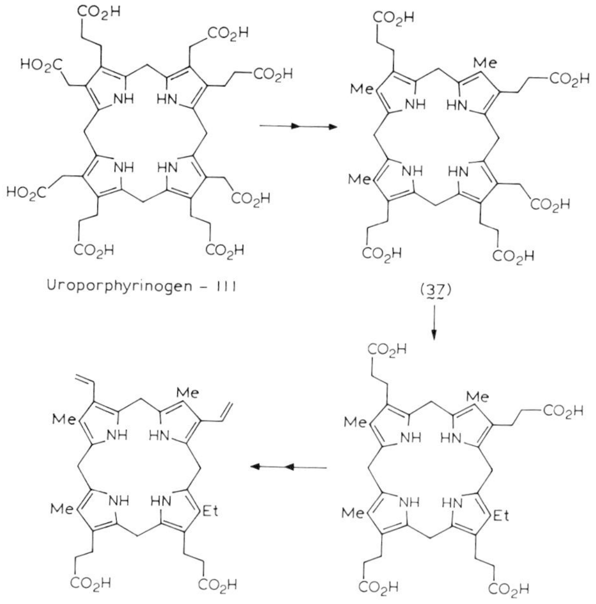

Scheme 5. Possible biosynthetic pathway for R^5 homologation in the BChl c, d, and e, (6–8) assuming intermediacy of the pentacarboxylic suburoporphyrin (37).

give the carbocation (31), which could again deprotonate to alkene (32) and be reduced to (26). Repetition of the same process would give the carbocation (33), which would need to be directly reduced to give the neopentyl homologue (27). This procedure seems less cumbersome than the decarboxylative methylation proposed in Scheme 3 for formation of (27); it is, of course, entirely possible that the carbocations (29) and (31) are also directly reduced without intermediacy of the alkenes (30) and (32). The pathways outlined in Scheme 4 are currently favored because of tentative identification [13] of the rearranged homologue (34) in small quantity from *C. vibrioforme* forma *thiosulphatophilum*; formation of (34) would have to result from rearrangement of the carbocation (33) to give (35), followed by deprotonation [giving (36)] and reduction, or simple one step carbocation reduction. Species such as (34) cannot be accounted for on the basis of the propionate pathway outlined in Scheme 3. It is possible that alkylation at the 5-position is the initial point of deviation from the Chl *a* biosynthetic pathway because activation of the 5a carbon in the form of an acetic acid side chain seems a good possibility [Compound (37); Scheme 5] and because the 5-acetic side chain in (37) is converted into methyl at the coproporphyrinogen-III formation step, [39a,b] and this is in itself several steps before the vinyl necessary for 4-side chain methylation is fashioned at protoporphyrinogen IX (38) (Scheme 6) or protoporphyrin-IX. A possible mechanism for transformation of acetate into ethyl is shown in Scheme 7; this mechanism is in agreement with the only experimental work [11] on the 5-position which, using labeling experiments with

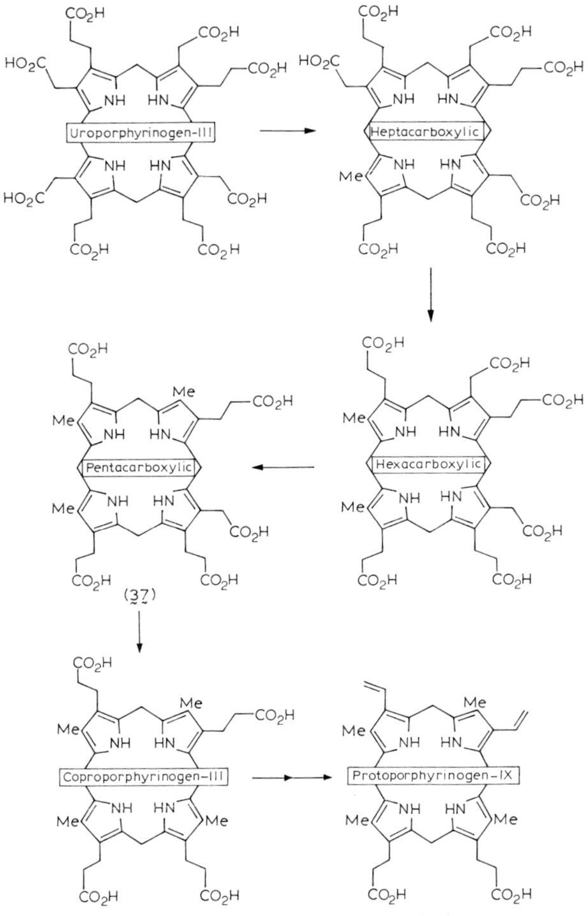

Scheme 6. Established heme biosynthetic pathway [39] between uroporphyrinogen III and protoporphyrinogen IX (38) showing the presence of pentacarboxylic compound (37).

$^{13}CD_3$ methionine, showed that the methyl group is transferred intact, and that no vinyl-type intermediates [e.g. (39)] are possible.

It has been possible to deduce some fundamental points about the biosynthesis of the BChl c, d, and e simply by consideration of structural facts. For example, a trend in all three series (Tables 1, 2, and 3) towards (R) absolute stereochemistry at the 2-(1-hydroxyethyl) when the R^4 substituent is small (e.g. ethyl), and towards (S) absolute stereochemistry when R^4 is a large group (e.g. isobutyl or neopentyl) has been interpreted [7] in terms of remote control of induced stereochemistry by the bulk of the R^4 substituent, possibly involving rotamers of the (presumed) 2-vinyl precursor (Figure 1). For such control to be possible, one must conclude that the R^4 substituent has been homologated (Scheme 3 or 4) prior to hydration of the 2-vinyl. No attempt

Scheme 7. Possible mechanism for the biosynthetic transformation of the 5-acetic side chain in (37) into 5-ethyl, mediated by S-adenosylmethionine (cf. Scheme 5).

has been made to isolate or purify enzymes involved in biosynthesis of the Chlorobiaceae pigments.

The problem of why the Chlorobiaceae use such a complex series of homologated antenna pigments has also been investigated. Optical spectra, in dichloromethane, of pure homologues show no R^4 side chain dependence on absorption maxima; however, in hexane [14] and in living cells there is a very great dependence on the size of

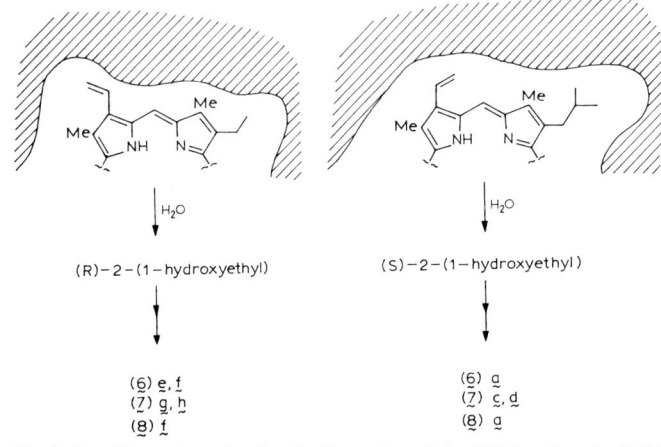

Fig. 1. Possible explanation for the formation of the BChl c, d, and e 2-(1-hydroxyethyl) groups in the (R) or (S) absolute configuration, based on rotation of the 2-vinyl group in a precursor.[7,14]

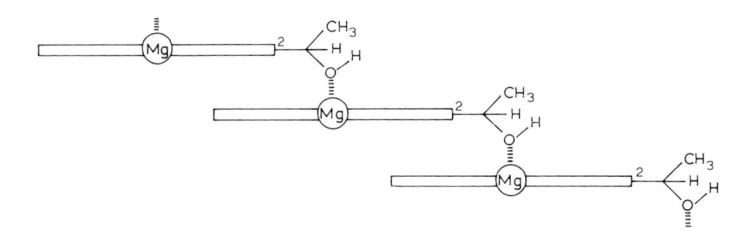

Fig. 2. Diagrammatic representation of the intermolecular interactions responsible for oligomeric stacked arrays in photosynthetic antenna systems of green and brown bacteria (Chlorobiaceae) [14]. An additional interaction involves the protein backbone (not shown).

the R^4 substituents, and concomitantly, upon the chirality of the 2-(1-hydroxyethyl).[8] This has led to the conclusion that the antenna system is composed of an overlapping BChl oligomer, supported by a protein backbone [14]. When the oligomer is small, the red-shift (monomer to oligomer) in the optical spectrum of the antenna system is small, and vice versa for large oligomers. The size of the antenna aggregate is believed [Smith, K.M., unpublished data] to be controlled indirectly by the size of the R^4 substituent, but directly by the chirality of the 2-(1-hydroxyethyl) (which is linked to size of R^4), because a major interaction holding the oligomer together involves the central magnesium ion of one BChl and the oxygen of the 2-(1-hydroxyethyl) of another (Figure 2). Thus, the bacteria are able to modify the absorption maxima of their antenna systems by inducing a methylating enzyme under conditions of subdued light. Alkylation in newly developing cells causes a red shift in the antenna system,

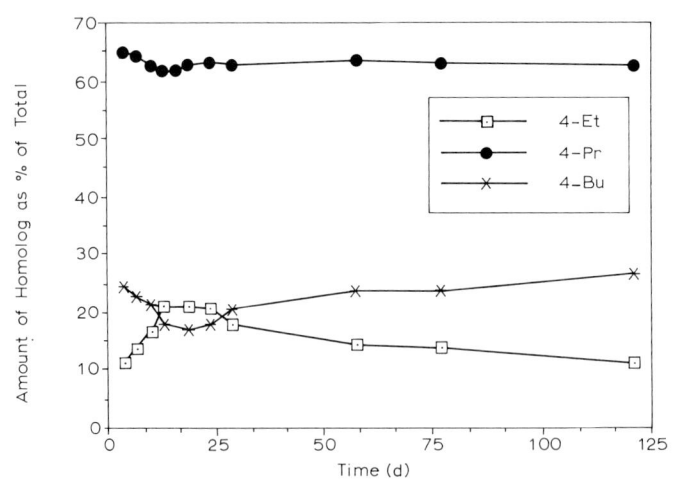

Fig. 3. Growth curves showing relative proportions of BChl c homologues produced, with time after innoculation into new medium, using the *C. vibrioforme* forma *thiosulfatophilum* (Davis) strain [40].

which makes more light available because of better penetration of low energy light through the medium, and because existing bacteria in the media are absorbing at higher energy.

This adaptation phenomenon takes place in colonies grown from single cells, and is reversible, as shown in Figure 3. Subculturing into new medium (transparent) results in an initial decrease in n-propyl and isobutyl containing pigments due to reduction (or complete termination) of methylation. With time, as the culture develops past its steady state (normally 5–10 days) reduced light conditions in the almost opaque medium trigger alkylation in newly developing cells which, at steady state, replace old ones which decompose and are re-assimilated into the medium; as time goes by (Figure 3) more and more alkylation takes place at the expense of R^4=ethyl pigments. The whole process can be reversed by inoculation into new (transparent)medium, and the bacteria re-adapt in newly developed cells.

The BChl d antenna system can change from a long wavelength absorption maximum in living cells at 714 nm for the R^4=ethyl pigment (B1–20 strain) to about 730 nm for the normal (*C. vibrioforme*) strain.[8] However, an even more remarkable transformation has been noted because this (BChl d producing) strain has been shown to convert to production of the BChl c; such an event had been noted earlier using only optical spectroscopy [41], but was subsequently fully confirmed by complete identification of the pigments. Thus, the bacteria turn to *meso* methylation rather than 4- or 5-side chain homologation as a means of further adapting to subdued light conditions. The functional effect of the transformation to a new (BChl c producing) strain (named *C. vibrioforme* Davis) is to move the absorption maximum in living cells from 730 nm to about 750 nm [36]. The original *C. vibrioforme* and the new Davis strain have been shown to be morphologically identical with each other (Pfennig, N., personal communication). Whether or not the δ-*meso* methylation by the BChl d to produce the BChl c strain is a mutation, or simply an adaptation to external environmental conditions (as is the case in the 4- and 5-side chain alkylations, Figure 3) has not yet been established. Much remains to be done, in particular, with enzymology in these systems.

Acknowledgment

This work was supported by a grant from the National Science Foundation (CHE-86-19034).

References

1 Scheer, H., Svec, W.A., Cope, B.T., Studier, M.H., Scott, R.G. and Katz, J.J. (1974) J. Am. Chem. Soc. 96, 3714–3716.
2 Pfennig, N. (1977) Ann. Rev. Microbiol. 31, 275–290.
3a Purdie, J.W. and Holt, A.S. (1965) Can. J. Chem., 1965, 43, 3347–3353.

3b Holt, A.S., Purdie, J.W. and Wasley, J.W.F. (1966) Can. J. Chem. 44, 88–93.
3c Holt, A.S. and Hughes, D.W. (1961) J. Am. Chem. Soc. 83, 499–500.
3d Hughes, D.W. and Holt, A.S. (1962) Can. J. Chem. 40, 171–176.
4   Brockmann, Jr., H. (1976) Philos. Trans. Roy. Soc. London, Ser. B 273, 277–285.
4   Brockmann, Jr., H. and Tacke-Karimdadian, R. (1979) Liebigs Ann. Chem. 419–430. See also ref. 4.
5   Brockmann, Jr., H., Gloe, A., Risch, N. and Trowitzsch, W. (1976) Liebigs Ann. Chem. 566–577. See also ref. 4.
7   Smith, K.M., Craig, G.W., Kehres, L.A. and Pfennig, N. (1983) J. Chromatog. 281, 209–223.
8   Smith, K.M. and Goff, D.A. (1985) J. Chem. Soc., Perkin Trans. 1 1099–1113.
9   Gloe, A., Pfennig, N., Brockmann, Jr., H. and Trowitzsch, W. (1975) Arch. Microbiol. 102, 103–109; Risch, N. and Brockmann, Jr., H. (1976) Liebigs Ann. Chem. 578–583. See also ref. 4.
10 Simpson, D.J. and Smith, K.M. (1988) J. Am. Chem. Soc. 110, 1753–1758.
11 Kenner, G.W., Rimmer, J., Smith, K.M. and Unsworth, J.F. (1978) J. Chem. Soc., Perkin Trans. 1 845–852.
12 Caple, M.B., Chow, H.C. and Strouse, C.E. (1978) J. Biol. Chem. 253, 6730–6737.
13 Goff, D.A. (1984) Ph.D. Dissertation, University of California, Davis.
14 Smith, K.M., Kehres, L.A. and Fajer, J. (1983) J. Am. Chem. Soc. 105, 1387–1389.
15 Gloe, A. and Risch, N. (1978) Arch. Microbiol. 118, 153–156; Risch, N., Brockmann, Jr., H. and Gloe, A. (1979) Liebigs Ann. Chem. 408–418.
16 Brockmann, Jr., H. and Lipinski, A. (1983) Arch. Microbiol. 136, 17–19; Gest, H. and Favinger, J.L. (1983) Arch. Microbiol. 36, 11–16.
17 Michalski, T.J., Hunt, J.E., Bowman, M.K., Smith, U., Bardeen, K., Gest, H., Norris, J.R. and Katz, J.J. (1987) Proc. Natl. Acad. Sci. USA 84, 2570–2574.
18 Andersen, T., Briseid, T., Nesbakken, T., Ormerod, J., Sirevag, R. and Thorud, M. (1983) FEMS Microbiol. Lett. 19, 303–306.
19 Oh-Hama, T., Seto, H. and Miyachi, S. (1985) Arch. Biochem. Biophys. 237, 72–79.
20 Oh-Hama, T., Seto, H. and Miyachi, S. (1986) Arch. Biochem. Biophys. 246, 192–198.
21a Walker, C.J., Mansfield, K.E., Smith, K.M. and Castelfranco, P.A. (1991) Biochem. J. (in press)
21b Walker, C.J., Mansfield, K.E., Rezzano, I.N., Hanamoto, C.M., Smith, K.M. and Castelfranco, P.A. (1988) Biochem. J. 255, 685–692.
22 Mori, M. and Sano, S. (1972) Biochim. Biophys. Acta 264, 252–258.
23 Tait, G.H. (1972) Biochem. J. 128, 1159–1169.
24 Jones, O.T.G. (1978) in: The Photosynthetic Bacteria, R.K. Clayton and W.R. Sistrom Eds. Plenum, New York, pp. 751–777.
25 Sistrom, W.R., Griffith, M. and Stanier, R.Y. (1956) J. Cell. Comp. Physiol. 48, 459–464; Richards, W.R. and Lascelles, J. (1969) Biochemistry 3, 3473–3482.
26 Pudek, M.R. and Richards, W.R. (1975) Biochemistry 14, 3132–3137.
27 Cohen-Bazire, G., Sistrom, W.R. and Stanier, R.Y. (1957) J. Comp. Cell Physiol. 49, 25–68.
28 Emery, V.C. and Akhtar, M. (1985) J. Chem. Soc., Chem. Commun. 1646–1647.
29 Simpson, D.J. and Smith, K.M. (1988) J. Am. Chem. Soc. 110, 2854–2861.
30 Smith, K.M. and Simpson, D.J. (1987) J. Chem. Soc., Chem. Commun. 613–614.
31 Brown, A.E. and Lascelles, J. (1972) Plant Physiol. 50, 747–749.
32a Ajaz, A.A., Corina, D.L. and Akhtar, M. (1980) J. Chem. Soc., Chem. Commun. 511–513.
32b Akhtar, M., Ajaz, A.A. and Corina, D.L. (1984) Biochem. J. 224, 187–194.
32c Emery, V.C. and Akhtar, M. (1985) J. Chem. Soc., Chem. Commun. 600–601.
33 Emery, V.C. and Akhtar, M. (1987) Biochemistry 26, 1200–1208.
34 Oh-Hama, T., Seto, H. and Miyachi, S. (1986) Eur. J. Biochem. 159, 189–194.
35 Smith, K.M. and Huster, M.S. (1987) J. Chem. Soc., Chem. Commun. 14–16; Huster, M.S. and Smith, K.M. (1990) Biochemistry 29, 4348–4355.

36 Smith, K.M. and Bobe, F.W. (1987) J. Chem. Soc., Chem. Commun. 276–277; Bobe, F.W., Pfennig, N., Swanson, K.L. and Smith, K.M. (1990) Biochemistry 29, 4340–4348.
37 Kondrat'eva, E.N. and Uspenskaya, V.E. (1968) Nauch. Dokl. Vyssh. Shk., Biol. Nauki, 97–110; CA 69:41862a.
38a Richards, W.R. and Rapoport, H. (1968) Biochemistry 6, 3830–3840.
38b Richards, W.R. and Rapoport, H. (1967) Biochemistry 5, 1079–1089.
39a Jackson, A.H., Sancovich, H.A. and Ferramola de Sancovich, A.M. (1980) Bioorg. Chem. 9, 71–120.
39b Jackson, A.H., Ferramola, A.M., Sancovich, H.A., Evans, N., Games, D.E., Matlin, S.A., Ryder, D.J. and Smith, S.G. (1976) Ann. Clin. Res., Suppl. 8, 64–69.
40 Huster, M.S. (1988) Ph.D. Dissertation, University of California, Davis.
41 Broch-Due, M. and Ormerod, J.G. (1978) FEMS Microbiol. Lett.3, 305–308.
42 Swanson, K.L. and Smith, K.M. (1990) J. Chem. Soc. Chem. Commun. 1696–1697.

CHAPTER 7

The genes of tetrapyrrole biosynthesis

PETER M. JORDAN AND BOB I.A. MGBEJE

School of Biological Sciences, Queen Mary and Westfield College, University of London, Mile End Road, London E1 4NS, U.K.

1. Genes of haem biosynthesis

1.1. Mapping of haem biosynthesis genes

The techniques used for the genetic analysis of *hem* mutants, and indeed mutants of other biosynthetic pathways, are very dependent on the organism, however, the principles involved and the conclusions drawn are very much the same. Wild-type strains do not accumulate porphyrins under normal growth conditions due to a very efficient regulatory mechanism, involving haem as the main effector [1], however, intermediate porphyrins accumulate when the synthesis of haem is deficient. Usually

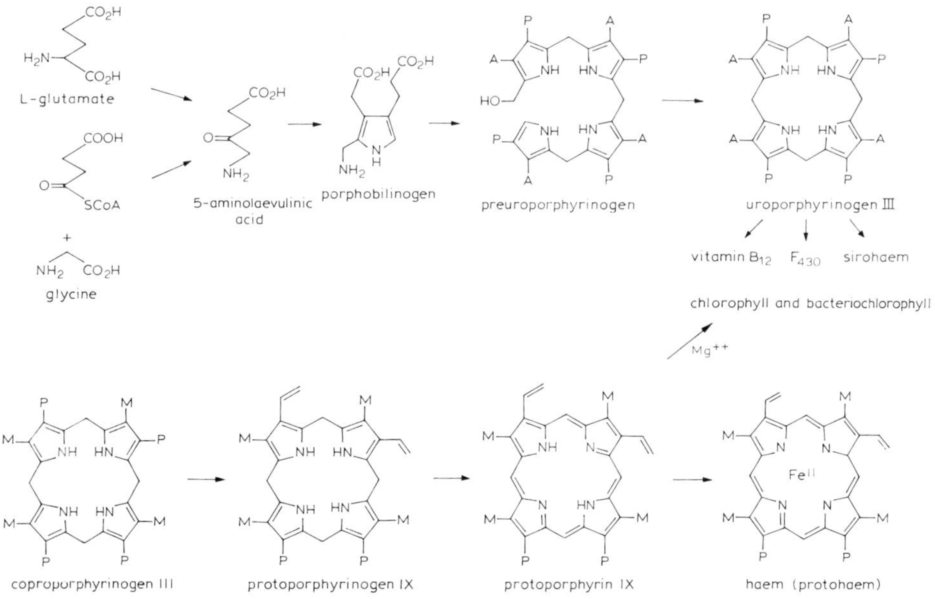

Scheme 1. The haem biosynthesis pathway

only the intermediate which precedes the mutation accumulates and hence the pattern of accumulation often reflects the site of mutation. Scheme 1 shows the haem biosynthesis pathway.

Haem-deficient mutants have, for obvious reasons, been studied most thoroughly in bacteria. With the availability of mutants and using the techniques of phage P1 mediated transduction [2] or conjugation [3] it has been possible to map the loci of the haem biosynthetic genes on several bacterial chromosomes. Mapping of haem genes in the Enterobacteriaceae, *Escherichia coli* and *Salmonella typhimurium* LT2, have lagged behind the genetic study of the other markers in these two species. This was mainly due to the impermeability of haem in these bacteria, which made the study of *hem* mutants difficult [4–6]. Haem-deficient mutants of bacteria are partially resistant to aminoglycoside antibiotics and this property was routinely used for their isolation [6]. The mechanism of resistance is not yet well established, but a diminished uptake and metabolism of the drug has been suggested [7].

1.1.1. The hem genes in Escherichia coli
The first haem-deficient mutant of *E. coli* was isolated by Beljanki and Beljanki [8] using a streptomycin selection procedure. Several years later, new haem-deficient mutants of *E. coli* were isolated and the first *hem* locus was mapped [9,10]. Since then, the genes of virtually the complete pathway have been mapped in *E. coli* [5,6,10–12]. The use of haem-permeable mutants in *E. coli* greatly facilitated the studies [12].

There was originally some controversy surrounding the locus of the *hemA* gene, coding for the synthesis of 5-aminolaevulinic acid (5-ALA) on the *E. coli* chromosome. The first 5-aminolaevulinic acid-requiring mutant was described by Wulff [13] who also showed a preliminary map location for the '5-aminolaevulinic acid synthase' gene. Subsequently, Sasarman et al. [10] mapped the *hemA* gene in *E. coli* K-12 and found a locus different from that reported by Wulff. Thus, the two mutants seemed to differ from each other [6]. Later, Powell et al. [12] also described a mutation different from that of Sasarman et al. [10] which was termed as *popC*. Wulff's *hemA$^-$* mutation was reported to map in the *pro-thr-leu* region. At that time it was thought [6] that the mutant of Wulff was likely to be a recurrence of that described by Powell et al. [12] although the information available was insufficient to define the precise position of the gene on the linkage map [12]. The *hemA$^-$* gene of Sasarman et al., [10] giving rise to 5-aminolaevulinic acid-requiring mutants was, in contrast, co-transducible with a *trp* gene. A 5-aminolaevulinic acid-requiring mutant which is co-transducible with the *trp* gene had also been isolated by Haddock and Schraider [14]. It has been confirmed that this mutant is also co-transducible with the *trp* locus [12]. It was thus clear, almost twenty years ago, that there are two independent genes in *E. coli* which can be mutated to give a specific requirement for 5-aminolaevulinic acid.

These findings are now easily explained by the recent discovery [15,16] that 5-aminolaevulinic acid synthesis in *E. coli* occurs via the C_5 pathway involving two dedicated reaction steps, glutamyl-tRNA reductase and glutamate 1-semialdehyde aminotransferase (see also Chapters 1 and 5). Thus the two 'hemA$^-$' mutations already described can be explained as mutations in these two structural genes. Earlier indirect support for a C_5 pathway for 5-aminolaevulinic acid synthesis in *E. coli*, was the failure to obtain 5-aminolaevulinic acid from succinate in *E. coli* [17,18] even if the *hemA* gene was present in more than one copy per cell [18]. The exclusive operation of the C_5 pathway in *E. coli* is also consistent with the placement of this species within the γ-subgroup of purple bacteria. Another member of the subgroup, *Chromatium vinosum*, was found to form 5-aminolaevulinic acid exclusively via the C_5 5-aminolaevulinic acid route [19,20]. It is of interest to note that two genes have also been identified for 5-aminolaevulinic acid synthesis in another member of the Enterobacteriaceae, *S. typhimurium* [21].

There was also some controversy regarding the positioning of the *hemC* and *hemD* loci on the *E. coli* chromosome map. The *hemD* locus had been mapped at minute 83 on the chromosome of *E. coli* K-12 between *ilv* and *cyaA* with *hemC* on the other side of *cyaA* [11]. From the recent work of Thomas and Jordan [22], Thomas (1986) [23], Jordan et al. [24], Sasarman et al. [25], the precise positioning of the *hemC* and *hemD* genes with respect to *cyaA* are now known to be in the order *cyaA*, *hemC*, *hemD*. The loci of the *hemC* and *hemD* genes which are neither linked nor in the correct sequence on the *E. coli* chromosome map [11,26] have now been officially repositioned [27]. The *hemC* and *hemD* genes appear to be under the control of a major promoter in a *hem* operon together with at least three other genes [28]. The position of the *hem* genes on the *E. coli* chromosome are summarized in Table 1.

1.1.2. The hem genes in S. typhimurium

A similar less complete map has also been obtained for the *hem* loci in *S. typhimurium* [29,7]. In this bacterium mutations have been described for the five genes *hemA*, *hemB*, *hemC*, *hemD* and *hemE*. As in *E. coli*, the loci are widely distributed over the

TABLE 1
Location of *hem* genes on the chromosomal map of *E. coli* K12

Locus	Mins	Enzyme
hemA	27	Glutamyl-tRNA reductase
hemL	–	Glutamate 1-semialdehyde aminotransferase
hemB	8	5-Aminolaevulinic acid dehydratase
hemC	86	Porphobilinogen deaminase
hemD	86	Uroporphyrinogen III synthase
hemE	90	Uroporphyrinogen decarboxylase
hemF	17	Coproporphyrinogen oxidase
hemG	87	Protoporphyrinogen oxidase
hemH	11	Ferrochelatase

Map positions may have been adjusted in line with current experimental literature [27].

TABLE 2
The positions of *hem* genes on the *Salmonella typhimurium* chromosomal map

Locus	Mins	Enzyme
hemA	35	Glutamyl-tRNA reductase
hemL	5	Glutamate 1 semialdehyde aminotransferase
hemB	8	5-aminolaevulinic acid dehydratase
hemC	84	Porphobilinogen deminase
hemD	84	Uroporphyrinogen III synthase
hemE	88	Uroporphyrinogen decarboxylase
hemF	18	Coproporphyrinogen oxidase
hemG	85	Protoporphyrinogen oxidase
hemH	12	Ferrochelatase

Only *hemA* through *hemE* are currently known in *S. typhimurium*. The positions of genes *hemF* through *hemH* are inferred from those reported for *E. coli* [21].

chromosome (Table 2) except for the *hemC* and *hemD* loci which are closely linked in both bacteria. Recently, an additional gene, *hemL*, has been identified for 5-aminolaevulinic acid synthesis in *S. typhimurium* [21]. It is thought that the *popC* gene of *E. coli* [12] is equivalent to *hemL* in *S. typhimurium* based on their similar map position and phenotypes [30]. The previously identified *hemA* gene [7] maps at 35 min while the *hemL* gene maps at 5 min on the *S. typhimurium* genetic map [21]. It is thus certain that the considerations discussed above for the existence of at least two genes for 5-aminolaevulinic acid synthesis in *E. coli* applies to *S. typhimurium* since the latter also synthesizes 5-aminolaevulinic acid from glutamate [30].

1.1.3. The hem genes in Staphylococcus aureus

Distribution of the *hem* loci in *Staphylococcus aureus* appears to be of a different pattern to that observed in *E. coli* and *S. typhimurium*. Investigations by Tien and White, [31] produced a set of protohaem mutants. These mutants were tested for their ability to grow rapidly in the presence of 5-aminolaevulinic acid, porphobilinogen, or protohaem in nutrient agar (the porphyrinogens are unstable in air and could not be used to test for growth response). Based on the test, the mutants were separated into five classes representing five enzymatic lesions in the pathway. Genetic analysis of the mutants showed that the genes specifying these enzymes are tightly linked, co-transducible and arranged in sequence as are the reactions in the biosynthetic pathway.

1.1.4. The hem genes in Bacillus subtilis

The first *hem* locus to be mapped on the *Bacillus subtilis* chromosome was the *hemA* gene [32]. Subsequently, using two- and three-factor transduction crosses, the same group mapped the *hemB* and *hemC* loci [33] and *hemE, hemF,* and *hemG* loci [34] on the *B. subtilis* chromosome. The *hemA, hemB* and *hemC* loci are closely linked to each other and located between the *leu-1* and *pheA1* loci arranged according to their order in the biosynthetic pathway. The *hemE, hemF* and *hemG*, on the other hand are located near the *argC4* and *metD1* loci, rather far from the first three genes [34]. It was not possible to determine the order of the last three loci because the porphyrin auxotrophs could utilize uniformly only the last product, haem, from the medium. It is interesting, considering the data of Hartford [35] on the bidirectional chromosome replication in *B. subtilis*, that the two groups of *hem* loci situated on each arm of the chromosome, replicate at about the same time [34].

1.1.5. The hem genes in Rhodobacter sphaeroides
Because of its ability to adapt from aerobic growth to photosynthetic semi-anaerobic growth *R. sphaeroides* (previously *Rhodopseudomonas sphaeroides*) has been used extensively for studies on the regulation of tetrapyrrole biosynthesis especially with respect to bacteriochlorophyll biosynthesis. However, because of difficulties in cloning, transformation and sequencing the exceptionally high GC rich DNA, studies on this important bacterium have lagged behind until relatively recently. Information about genes from this bacterium are included in the molecular biology section p. 265.

1.1.6. The hem genes in yeast
Haem-deficient mutants of yeast, *Saccharomyces cerevisiae* have been isolated and studied by several groups [36–38]. Genetic and biochemical analysis was carried out for seven defined mutants which were deficient in six of the eight enzymic steps of the haem biosynthetic pathway. The six mutants were blocked at 5-aminolaevulinic acid synthase, porphobilinogen synthase, porphobilinogen deaminase, uroporphyrinogen decarboxylase, coproporphyrinogen III oxidase, and ferrochelatase. The seventh mutant had the same phenotype as the mutant deficient in ferrochelatase activity [38], however, it possessed a normal ferrochelatase activity when measured in vitro. The mutant was therefore assumed to be deficient in protoporphyrinogen oxidase activity and/or the capability to carry out the reduction of iron. A similar assumption was made in the mapping of the *hemG* locus encoding protoporphyrinogen oxidase in *E. coli* K-12 [39]. The auto-oxidation of protoporphyrinogen IX to protoporphyrin IX during extraction of the porphyrins might explain the apparent non-accumulation of prototoporphyrinogen IX and the accumulation of protoporphyrin IX [39]. There is no report of the isolation of a mutant defective in uroporphyrinogen III cosynthetase in yeast.

Although many haem mutants of yeast have been isolated, the genetic features of haem synthesis in yeast are only just becoming clear. It has been reported that there is no apparent link between the *hem* loci for 5-aminolaevulinic acid synthase, porphobilinogen deaminase and coproporphyrinogen III oxidase in yeast [37]. Urban-Grimal and Labbe-Bois [38] mapped the first *hem* locus in yeast. They showed that the allele controlling 5-aminolaevulinic acid dehydratase activity was linked to *leu1*, a centromere marker on chromosome VII. Recently, the *hem13* locus, encoding coproporphyrinogen oxidase, was located on the right arm of chromosome IV, at a map distance of 23 centrimorgans (cM) from *trpI* [40]. Subsequently, it was established by the same group [41] that the *hem12* locus, encoding uroporphyrinogen decarboxylase, was linked to the *trpI* locus (centromere-linked marker of chromosome IV) with a map distance of 35 cM, and to *hem13* with an average distance of 10 cM.

1.1.7. The hem genes in mammalian species
The *hem* genes in mammals have been mapped mainly on the human chromosome using classical family linkage studies, Southern blot and in situ hybridization techniques. The first report of the mapping of *hem* genes on the human chromosome was by Meisler et al. [42]. They assigned the gene for human porphobilinogen deaminase to chromosome 11, using hybrids derived from human fibroblast and mouse RAG, LM/TK, or A9 cells and using isoelectric focussing for detection of human porphobilinogen deaminase. The result was confirmed by Wang et al. [43] who further localized the gene to the region 11q23 → 11qter. Genes for other enzymes of the haem biosynthetic pathway examined so far are localized on different chromosomes; 5-aminolaevulinic acid synthase has been localized to 3pter → 3q13.2 [44], 5-aminolaevulinic acid dehydratase to 9q34 [45], uroporphyrinogen decarboxylase to 1p34 [46], and coproporphyrinogen oxidase to chromosome 9 [47]. The loci for the human cosynthase has been mapped at 10q25.2 → q26.3 [48]. In mice, the locus encoding 5-aminolaevulinic acid dehydratase (*lv*), is located on chromosome 4 [49]. The genes specifying protoporphyrinogen oxidase and the ferrochelatase have not yet been mapped on human chromosomes.

In humans, the deficiency of haem synthesis enzymes lead to diseases collectively termed the porphyrias. These diseases are typified by the accumulation of intermediates prior to the enzymic lesion. A detailed survey of the porphyrias is outside the scope of this chapter but most readable accounts are given elsewhere [50,51]. The porphyrias of most importance are summarized in Table 3. Those porphyrias with a lesion either in 5-aminolaevulinic acid dehydratase or porphobilinogen deaminase cause the 'acute' porphyrias characterised by episodes of acute abdominal pain and associated with the accumulation of 5-aminolaevulinic acid alone or with porphobilinogen. There is much evidence to suggest that 5-aminolaevulinic acid, acting as a 4-aminobutyric acid analogue, is the causative agent in the symptoms which affect the central nervous system [52]. The remaining porphyrias, the 'non-acute porphyrias' are characterized by photosensitivity, sometimes of an extreme kind, with severe scarring and dermatological involvement. These effects are caused by the over production of porphyrinogens which are then oxidized to photosensitizing porphyrins.

As a general rule, the 'acute' porphyrias are caused by lesions in the enzymes responsible for the biosynthesis of uroporphyrinogen III from 5-aminolaevulinic acid and the 'non acute' porphyrias by lesions in enzymes involved in the further transformation of uroporphyrinogen III into haem. However, in variegate porphyria the

TABLE 3
Enzymic lesions in the haem biosynthesis pathway and the associated human disease

Enzyme affected	Disease
5-Aminolaevulinic acid synthase	various unassigned anaemias
5-Aminolaevulinic acid dehydratase	5-aminolaevulinate porphyria (Doss porphyria)
Porphobilinogen deaminase	acute intermittent porphyria (AIP)
Uroporphyrinogen III cosynthase	congenital erythropoietic porphyria (CEP)
Uroporphyrinogen decarboxylase	hepatoerythroporphyria (HEP)
	porphyria cutanea tarda (PCT)
Coproporphyrinogen oxidase	coproporphyria
Protoporphyrinogen oxidase	variegate porphyria (VP)
Ferrochelatase	erythropoietic protoporphyria

low levels of protoporphyrinogen oxidase causes the accumulation of protoporphyrinogen which leads to the inhibition of porphobilinogen deaminase, the accumulation of 5-aminolaevulinic acid and porphobilinogen and the development of 'acute' symptoms [53].

The study of porphyrias has greatly assisted our understanding of the regulation of haem biosynthesis and played an important role in the identification and sequencing of genes encoding the enzymes of the pathway.

1.2. Molecular biology of haem biosynthesis

During the past decade, tremendous progress has been made in the isolation and purification of the enzymes of the haem biosynthetic pathway. This has made it possible to obtain information on some of their physicochemical and catalytic properties and to raise enzyme-specific antibodies. In recent years, the availability of these antibodies and the use of molecular biology techniques has permitted the cloning of genes and cDNAs encoding many enzymes of the pathway from various sources. Practically all the cloned genes have been sequenced and the primary structures of the proteins have been deduced. The data obtained has also enabled three porphyrias, acute intermittent porphyria, congenital erythropoietic porphyria and hepatoerythropoietic porphyria to be investigated at the nucleic acid level (see [50,51,54] for reviews).

1.2.1. 5-Aminolaevulinic acid synthesis
1.2.1.1. 5-Aminolaevulinic acid synthase (the glycine pathway). Genes encoding 5-aminolaevulinic acid synthase, *hemA*, have been isolated from several bacterial sources, *Rhizobium meliloti* [55], *Rhizobium* strain *NGR 234* [56], *Bradyrhizobium japonicum* [57] and *R. sphaeroides* [58]. The *R. meliloti* gene has been shown to be

transcribed from two highly homologous promoters [55]. In addition, nuclear genes or mRNA encoding the equivalent mitochondrial enzyme, have been isolated from yeast [59–61], chicken [62–64], mouse [65], rat [66] and humans [67].

The complete nucleotide sequence coding for 5-aminolaevulinic acid synthase and the derived amino acid sequence have been reported for human [67], rat [66], mouse [65], chicken [62,64], yeast [68], *B. japonicum* [69] and *R. meliloti* [55]. Two genes for 5-aminolaevulinic acid synthase, *hemA* and *hemT*, in *R. sphaeroides* have been described [58,70] and sequenced [71]. Two species of enzyme had been isolated previously [72,73] from *R. sphaeroides* and it is tempting to suggest that these two enzymes are encoded by the *hemA* and *hemT* genes, although there is, as yet, no direct evidence to support this view.

In vitro translation of mRNA selected by hybridization to the cloned 5-aminolaevulinic acid synthase gene or of total RNA followed by immunoprecipitation with anti-5-aminolaevulinic acid synthase antibody and/or cDNA sequencing, has shown that the 5-aminolaevulinic acid synthase in yeast and mammalian cells [54,68] are both synthesized as larger precursors. These precursors are subsequently processed post-translationally on import from the cytoplasm into the mitochondrial matrix. In chicken, an erythroid form of 5-aminolaevulinic acid synthase has been characterized [74] which differs in size from the liver enzyme and which is regulated in a cell specific manner. Yamamoto et al. [63] who isolated cDNA specifically encoding the erythroid form suggested that these isoenzymes of 5-aminolaevulinic acid synthase might be encoded for by two different genes. However, this suggestion was controversial for a while. Elferink et al. [75] demonstrated that in chicken reticulocytes, 5-aminolaevulinic acid synthase mRNA species identical to the liver mRNA can be identified and that the 5-aminolaevulinic acid synthase from these tissues displays the same apparent molecular weight as judged by immunoblot analysis. Other studies by Schoenhaut and Curtis [65] also suggested that 5-aminolaevulinic acid synthase mRNA is the same in the liver and erythroid spleen of mice. More recently, Srivastava et al. [76], studying the regulation of 5-aminolaevulinic acid synthase mRNA in different rat tissues, showed that the mRNA in all rat tissues examined was the same and that only a single species existed. These results taken together indicated that the same 5-aminolaevulinic acid synthase gene expressed in liver cells is also expressed in all cell types. These studies did not, however, exclude the possibility that an additional 5-aminolaevulinic acid synthase gene is expressed in erythroid cells [54,77]. This has indeed been found to be the case. By analysing chicken 5-aminolaevulinic acid synthase cDNA clones isolated from both liver and erythroid cells, Riddle et al. [78] have demonstrated unequivocally that at least two separate genes encode the 5-aminolaevulinic acid synthase isoenzymes found in liver and erythroid tissues. The experiments showed that while the product of the erythroid gene is expressed exclusively in erythroid cells, the hepatic form is expressed ubiquitously suggesting that this is the non-specific or house-keeping form found in all tissues by other workers. It was proposed that the failure to detect heterologous RNA transcripts by other workers, may have been due to the clear lack of nucleotide identity between the two species. There were also speculations that other RNA species detected in different chicken tissues [63] might represent new and uncharacterized 5-aminolaevulinic acid synthase (or 5-aminolaevulinic acid synthase-related) transcripts in this gene family.

Sequencing of the chicken 5-aminolaevulinic acid synthase gene [64] obtained from a chicken genomic library by probing with a chicken liver 5-aminolaevulinic acid synthase cDNA clone, revealed that the structural gene is 6.9 kb in length and contains 10 exons. The 10 exons encode a 5-aminolaevulinic acid synthase precursor protein of 638 amino acids [62]. A fragment of 291bp from the 5'-flanking region, including 34bp of the first exon, showed promoter activity when introduced upstream of a chicken histone H2B gene injected into the nuclei of *Xenopus laevis* oocytes [64]. This promoter region of the gene contains a TATA box and a CAAT box as well as four copies of the Sp1 factor binding site CCGCCC. Such GC boxes are present in the control regions of many eukaryotic house-keeping genes. Recently, the role of this sequence in the regulation of the chicken 5-aminolaevulinic acid synthase has been studied and the *cis*-acting sequences required for the expression of the gene have been characterized [79]. By constructing deletion and deletion-insertion mutations of the chicken 5-aminolaevulinic acid 5'-flanking region and examining the expression of this construct in micro-injected *Xenopus* oocytes, the 5'-flanking region required for expression has been delineated to 80bp upstream from the transcriptional initiation site. Expression of the insertion-deletion mutants demonstrated that only a TATA box at position -28, and a single GC box at position -78 was necessary for expression of the 5-aminolaevulinic acid synthase gene in *Xenopus* oocytes. This result was regarded as unusual in view of the current knowledge of the function of *cis*-acting elements in transcriptional control. It was noted that whereas most house-keeping genes do not have TATA boxes and initiate transcription from multiple start sites, the 5-aminolaevulinic acid synthase gene required a TATA box for transcriptional initiation.

Comparison of the two chicken (ubiquitous and erythroid) 5-aminolaevulinic acid synthase proteins showed that their primary sequence, as deduced from their gene sequence, were similar [78] despite the apparent dissimilarity in their gene nucleotide sequence. This similarity extends relatively evenly throughout the carboxyl-terminal segments of the two proteins but ends abruptly within the amino termini. Of interest [78] is the observation that the position at which the similarity between the two proteins begins is very close to a splice sequence [64]. Studies of *B. japonicum* [69] and *R. meliloti* [55] 5-aminolaevulinic acid synthase genes have shown that their deduced protein sequences lack the amino acids corresponding to the first three exons of the chicken liver 5-aminolaevulinic acid synthase gene. Furthermore, it has been shown [62,63] that a degraded form of liver 5-aminolaevulinic acid synthase (corresponding to a molecular weight of approximately 50 kDa) is fully enzymatically active. These observations, taken together, suggest that the amino termini of the two chicken 5-

aminolaevulinic acid synthase enzymes are not necessary for enzymatic activity and that the amino termini of the two isozymes may have evolved independently, perhaps bringing, or acquiring, additional regulatory function(s) [78].

Comparison of the amino acid sequences predicted for 5-aminolaevulinic acid synthase, from the gene and cDNA sequences, indicates a high degree of similarity between the 5-aminolaevulinic acid synthase from various sources. There was 53% amino acid identity between the *B. japonicum* and *R. meliloti* protein over the 76 amino acid sequence available for *R. meliloti* while the *B. japonicum* and chicken protein showed 48.8% identity over the entire length of the *B. japonicum* protein [69]. The amino acid sequences of the mature chicken and yeast proteins were found to have 41% similarity [68]. Similarly, strong similarity was observed between the mature protein sequences of human and rat (83%), chicken (78%) but somewhat less (45%) with that of mouse [67]. Comparison of the amino acid sequence of the *hemA* gene product of *E. coli* K-12 (mapping at 26 min) with the other known 5-aminolaevulinic acid synthase sequences revealed no similarity at all [18,80,81]. As should be obvious from earlier discussions, there was also a lack of similarity between the *S. typhimurium* protein and the 5-aminolaevulinic acid sequences. This is expected in view of the existence in *E. coli* [16,81], and in *S. typhimurium* [30], of the glutamate pathway for the synthesis of 5-aminolaevulinic acid.

The high degree of amino acid sequence conservation of 5-aminolaevulinic acid synthase between such evolutionarily distant organisms as *Bradyrhizobium* and chicken suggests that major tertiary structural similarity exists for the enzymes [69]. Although the amino-terminal presequences of the eukaryotic 5-aminolaevulinic acid synthase genes are characteristic of those for imported mitochondrial matrix proteins, there seems to be a total lack of similarity between the yeast and chicken presequence [68] and between the human and mouse erythropoietic presequence [67]. There was, however, a high degree of similarity between the presequences of human, rat and chicken [67]. The rather surprising finding [76] that the rat liver protein showed less similarity to that of the mouse erythropoietic spleen than to the human and chicken liver enzymes was the first indication for the existence of two distinct genes for the liver and erythroid 5-aminolaevulinic acid synthases. The lack of similarity between the presequences of the rat and mouse enzyme also suggested that different genes may be responsible for producing proteins destined for different metabolic roles. It thus appears that greater evolutionary constraints have been placed on the maintenance of structure in catalytically functional domains than in the presequence which is required solely for import and which is species specific [68].

1.2.1.2. The genes of the glutamate pathway (C_5) pathway. The discovery that 5-aminolaevulinic acid is biosynthesised from glutamate via the C_5 pathway has resulted in the search for the requisite enzymes in a wide range of organisms. It appears that this pathway is far more widespread than previously thought and that it may be more significant in the biosphere than the glycine pathway [20] (see also Chapters 1 and 5 of this volume).

Before it was realised that the C_5 pathway operated in *E. coli* [16], it was clear that there were at least two genes for 5-aminolaevulinic acid synthesis in *E. coli* [6,12], (see Mapping of *hem* genes above). Two genes, *hemA* and *hemL*, were also identified for *S. typhimurium* [21]. Mutants defective in *hemA* had a more severe auxotrophic phenotype than those lacking *hemL*. In the context of a C_4 (glycine) route, it was suggested that *hemA* might encode 5-aminolaevulinic acid synthase whose action, or synthesis, is facilitated by *hemL*. Based on the observation that a large number of *hemA*⁻ mutants, including nonsense mutants, could synthesize sirohaem and that *hemA*⁻ and *hemL*⁻ mutants were capable of synthesizing vitamin B_{12}, it was further postulated that a secondary route of 5-aminolaevulinic acid synthesis operates in *S. typhimurium*; the second route appeared to function anaerobically and was thought to be independent of the *hemA* and *hemL* genes. In view of the discovery of the existence of a C_5 pathway, for 5-aminolaevulinic acid synthesis in *S. typhimurium* [30] a more tenable explanation is that these genes represent two of the three putative genes for 5-aminolaevulic acid synthesis via the glutamate pathway. The ability of *hemL*⁻ mutants to grow, albeit poorly, implies that additional mechanisms for the transformation of glutamate semialdehyde into 5-aminolaevulinic acid may exist. One possibility is that the transamination of the aminoaldehyde may be catalysed at a low rate by other transaminases as a side reaction. Slow non-enzymic conversion of glutamate semialdehyde into 5-aminolaevulinic acid may also be possible. The *hemA* gene of *E. coli* [18,81,82] and *S. typhimrium* [83] has been cloned and sequenced. The two nucleotide sequences and their deduced amino acid sequences show a very high degree of similarity. The *hemA* gene of *B. subtilis* has also been cloned and sequenced [84]. It is the first member of an operon of at least four genes containing *hemC* and possibly *hemD*. The predicted amino acid sequence of the *B. subtilis hemA* protein showed 34% similarity with the deduced amino acid sequence of the protein from Enterobacteriaceae. *B. subtilis* has recently been shown to form 5-aminolaevulinic acid from glutamate [85]. The *S. typhimurium hemA* gene [83], like the *E. coli* gene [81,82] is co-transcribed with the *prfA* gene, encoding polypeptide chain release factor 1 (RF-1). It may be relevant, as suggested by Elliott [83] that the C_5 pathway utilizes two elements of the protein synthetic machinery, $tRNA_{glu}$ and glutamyl-tRNA synthase. This might provide, at least, an evolutionary rationale for co-transcription of *hemA* and the gene encoding a peptide release factor.

It has been proposed [16] that the *hemA* gene of *E. coli* encodes glutamyl-tRNA dehydrogenase or a component of it, although the possibility of the gene being a regulatory gene affecting the expression of the dehydrogenase was not ruled out. More recently, Avissar and Beale [86] have cloned and expressed a structural gene from *Chlorobuim vibrioforme* that complements in *trans hemA* mutation in *E. coli*. More evidence was provided to show that the *hemA* genes encode the dehydrogenase.

The *hemL* gene for *S. typhimurium* has also been sequenced and shown to encode glutamate 1-semialdehyde amino transferase [30]. Also cloned and sequenced is a cDNA sequence encoding barley glutamate 1-semialdehyde amino transferase [87].

The predicted amino acid sequences for the two amino transferases showed more than 50% identity [30].

The RNA required for the biosynthesis of 5-aminolaevulinic acid from glutamate has been characterized [88] and in all cases, the effective RNA contained the UUC glutamate anticodon, as determined by its specific retention on an affinity resin containing an affinity ligand directed towards this anticodon (See a review by O'Neill and Soll, 1990) [89]. Detailed investigations with *Synechocystis* have shown that the levels of $tRNA_{glu}$ are not altered if the rate of chlorophyll production is changed [90].

1.2.2. 5-Aminolaevulinic acid dehydratase
The structural gene for 5-aminolaevulinic acid dehydratase, *hemB* (*hem2* in yeast) has been cloned and sequenced in *E. coli* K-12 [91–93], yeast [94], rat liver [95], and human liver [96,97]. Analysis of the sequence data for the *E. coli* gene revealed the presence of two promoter regions, two Shine-Dalgarno sequences and two 'start' sites [91,93]. The two possible start sites would encode proteins of 324 and 335 amino acids with M_rs of 35,505 and 36,763 respectively [93]. Mapping of the 5'-end of *hemB* mRNA by primer extension analysis showed that the transcription initiates at the second promoter, immediately downstream from the first ATG codon [91]. Hence, only the second ATG is located inside the *hemB* gene and can represent the start codon for the 5-aminolaevulinic acid dehydratase. This was confirmed by results obtained in vitro DNA-directed translation experiments followed by electrophoresis on SDS-polyacrylamide gel [91]. The subunit molecular weight of the protein was 35KDa in close agreement to the 35.6KDa value expected if the mRNA was translated from the second ATG. As with the *E. coli* gene, the ATG initiation codon of the human cDNA was preceded by an in phase ATG with no intervening termination codon [97].

Comparison of the amino acid sequence of *E. coli* 5-aminolaevulinic acid dehydratase, deduced from its gene sequence, with the other known 5-aminolaevulinic acid dehydratase sequences, showed a fairly extensive similarity between the prokaryotic protein and the eukaryotic proteins, 37% with humans [91], 40% with humans, 40% with rat, and 36% with yeast [93]. Local regions of exceptionally strong similarity were observed, including two regions with known functional significance [91,93], where twelve out of the sixteen (75%) amino acids were identical for all the proteins [93]. The first region corresponds to the putative zinc binding site [98–100] and contains the two active cysteine residues and also histidine. The second region is the active site lysine identified by Gibbs and Jordan [101] for the human and bovine 5-aminolaevulinic acid dehydratases. The result of these studies show that the gene for 5-aminolaevulinic acid dehydratase is indeed highly conserved. This is significant because 5-aminolaevulinic acid dehydratase plays a part in the regulation of an important enzymic step in the haem biosynthetic pathway. Porphobilinogen (PBG) availability controls the activity of porphobilinogen deaminase in *E. coli* [102]. The rationale for the dependence of porphobilinogen deaminase activity on porphobilinogen availability had earlier been explained by the fact that porphobilinogen is needed for the synthesis of the dipyrromethane cofactor of porphobilinogen deaminase [103,104]. This explains why *hemA*⁻ and *hemB*⁻ mutants also appear to be 'mutants' for *hemC*. The dipyrromethane cofactor [103] requirement appears to be ubiquitous [105].

1.2.3. Porphobilinogen deaminase
The first reported cloning of a cDNA for porphobilinogen deaminase was for the rat enzyme [106], however the first nucleotide sequence for a porphobilinogen deaminase gene was reported by Thomas and Jordan [22] for the *E. coli* enzyme. Subsequently, the porphobilinogen deaminase gene in yeast, *S. cerevisiae*, was cloned [107] and cDNA clones coding for the enzyme in human erythrocytes [108], human lymphoblastoid cell lines [109] and rat erythropoietic spleen [110] were cloned and/or sequenced. More recently the porphobilinogen deaminase gene in *B. subtilis* has been cloned and sequenced [84]. The cDNA from *Euglena gracilis* has also been identified and sequenced [111] as has the mouse porphobilinogen deaminase gene [112].

Sequencing of the *E. coli* gene and examination of the sequence data [22] revealed the presence of five in-frame methionine codons, at the 5'-end of the open reading frame, in close proximity to each other at positions 99, 105, 120, 201 and 240. However, only two of the ATG codons at 99 and 119 were preceeded by putative Shine-Dalgarno ribosome binding sequences. N-Terminal analysis of the purified protein defined unambiguously the translational start codon as the ATG at position 119. The putative *hemC* promoter was also identified as the TAGGAT at the −10 region and CTGACA at the −35 region. A comparison of this sequence and the rest of the 5'-sequence with that reported for the 5'-flank of the *cyaA* locus [113] revealed similarities between the sequences indicating that the two genes are linked. It is now known that the two genes are transcribed from independent promoters in a divergent manner [22]. Alignment of the derived amino acid sequence of the *E. coli* porphobilinogen deaminase with that predicted for the *B. subtilis* protein revealed 46% identity between the two proteins [84]. Both bacterial proteins are also similar to human porphobilinogen deaminase. The cysteine residue (cysteine 252 in *E. coli*) required for binding the dipyromethane cofactor [103–105] is conserved in all the porphobilinogen deaminases that have been sequenced. A second cysteine (cysteine 99 in *E. coli*) is conserved in *B. subtilis*, *E. coli* and in the human protein.

The determination of the cDNA sequence specifying the porphobilinogen deaminase from *Euglena gracilis* [111] is particularly significant on two counts. Firstly, it indicates a derived protein sequence which has substantial similarity with other porphobilinogen deaminases, especially that from *E. coli* (80% similarity) and secondly, it has yielded valuable information about the recognition signals required for import of the pre-protein across the chloroplast membranes.

In humans, two isoforms of porphobilinogen deaminase, one found in all cells and the other found only in erythroid cells, can be distinguished [114]. Comparison of the nucleotide sequence of the cDNA complementary to the erythroid [115] and non-erythroid [109] mRNA revealed that these two isoforms are translated from two mRNAs that differ solely in their 5'-termini [109]. Analysis of porphobilinogen deaminase mRNA from different tissues using a RNAse mapping technique demonstrated that both mRNA species are distributed according to an absolute tissue-specificity, the erythroid form being restricted to erythropoietic cells. Southern analysis of the human genomic DNA, however, shows that the gene is present as a single copy in the human genome [115]. The gene, encoding human porphobilinogen deaminase, is split into 15 exons, spread over 10 kb of DNA and has two promoters [116]. The two distinct mRNAs are produced through alternative splicing of two primary transcripts arising from the two promoters [116]. The upstream promoter is active in all tissues and has some of the structural features of a house-keeping promoter. The second promoter, located 3 kb downstream, is active only in erythroid cells and displays structural similarity with the β-globin gene promoter. In addition, the erythroid porphobilinogen deaminase and α-globin gene are induced during erythroid differentiation. This suggests that some common *trans*-acting factor may co-regulate the transcription of these two genes during erythropoiesis [116]. The erythroid promoter fused to herpes simplex virus thymidine kinase coding sequence is correctly expressed and regulated during murine erythroleukemia (MEL) cell differentiation [117]. This same construct was inactive in non-erythroid cells and only weakly stimulated by a heterologous non-tissue-specific enhancer. These results suggest that structural features within the porphobilinogen deaminase erythroid promoter are responsible for its strict erythroid-specific expression. Using DNAse 1 footprinting, gel retardation and methylation interference assays, it has been shown that the erythroid-specific promoter contains three binding sites for the erythroid-specific factor I (NF-E1) and one site for a second erythroid specific factor, NF-E2 [118]. Although NF-E2 binds to a sequence containing the Ap1 consensus, it appears to be different from Ap1. The NF-E1 factor also binds the β-globin promoter [119] and the β-globin 3' enhancer [120]. Both the NF-E1 [121,122] and NF-E2 factor [121] have been implicated in the erythroid-specific transcription of the porphobilinogen deaminase gene. Also implicated in the activity of the porphobilinogen deaminase erythroid promoter is a CACCC motif, located upstream of the NF-E1 binding site, which matches the consensus CAC box found upstream of the β-globin gene cap site in multiple species [121,123]. It appears that there is some cooperation between the CAC motif and the NF-E1 binding site in the initiation of the porphobilinogen deaminase erythroid promoter [121,123].

The structural organization of the mouse porphobilinogen deaminase gene has also been studied [112]. The overall organization is similar to that of the human gene. In the housekeeping promoter, only a short stretch of homology is found including two potential Sp1 binding sites. In contrast, more extensive similarity appears in the erythroid-specific promoter including the NF-E1 consensus binding sequence and the CACCC motif, but the CACCC motif occurs as a tandem duplication in the mouse genome. An important difference between the human and mouse porphobilinogen deaminase erythroid promoter is that the NF-E2 binding site is not conserved in the mouse genome. Considering that NF-E2 has been implicated in porphobilinogen deaminase gene transcription during late induction of MEL cell differentiation [121], it is probable that an NF-E2 site exists elsewhere in the mouse porphobilinogen deaminase gene or that its role can be replaced by that of another control region, possibly the tandem duplication of the CACCC motif [123].

Recently, some progress has been made, at the DNA level, in unravelling the mutations that lead to acute intermittent porphyria (AIP), an autosomal dominant disorder, resulting from a partial deficiency in porphobilinogen deaminase [51,54]. Different classes of mutation, as indicated by immunological studies, have been described suggesting that the porphobilinogen deaminase defect in AIP is heterogenous [54]. The majority of AIP patients show a cross-reactive immunological material (CRIM)-negative mutation, while some patients show moderate to marked CRIM-positive mutations [114,124]. Two types or subclass of the CRIM-negative mutation of the disease have been described based on the erythrocyte deaminase level. The first is characterized by porphobilinogen deaminase deficiency in the erythrocytes whereas the second, while showing clinical and biochemical criteria indicating AIP, does not show a deficiency in the erythrocyte porphobilinogen deaminase (see [54] and references therein). It has been suggested that the latter defect might result from a mutation either in the nonerythroid sequences of the gene or in a distinct gene, the product of which is important in the *trans*-regulation of porphobilinogen deaminase expression in non-erythroid cells. Investigation of a large family with this subclass of AIP by using restriction fragment length polymorphisms (RFLPs) of the porphobilinogen deaminase gene, established the linkage between the porphyria and the porphobilinogen deaminase locus in this family [125]. Upon cloning and sequencing the nonerythroid part of the mutated porphobilinogen deaminase gene from an affected individual, a G → A transition was observed at the first position of the first intron [54,125,126]. This modified the normal splice consensus sequence CG*G*TGAGT to CG*A*TGAGT. A similar mutation has previously been shown to abolish completely the normal splicing process [127,128]. Since the first intron interrupts the sequence for coding the non-erythroid iso-form of porphobilinogen deaminase [116], it is to be expected that abnormal splicing leads to detrimental effects upon the gene product. In contrast, in erythroid cells, transcription of the gene starts 2.8 kb downstream from the identified mutation and it is therefore logical that this mutation has no consequence for the expression of the porphobilinogen deaminase gene in these cells. This is in keeping with the normal porphobilinogen deaminase activity in erythrocytes from patients within these affected families.

A mutation has been reported to account for the enzymatic abnormality in CRIM positive acute intermittent porphyria. Using PCR techniques [129] a point mutation G → A in exon 12 of the porphobilinogen deaminase gene has been shown to lead to exon skipping. The resultant shortening of the mRNA produces an abnormal protein resulting in AIP. An earlier report [130] had indicated that a post-transcriptional splitting abnormality may be responsible for the enzyme defect. From the size of the RNA 'deletion' it is likely that a different exon from exon 12 was absent from the mutant RNA [129]. These results emphasize the fact that even in phenotypically homogenous subtypes of AIP, different mutations may be responsible for the disease.

Three intragenic restriction fragment length polymorphisms (RFLPs) of human porphobilinogen deaminase have been described with Mspl, PstI and Bstn1 [131–133]. The polymorphic sites map to the same region of the porphobilinogen deaminase gene within the first intron [54]. The RFLPs can be used for linkage analysis and have been useful for presymptomatic carrier detection in AIP families. Haplotypes have been obtained for all three RFLPs by a single PCR [134].

1.2.4. Uroporphyrinogen III synthase (cosynthase)
As with other aspects of haem biosynthesis, the study of the molecular biology of the cosynthase has lagged behind that of the other haem biosynthetic enzymes because of the constraints placed by the difficulty in purifying the enzyme and the unavailability of a reliable assay for the enzyme. With the identification of the hydroxymethylbilane, preuroporphyrinogen, as the substrate for the enzyme [135;137] and the advent of a rapid assay method [138] and improved h.p.l.c. separation of uroporphyrins [139], considerable progress has been made in the study of this enzyme. The *E. coli* gene, *hemD*, has previously been cloned in our laboratory [23] and has been sequenced [24,140,25].

The complete nucleotide sequence of a full length cDNA encoding the human uroporphyrinogen III synthase has also been reported [141]. Comparison of this sequence with the published *E. coli* sequence from our laboratory [24,140] and elsewhere [25] showed 44% nucleotide (with 42 gaps of 1–4 nucleotides) and 22% amino acid (with 14 gaps of 1–3 residues) sequence similarity between the two synthases. With no gap penalty there is little similarity between the deduced amino acid sequences of the two proteins. Only a short portion of sequence AIGPTTARAL at position 223 in the human sequence and AIGRTTALAL at position 86 of the *E. coli* sequence have considerable similarity with eight out of ten residues identical. Whether any significance can be attached to the finding requires elucidation of the tertiary structure of the proteins. Analysis of the secondary structures of the *E. coli* and human uroporphyrinogen III synthase, predicted by the method of Garnier also failed to reveal any similarities in their secondary structures. These findings are surprising in view of the similarities in the protein structures of the human and *E. coli* dehydratases and porphobilinogen deaminases (see above). However, a hydropathy plot of the human *hemD* showed a high degree of agreement with the *E. coli* plot (Jordan and Mgbeje, unpublished). More recently a fragment of an ORF that may encode uroporphyrinogen III synthase in *B. subtilis* has been sequenced [84]. It was the fourth ORF of a 'hem' operon containing *hemA* and *hemC*. Assignment of this ORF to *hemD* was based on the mapping of *hemD* in *B. subtilis* [142] and needs confirmation. The section so far sequenced showed 19% similarity with the *E. coli* sequence.

A defect in uroporphyrinogen III synthase results in congenital erythropoietic porphyria (Gunther's disease) a rare disorder inherited as an autosomal recessive trait [143]. The molecular abnormality responsible for the characteristic defect in uroporphyrinogen III synthase has been investigated in two patients [144]. In the first case, data obtained revealed the coexistence of two distinct point mutations: a T to C change in codon 73 (arginine in place of a cysteine) and a C to T change in codon 53 (leucine in place of a proline). The second patient was found to be homozygous for the same mutation in codon 53. This was in agreement with the fact that the patient was born from consanguinous parents.

Interestingly the sequence of the human uroporphyrinogen III synthase gene [141] revealed the presence of a short segment with significant homology to a site in the human α-globin gene which binds the D ring of hydroxymethylbilane, i.e. the pyrrole that is specifically rearranged by uroporphyrinogen III synthase. It may be significant that the two mutations reported by Deybach et al. [144] are located in the beginning of the coding sequence and upstream and downstream, respectively, of this conserved sequence. Studies using expression of mutated cDNAs are needed to clarify the consequences on the enzymatic activity.

1.2.5. Uroporphyrinogen decarboxylase
The first uroporphyrinogen decarboxylase gene to be cloned was as a cDNA sequence complementary to the uroporphyrinogen decarboxylase mRNA from rat [145]. The cDNA was used subsequently as a probe to isolate the human cDNA [146]. Both cDNAs have been sequenced [146,147]. The nucleotide and amino acid sequence similarities between the human [146] and rat uroporphyrinogen decarboxylase were 85% and 90% respectively [147]. In the human genome the gene is present as a single copy [146] providing further evidence that one enzyme is able to catalyse all four decarboxylations during the conversion of uroporphyrinogen III to coproporphyrinogen III. Northern analysis [146] and sequencing of isolated cDNA clones from human erythroid and non-erythroid libraries [146,148] suggests that the uroporphyrinogen decarboxylase mRNA is identical in the different cell types.

As with the human 5-aminolaevulinic acid synthase and porphobilinogen deaminase, uroporphyrinogen decarboxylase is a house-keeping enzyme whose activity is enhanced during erythropoietic differentiation [149]. This enhancement has been correlated with an increase in the amount of the corresponding mRNA [150] resulting, in part, from an enhanced transcription of the uroporphyrinogen decarboxylase gene [146]. The mechanism of this regulation is not understood.

The human uroporphyrinogen decarboxylase gene has been isolated and its organization studied [149]. It is composed of ten exons spread over 3 kb of DNA. Two transcription start points, the major one and a minor, separated by 6bp, were mapped by primer extention and RNase mapping in several tissues or cell lines. The two initiation sites were used in all tissues tested in identical proportions. The major downstream initiation site is 15 base pairs from the ATG codon for the N-terminal methionine. Sequencing of the 900bp located upstream from the initiation sites revealed two modules typical of house-keeping genes i.e. a pseudo TATA box and a unique SpI box clustered 70bp immediately upstream from the start site.

Uroporphyrinogen decarboxylase deficiency in man is responsible for familial porphyria cutanea tarda (PCT) and hepatoerythropoietic porphyria (HEP). The molecular genetics and pathogenesis of human uroporphyrinogen decarboxylase has been reviewed [51,54,151–153]. A study of a family with a HEP showed that the enzyme defect resulted in rapid degradation of the protein in vivo [148]. Cloning and sequencing of a cDNA for mRNA from lymphoblastioid cells, from a homozygous patient with HEP, revealed that the mutation was due to the replacement of a glycine residue by a glutamic acid residue at position 281. The mutation appears to make the enzyme unstable and susceptible to digestion with proteases. Further studies [154] have demonstrated the prevalence of the 281 (Gly>Glu) mutation in HEP and PCT. However, the failure to detect the mutation in some of the cases tested [154] indicates a molecular heterogeneity of the mutation among these groups of patients [54,151] as has been found for porphobilinogen deaminase. In a family with PCT, it has also been shown that a glycine → valine substitution at amino acid 281 is the specific mutation responsible for the uroporphyrinogen decarboxylase deficiency [155]. This gives rise to a 4 hour half-life protein as compared to 104 hours for the normal protein.

Studies on uroporphyrinogen decarboxylase from the yeast, *S. cerevisiae*, have demonstrated that the functioning of the enzyme in this organism is very similar to that in mammalian cells [156]. A yeast mutant with biochemical characteristics mimicking those found in PCT was described. Recently, the effects, in vivo, of mutationally modified uroporphyrinogen decarboxylase in different *hem12* (the gene encoding the yeast decarboxylase) mutants have been studied [41]. From the results presented it appears that, irrespective of the extent of the deficiency and of the difference in the functioning of the mutated enzyme in vivo, the proportion of accumulated and excreted porphyrins remain similar. Similar results have been observed in mammals suggesting the same mode of action for the enzyme in yeast and mammals [41]. Most recently in our laboratory Jones and Jordan (unpublished) have shown that the N-terminus of uroporphyrinogen decarboxylase of *R. sphaeroides* has a remarkable similarity to the mammalian and yeast enzymes suggesting that the uroporphyrinogen decarboxylases may have similar three dimensional structures and evolutionary origins.

1.2.6. Coproporphyrinogen oxidase
Only the coproporphyrinogen oxidase gene (*hem13*) from yeast has so far been cloned and sequenced [40]. The open reading frame encodes a protein of 328 amino acids. The sequence data confirmed the high content of aromatic amino acid residues

previously reported for the yeast [157] and the bovine liver [158] enzyme. This amounts to 11.3% (15% if the histidine is taken into account), which is unusually higher than the 5–7% usually found in enzymes generally. These aromatic residues were found clustered and close to the amino acid residues known as potential iron-ligands (ie cysteine, methionine, glutamate, aspartate, histidine, and tyrosine) and it was suggested that these aromatic residues might be important for providing a suitable environment to the substrate molecule, or to the iron atom and its ligand [40].

The synthesis of coproporphyrinogen oxidase in yeast is subject to negative control by oxygen and haem and this regulation operates at the pre-translational level [159]. 5'-Deletion analysis of the coproporphyrinogen oxidase gene [40] revealed that the DNA sequence located 409 nucleotide from the translation-initiation start codon was needed for depression under oxygen limitation. The significance of this is not understood, however.

The gene encoding coproporphyrinogen oxidase has been cloned and sequenced in *R. sphaeroides* [160]. The protein sequence derived from the DNA sequence shows less similarity to the yeast enzyme than in the case of uroporphyrinogen decarboxylase.

1.2.7. Protoporphyrinogen oxidase
There is no report on the cloning or sequencing of the protoporphyrinogen oxidase gene from mammals. Recently, a DNA region from *B. japonicum* encoding pleiotropic functions in haem metabolism and respiration has been cloned [161]. Data were presented to show that the DNA region contained a structural gene for the protoporphyrinogen oxidase. The possibility that this was a regulatory gene governing the expression of protoporphyrinogen oxidase was, however, not ruled out.

1.2.8. Ferrochelatase
The gene encoding ferrochelatase, *hem15*, has been cloned and sequenced from yeast [162,163]. A cDNA encoding mouse ferrochelatase, isolated from a mouse erythroleukemia (MEL) cell cDNA library, has also been sequenced [164].

The yeast ferrochelatase gene is 1179 kb long and encodes a precursor protein containing 31 amino acid presequence. The mature enzyme contains 362 amino acids. Analysis of the 3' flanking sequence revealed an oppositely oriented 74 bp long tRNA^{val} (GUU) coding sequence. The ferrochelatase is relatively abundant in lysine (9%) and contains no apparent transmembrane segment [162]. Northern analysis showed a slight (1.5–2 fold) repression of *hem15* expression by glucose.

The DNA region encoding the mouse ferrochelatase specifies a precursor protein of 420 amino acids (Mr 47,130) consisting of a putative leader sequence of 53 amino acids and a mature protein of 367 amino acid residues (Mr 41,692). The isolated cDNA allowed for the expression of active ferrochelatase by transfected culture cells. RNA blot analysis revealed two species of ferrochelatase mRNA. The band pattern of the RNA was the same as that of the mouse liver implying that the ferrochelatase in erythroid and hepatic cells are the same. The level of the ferrochelatase mRNA in MEL cells increased when the cells were induced to differentiate by treatment with dimethyl sulfoxide.

Comparison of the deduced amino acid sequence of the yeast ferrochelatase to that of the mouse ferrochelatase [164] revealed that two regions (residues 182–199 and 253–309 of mouse ferrochelatase) were particularly well conserved.

A reduction in ferrochelatase activity results in erythropoietic protoporphyria, characterized by excessive accumulation and excretion of protoporphyrin [143]. The mutational event(s) causing the enzyme defect is not known, although a study of the defective enzyme in recessive bovine protoporphyria suggested that a point gene mutation was probably responsible for reduced catalytic activity by changing the conformation of the enzyme without altering its kinetic parameters [165].

A summary of the haem synthesis genes or cDNAs cloned is shown in Table 4.

2. Genes of cobalamin (vitamin B_{12}) biosynthesis

Although the metabolic role of cobalamin (vitamin B_{12}) is fairly well characterized, the biosynthetic pathway from uroporphyrinogen III to vitamin B_{12} is poorly understood and the precise number of enzymatic steps is not known. Progress has been slow due to the lack of a suitable genetic model, until recently, with which to examine cobalamin biosynthesis systematically. Unfortunately, E. coli K-12 appears unable to form vitamin B_{12} de novo from uroporphyrinogen III (or does so at a remarkably slow rate), although it has retained the ability to convert cobinamide into cobalamin and into its two coenzyme forms, adenosylcobalamin and methylcobalamin [166]. Most organisms that have hitherto been known to possess the vitamin B_{12} pathway are either difficult to grow in the laboratory or have poorly developed methods for genetic studies. The detection of cobalamin synthesis in anaerobically grown cultures of S. typhimurium by Jeter et al. [167] was therefore a major breakthrough presenting a means to dissect the genetic details of the B_{12} pathway in a well characterised organism. An outline of the pathway from uroporphyrinogen III to cobalamin is shown in Scheme 2. A detailed discussion of the pathway is to be found in Chapter 3.

Since the genes encoding the enzymes for the biosynthesis of uroporphyrinogen III, *hemA*, *hemL*, *hemB*, *hemC* and *hemD* have been discussed in the previous section, only the genes specifying the later stages of the corrin synthesis pathway from uroporphyrinogen III will be discussed in this section. The first steps in the formation of vitamin B_{12} from uroporphyrinogen III involve the methylation of rings A and B of uroporphyrinogen III by S-adenosyl methionine dependent methylases. Although *E. coli* does not appear to synthesize vitamin B_{12} (except perhaps under strictly anaerobic conditions) this bacterium does elaborate sirohaem, a prosthetic group required for sulphite and nitrite reductases in a pathway which is common to the early stages of the B_{12} pathway. A group of mutants of *E. coli* which are unable to grow on

TABLE 4
Enzymes of haem biosynthesis: cDNAs and/or genes cloned

Enzyme	Source	Reference
5 Aminolaevulinate synthase	Human liver	67
	Rat liver	66,76
	Mouse liver & erythropoietic spleen	65
	Chicken liver	62,66,78
	Chicken erythrocyte	78
	Chicken gene	64
	Yeast	59,60,61
	B. japonicum	57,69
	Rhizobium NGR 234	56
	R. melilotti	55
	R. sphaeroides	58,71
Glutamyl-tRNA dehydrogenase	E. coli	18,81,82
	S. typhimurium	83
	B. substilis	84
Glutamate 1 semialdehyde amino transferase	S. typhimurium	30
	Barley	87
5-Aminolevulinate dehydratase	Human liver	96,97
	Rat liver	95
	Yeast	94
	E. coli	91,92,93
Porphobilinogen deaminase	Human lymphocyte	109
	Human erythrocyte	115
	Human erythropoietic spleen	108
	Human gene	116
	Rat erythropoietic spleen	47, 110
	Mouse	112
	Yeast	107
	Euglena gracilis	111
	E. coli	22
	B. subtilis	84
Uroporphyrinogen III synthase	Human liver	141
	E. coli	24,25,140
	B. subtilis	84
Uroporphyrinogen decarboxylase	Human erythropoietic spleen	108,146
	Human lymphocyte	148
	Human gene	149
	Rat erythropoietic spleen	145,147
Coproporphyrinogen oxidase	Yeast	40
Protoporphyrinogen oxidase	B. japonicum	161
Ferrochelatase	Yeast	162,163
	Mouse	164

nitrate or sulphate and which lack the capability of sirohaem synthesis may be complemented by the *cysG* locus [168]. This locus has been sequenced and investigated in detail by Peakman et al. [169] and more recently the gene has been expressed and the gene product identified as S-adenosylmethionine: uroporphyrinogen III methylase [170]. A similar locus is present in *S. typhimurium* mapping at around minute 41 [167].

In the initial experiments of Jeter et al. [167], *cob* mutants of *S. typhimurium* were isolated and characterized. Three classes of mutants designated *cobI*, *cobII* and *cobIII* mutants were isolated. Further genetic analysis of the cobalamin biosynthetic (*cob*) genes in this bacterium [171] revealed that many of the genes are located in a cluster at 41 map units on the bacterial chromosome and near the *his* operon. In fact, no essential genes appear to lie between the *his* and *cob* operons. Transcriptional polarity studies [171] showed that the *cob* genes responsible for synthesis of the corrinoid intermediate, cobinamide (branch I of the pathway; see Scheme 2) were organised into a single operon. Genes for the synthesis of 5,6-dimethylbenzimidazole (branch II) and the final assembly of the complete cobalamin molecule (branch III) were organised into two or more additional operons. All of the known *cob* genes (in branches I, II and III) were transcribed in a 'counter clockwise' direction opposite to the direction of transcription of the *his* operon. Analysis of mutants carrying deletions from the *his* genes into the *cob* region established the order of loci as *his - cobI - cobIII - cobII*.

Transcription of cobalamin biosynthetic genes in *S. typhimurium* is repressed by cobalamin and by molecular oxygen [172]. This regulation is primarily exerted on

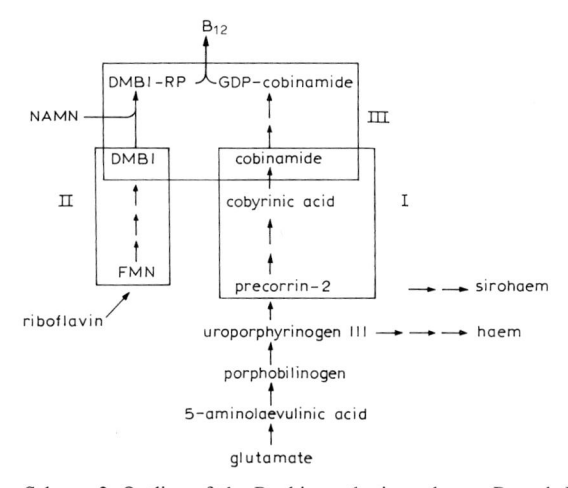

Scheme 2. Outline of the B_{12} biosynthetic pathway. Branch I represents cobinamide biosynthesis from precorrin-2 (dihydrosirohydrochlorin), branch II represents dimethylbenzimidazole (DMBI) biosynthesis and branch III represents cobalamin biosynthesis from these two precursors.
Abbreviations: DMBI, dimethylbenzimidazole; DMBI-RP, 1-α-D-ribofuranoside-DMBI; NAMN, nicotinic acid mononucleotide; SAM, *S*-adenosylmethionine

cobI i.e. the biosynthesis of the corrin ring. Mutation in the *oxrA*, *oxrB*, or *oxrC* genes (which affect other oxygen controlled genes in *S. typhimurium*) have no effect on *cobI* transcription either aerobically or anaerobically [173]. This implies that expression of the cobalamin biosynthesis genes is controlled by a mechanism distinct from those identified for other genes regulated in response to oxygen. Further work [174] suggested that molecular oxygen per se does not signal repression of *cobI* transcription, rather the *cobI* operon is induced in response to a reducing environment within the cell (i.e. regulatory mechanisms that involve direct interaction of molecular oxygen with a regulatory protein are unlikely). The *cob* genes seem to be subject to catabolite repression [172]. As in most cases of catabolite repression in bacteria, the effect is reversed by exogenous cAMP. It was further suggested that although cAMP may contribute to establishing the physiological conditions that favour cobalamin biosynthesis, it may not participate in the process directly. In other words, it seems that the stimulatory effect of cAMP on the biosynthesis of cobalamin under anaerobic condition is a secondary consequence of the global effect on metabolism caused by this nucleotide under anaerobic conditions. Interestingly the promoter of the *hem* operon in *E. coli* is immediately adjacent to, and divergent from, the *cyaA* promoter which regulates the formation of adenyl cyclase [22]. The *hemC* and *hemD* genes specify the enzymes responsible for the biosynthesis of uroporphyrinogen III. The juxtaposition of these promoters and the possible relationship between gene tetrapyrrole synthesis and cAMP formation may not be coincidental.

Recently, *cis*-dominant mutations (*cobRI* through *cobRIV*), mapping near the promoter-end of the *cobI* operon, that increased *cobI* gene expression in the presence of oxygen have been described [173]. These mutations increased aerobic *cobI* expression from 4- to 90-fold depending on the mutant employed and the carbon source used for growth. Also described was a recessive mutation (*cobF*) mapping far from the *cobI* operon, that caused increased expression of *cobI* under both aerobic and anaerobic conditions. The *cobF* mutation mapped near the *pyrC* gene around minute 24 on the *S. typhimurium* chromosome.

More recently, a new class of mutant (*cobIV*) defective in B_{12} biosynthesis and phenotypically distinct from those described above, has been isolated and characterised [175]. These mutations map between the *cysB* and *trp* loci (minute 34) and defined a new genetic locus, *cobA*. On the basis of the observed phenotype of the *cobA* mutants, it was proposed that the *cobA* gene product catalyses adenosylation of an early intermediate in the de novo B_{12} pathway and also adenosylates exogenous corrinoids.

It was also suggested that adenosylation is an obligatory, early step in de novo B_{12} synthesis and that exogenous corrinoids like cobyric acid and cobinamide are not true synthetic intermediates but must be adenosylated before they can be joined to 5,6-dimethylbenzimidazole (DMB) to form the B_{12} coenzyme (Ado-B_{12}). Under aerobic conditions mutants can assimilate adenosylated precursors but cannot make use of the non-adenosylated precursors. Under anaerobic conditions, alternative assimilatory functions to *cobA* are activated enabling *cobA* mutants to use non-adenosylated precursors. Two classes of mutants, *cobB* and *cobX*, that lack the anaerobic alternative adenosylation function have been isolated [176]. The *cobB* gene maps within the *cobIII* region of the main gene cluster whereas *cobX* lies just outside this region and may be within the operon encoding propandiol utilisation (*pdu*). The biosynthetic operons, especially the *cobI* operon which is involved in synthesis of the corrinoid ring, are repressed in the presence of O_2 [172]. The fact that the *cobA* locus at minute 34 maps outside the reported main *cob* clusters at minute 41 may reflect the need to express *cobA* under aerobic conditions to permit assimilation of exogenous corrinoids [175].

The *cobA* gene of *S. typhimurium* appears to be functionally equivalent to the *btuR* gene of *E. coli* [175]. The *btuR* gene maps at a point on the *E. coli* chromosome corresponding to the position of *cobA* in *S. typhimuruim*. Previous studies localised the *btuR* gene to 27.9 minutes on the *E. coli* genetic map between *cysB* and *trp* [177] and it was proposed that *btuR* might encode a repressor of transcription of the *btuB* locus. The *btuB* gene maps at minute 88 on the *E. coli* chromosome and encodes the outer membrane transport protein for vitamin B_{12} and other related compounds [26,178–180]. Vitamin B_{12} is not thought to be required for growth of *E. coli*. The *btuR* gene of *E. coli* has recently been sequenced [181]. It is 588 nucleotides long and encodes a 196 amino acid protein of 21,979 daltons in agreement with the size of the product made in maxicells. The gene is transcribed 'counter clockwise' with respect to the genetic map. Further results indicated that the *btuR* product was involved in the metabolism of adenosylcobalamin (as is *cobA* in *S. typhimirium*) and that this cofactor, or some derivative, controls *btuB* expression. The *btuB* gene has also been cloned [179] and sequenced [178]. The 1842 bp open reading frame encodes a polypeptide with 614 amino acids, the first 20 of which are very similar to the leader peptide signal sequence of outer membrane proteins.

Along the same lines as that carried out with *S. typhimurium*, a systematic genetic study of cobalamin biosynthesis has been undertaken for a strictly aerobic organism, *Bacillus megaterium* [182], in which the regulation of cobalamin biosynthesis was fundamentally different from that in *S. typhimurium*. *B. megaterium* was found convenient for the genetic study of cobalamin biosynthesis because it is capable of vitamin B_{12} synthesis under aerobic growth conditions and has a simple phenotype which may be used for monitoring vitamin B_{12} biosynthesis - dependence upon vitamin B_{12} for utilisation of ethanolamine as a sole nitrogen source under aerobic growth conditions [182]. Ethanolamine is deaminated by the action of ethanolamine ammonia-lyase (E.C. 4.3.1.7), an adenosylcobalamin-dependent enzyme. Consequently, to grow on ethanolamine as a sole nitrogen source, *B. megaterium* requires vitamin B_{12}. Identification of *B. megaterium* mutants deficient for growth on ethanolamine as the sole nitrogen source yielded a total of 34 vitamin B_{12} auxotrophs. These auxotrophs were divided into two major phenotypic groups - *cob* mutants and *cbl* mutants. The *cob* mutants contained lesions in biosynthetic steps before the synthesis of cobinamide, while *cbl* mutants were defective in the conversion of cobinamide into cobalamin. Analysis of phage-mediated transduction experiments revealed tight genetic linkage within the *cob* class and within the *cbl* class. Similar transduction analysis indicated that the *cob* and *cbl* classes were weakly linked. Cross-feeding experiments in which extracts prepared from mutants were examined for their effect on growth of various other mutants allowed a partial ordering to be made of the mutations within the cobalamin biosynthesis pathway. By complementation criteria, at least 11 genes of the cobalamin biosynthesis pathway of *B. megaterium* have been isolated and defined [183]. This figure might represent a reasonable proportion of the genes in the pathway, which may require as many as 30 enzymes. At least 6 genes involved in the biosynthesis of cobinamide are carried on a fragment of DNA approximately 2.7 kb in length.

Recently, *Pseudomonas putida* and *Agrobacterium tumefaciens* mutants deficient in cobalamin synthesis (*cob* mutants) have been isolated [184]. As in *B. megaterium*, *cob* mutants of *P. putida* were identified as being unable to use ethanolamine as a source of nitrogen in the absence of added cobalamin. In *A. tumefaciens*, *cob* mutants were simply screened for their reduced cobalamin synthesis. Furthermore, a genomic library of *Pseudomonas denitrificans* constructed on a mobilisable wide-host-range vector was used to complement the *P. putida* and *A. tumefaciens* cob mutants in order to clone genes for the cobalamin biosynthetic pathway from *P. denitrificans*. Based on the complementation data, 14 different genes involved in cobalamin biosynthesis in *Pseudomonas dentrificans* were cloned. Of these 14 genes, 12 are involved in the transformation of uroporphyrinogen III into cobinamide. The other two mutants complement *cob* mutants blocked in the conversion of cobinamide into cobalamin and are implicated in the last four steps of the cobalamin biosynthetic pathway. The 14 genes were grouped into four different complementation clusters (designated A-D) on the chromosome (Scheme 3).

Group A contains *cobF, cobG, cobH, cobI, cobJ, cobK, cobL,* and *cobM*
Group B contains *cobB,* and *cobE*
Group C contains *cobE, cobA, cobB, cobC* and *cobD*
Group D contains genes specifying cobinamide conversion to coenzyme B_{12}.

A 5.4 kb DNA for complementation group C containing *cob A-E* gene has been

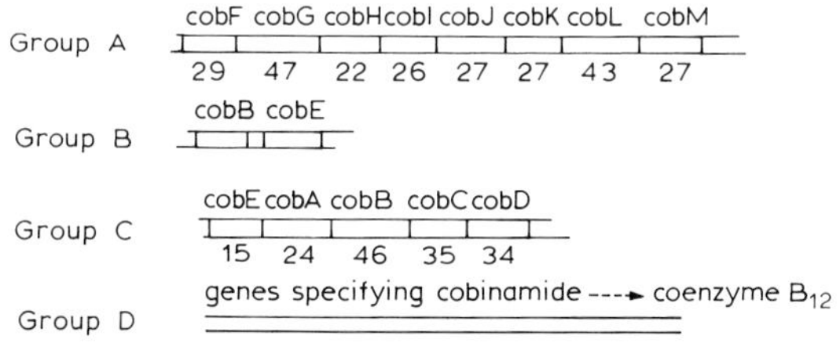

Group A cobF cobG cobH cobI cobJ cobK cobL cobM
          29   47   22   26   27   27   43   27

Group B cobB cobE

Group C cobE cobA cobB cobC cobD
          15   24   46   35   34

Group D genes specifying cobinamide ---→ coenzyme B_{12}

Scheme 3. Arrangement of the genes encoding the enzymes of B_{12} biosynthesis in *Pseudomonas denitrificans* sequenced [185]. Four of these genes (*cobA - D*) show characteristics of translationally coupled genes although an effective translational coupling has not been established. Also sequenced is a 8.7 kb DNA fragment from complementation group A containing *cobF - cobL* [186]. Six of these genes (*cobG-L*) have the characteristics of translationally coupled genes. Several of the genes have been identified [185,186]. The *cobA, cobB* and *cobI* genes have been assigned to the enzymes S-adenosylmethionine:uroporphyrinogen III methyl transferase (SUMT), cobyrinic acid *a,c* diamide synthase and S-adenosylmethionine:precorrin-3 methyltransferase respectively. The *cobA* protein shows extensive homology with the *E. coli cysG* protein.

The *cobC* and *cobD* genes are thought to encode proteins involved in the transformation of cobyrinic acid into cobinamide although the role of *cobE* is not yet known. Proteins encoded by the seven genes *cobF, cobG, cobH, cobJ, cobK, cobL, cobM*, have been shown to specify enzymes with a role in the pathway between precorrin-3 and cobyrinic acid. (See chapter 3 for details of these intermediates.) Of these, *cobF, cobJ, cobL* and *cobM* are thought to be S-adenosylmethionine-dependent methyltransferases involved in the remaining methylations of precorrin-2. The roles of *cobG, cobH* and *cobK* are being investigated.

It can be deduced from several publications that some 20 intermediates are required in the pathway from uroporphyrinogen III to cobalamin [184] (and references therein). However, the number of enzymes involved in this pathway could be smaller if methylases and amide synthases recognise several intermediates. It is certain that molecular biology techniques will bring about a better understanding of the cobalamin pathway and assist in the identification and characterization of the enzymes involved.

3. Genes of bacteriochlorophyll and chlorophyll biosynthesis

3.1. Bacteriochlorophyll biosynthesis genes
The purple photosynthetic bacteria *R. sphaeroides* and *Rhodopseudomonas capsulatus* are useful microorganisms in which to study the genetics of bacteriochlorophyll synthesis as well as the inter-relationship between the biosynthesis of bacteriochlorophyll and photosynthetic membrane assembly. The ability of these bacteria to grow aerobically in the dark as well as photosynthetically, has permitted the isolation of stable non-photosynthetic mutants that are unable to synthesize bacteriochlorophyll. Characterisation of mutants blocked at various steps in the pathway has enabled most of the steps in bacteriochlorophyll biosynthesis to be defined [187–191]. Using chromosome and plasmid mobilizing techniques, Marrs and coworkers have isolated a 45 kb gene cluster in *Rhodopseudomonas capsulatus* which contains genes, not only for bacteriochlorophyll (*bch*) and carotenoid (*crt*) biosynthetic enzymes, but also for the reaction centre and light harvesting polypeptides (*puf, puh* and *puc*) [192–194] (see Kiley and Kaplan [195] for a review on *crt, puf, puh* and *puc* genes. See also Young

et al. [196]). Complementation of the *bch* mutants with *R'*-plasmids bearing the genes specifying various bacteriochlorophyll biosynthetic enzymes [193] enabled identification of most of the genes required for bacteriochlorophyll biosynthesis in *R. capsulatus*. A physical genetic map was subsequently generated for this gene cluster [197].

It has been known for some time that oxygen represses bacteriochlorophyll synthesis in the purple photosynthetic bacteria [198]. To elucidate the mechanism by which this regulation takes place, it is necessary to be able to measure all the enzymes of the bacteriochlorophyll biosynthetic pathway. This has proven impossible, largely because of the lack of assays for most of the enzymes, especially those involved in the latter stages in the pathway from protoporphyrin IX to bacteriochlorophyll. To circumvent this problem Biel and Marrs [199] constructed fusions of the *lacZ* gene to various *bch* genes of *R. capsulatus* and used these fusions to examine the activity of the *bch* genes at the transcriptional level. In these fusions the *lacZ* gene of the bacteriophage was under the control of the promoters of the *bch* genes. The activities of the promoters could then be followed under different environmental conditions by measuring the β-galactosidase activity. Mutants in *bch* genes were identified by products which accumulated. Using this approach, it was shown that transcription of *bchA*, *bchB*, *bchC*, *bchG* and *bchH* genes were regulated in response to oxygen. Later studies have also shown that the levels of mRNA for the *bch* genes of *R. capsulatus* are affected (repressed) by light and oxygen [200–202]. It has been proposed that the expression of the *bch* genes (as with other genes involved in photosynthesis) in *R. capsulatus* in response to oxygen is mediated by the supercoiling of DNA under the influence of DNA gyrase [203]. However, the molecular mechanism by which supercoiling of DNA may lead to enhanced transcription during photosynthetic growth in *R. capsulatus* is unknown.

It might be expected that since most of the *bch* genes are clustered in one major region of the large chromosome, and since all those studied are regulated in response to oxygen, that the genes would be grouped into operons. Biel and Marrs [199] found no evidence to support this in *R. capsulatus* except for the case of the *bchA* and *bchC* genes which occur in a small operon and are transcribed in the order *bchC* towards *bchA*. However, since insertion mutations were not obtained in all the genes, it is possible that other operons exist.

Studies by Zsebo and Hearst [204] also suggested the *bchC* and *bchA* genes comprise an operon, but a contradictory conclusion to that of Biel and Marrs [199] was reached regarding the direction of transcription of the proposed operon. The disagreement has recently been resolved by Young et al. [196] who confirmed that *bchC* and *bchA* do comprise an operon and demonstrated that the direction of transcription was from *bchC* towards *bchA*. Furthermore a 134 bp region of DNA upstream of *bchC* was shown to contain the promoter function for this operon. Genetic mapping showed that the operon is located between one of the genes encoding enzymes for carotenoid biosynthesis (*crtF*) and genes for the bacteriochlorophyll reaction center and light harvesting (LH -1) antenna proteins and the first member of *puf*

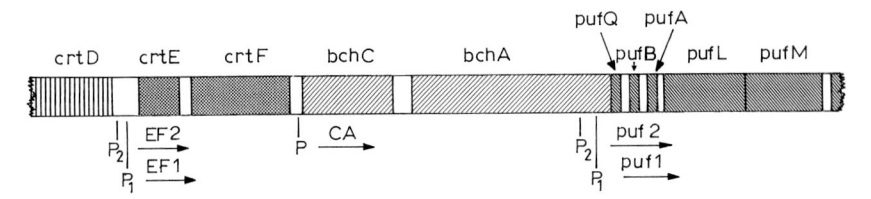

Scheme 4. Genetic map of *Rhodobacter capsulatus* in the *crt*, *bch* and *puf* regions. The *crt*, *bch* and *puf* operons are indicated by differential shading. The promoters are represented by P.

operon (*pufQ*) as shown in Scheme 4 [197,205,206]. A region of 147 bp separates the end of the *crtF* gene from the start of the *bchC* gene, the first gene in the subsequent *bchCA* operon while one base pair separates the end of the *bchA* gene from the first gene (*pufQ*) of the following *puf* operon [196]. Both promoters for the *puf* operon are located within the *bchA* gene. This region of DNA thus serves a dual function, both as part of the structural gene coding for the *bchA* polypeptide and also as the initiation sites for transcription of the downstream *puf* operon [196]. From both sequence data and functional analysis, it was concluded that transcription from the *bchCA* operon cannot be terminated before entering the *puf* operon, and therefore these operons 'overlap'. All three operons, *crtEF*, *bchCA*, and *pufQBALM* are transcribed in the same direction [196]. While the utility of the transcriptional linkage of the *crtEF*, *bchCA*, and *puf* operons is not fully understood, the genetic evidence clearly indicates that significant read-through occurs and that the super-operonal organisation of these genes is necessary for such read through [196].

The first structural gene in the *puf* operon, *pufQ*, of *R. capsulatus* [206–208] encodes a protein that is required for regulating the flow of tetrapyrroles through the bacteriochlorophyll biosynthesis pathway, the level of bacteriochlorophyll biosynthesis being directly dependent on the extent of *pufQ* expression [207]. The existence of this protein within the *puf* operon appears to provide the cell with a way of coupling the amount of bacteriochlorophyll biosynthesis to the synthesis of polypeptides of the reaction center (RC) and light harvesting complex (LH-1). Furthermore, the similarity between *pufQ* and the RC polypeptides, that are known to bind bacteriochlorophyll and quinone, suggest that the role of *pufQ* in bacteriochlorophyll biosynthesis could involve a tetrapyrrole-protein interaction similar to that proposed for the carrier polypeptide by Lascelles [187,209] (see Scheme 5) and that *pufQ* may interact with the quinone pool [207].

A similar arrangement for genes encoding bacteriochlorophyll biosynthesis enzymes and other components of the photosynthetic apparatus in *R. capsulatus* has been observed in *R. sphaeroides* [210]. Using a related technique to that of Biel and Marrs [199], a highly efficient mobilization system which employs plasmid pRK2073 [211] has been used to transfer a bank of *R. sphaeroides* genes into mutants unable to synthesize bacteriochlorophyll [212] (see Scheme 6). In this way, clones carrying genes designated *bchE*, *bclB*, *bclA* and *bclC* have been isolated. These clones were

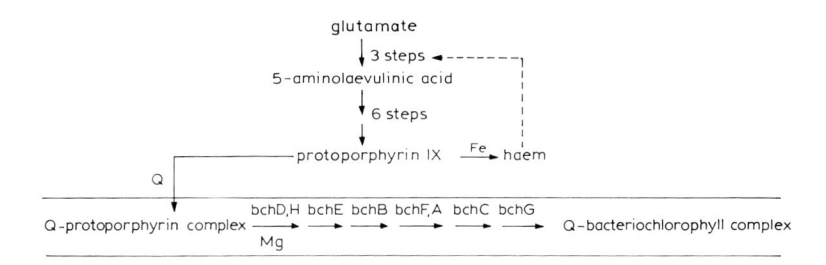

```
                        glutamate
                          │ 3 steps ◄─────┐
                  5-aminolaevulinic acid  │
                          │ 6 steps       │
                          ▼               │
            ┌──────── protoporphyrin IX ──Fe──► haem
         Q  │
────────────▼──────────────────────────────────────────────────────
                       bchD,H  bchE  bchB  bchF,A  bchC  bchG
Q-protoporphyrin complex ───► ───► ───► ───► ───► ───► Q-bacteriochlorophyll complex
                         Mg
───────────────────────────────────────────────────────────────────
```

Scheme 5. Carrier model of *PufQ* function. The polypeptide (Q) encoded by the *pufQ* locus interacts with protoporphyrin IX to form a membrane-bound complex which is an obligatory intermediate for the initiation of bacteriochlorophyll biosynthesis. It is suggested that all the steps to bacteriochlorophyll also involve similar interactions with the membrane Q carrier protein complex.

used as probes to estimate levels of specific transcripts in cells at various stages of pigmentation, after the oxygen level was lowered. Over the time course studied, the level of cellular bacteriochlorophyll increased about 100-fold [212]. Maximum levels of transcripts were observed at an early stage of photosynthetic membrane synthesis (within 1 hour of lowering of 0_2 content) when only 7% of the eventual level of pigment had been synthesized. A comparison of the timing of the appearance of mRNA specifying the enzymes of the bacteriochlorophyll biosynthesis pathway and the light harvesting and reaction center polypeptides [213] led to the conclusion that the rise in mRNA species specifying the enzymes of bacteriochlorophyll biosynthesis is a primary event in photosynthetic membrane assembly. It is interesting that the products of this pathway are needed to stabilize the polypeptide component of photosynthetic complexes in *R. capsulatus* [214]. While the studies of Hunter and Coomber [212] showed that oxygen tension affects the amount of transcripts encoded by clones carrying *bch* genes, they acknowledge that these studies do not differentiate between increase in transcription or decrease in degradation of these mRNA species. Subsequently nine genes for bacteriochlorophyll biosynthesis have been identified within a 45 kb photosynthetic gene cluster on the *R. sphaeroides* genome [215,216]. The organization of these genes (together with that of the other photosynthetic genes within the cluster) is similar to that of *R. capsulatus*.

Progress has been made in determining the DNA sequence for the proposed bacteriochlorophyll genes. The first of the bacteriochlorophyll genes to be sequenced was the *bchC* gene from *R. capsulatus* [217]. The gene encodes 2-desacetyl-2-hydroxyethyl bacteriochlorophyllide *a* dehydrogenase, an enzyme that catalyses the penultimate step in bacteriochlorophyll *a* biosynthesis. The deduced amino acid sequence shows that the *bchC* gene encodes a 33 kDa protein that is less hydrophobic than integral membrane proteins of *R. capsulatus*, although there are hydrophobic segments that could in principle interact with a lipid membrane. S1 nuclease protection and primer extension experiments revealed that the 5' end has significant similarity to the previously characterized *puf* operon promoter region [206,208].

```
              5-aminolaevulinic acid
                        ↓ 6 genes
                  protoporphyrin IX
                        ↓ bchD
                 Mg-protoporphyrin IX
                        ↓ bchH
         Mg-protoporphyrin IX monomethyl ester
                    N6  ↓ bchE
          Mg-2,4-divinylpheoporphyrin a₅ MME
                    N5  ↓ bchB
                  protochlorophyllide
                        ↓ bchL
                   chlorophyllide a
                        ↓ bchF
       2-desvinyl-2-hydroxyethyl chlorophyllide a
                   N22  ↓ bchA
   2-desacetyl-2-hydroxyethyl bacteriochlorophyllide a
                  T 127 ↓ bchC
                 bacteriochlorophyllide a
                        ↓ bchG
                  bacteriochlorophyll a
```

Scheme 6. The bacteriochlorophyll biosynthesis pathway showing the positions of lesions in mutant strains. The genes are located essentially according to Biel and Marrs [199] although *bchL* has been added. The mutants have been assigned according to Jones [190].

Three transcribed ORFs in the *R. capsulatus* gene cluster [197] originally designated *f0, f108* and *f1025* [218] have been identified [219]. Using site directed mutagenesis and complementation analysis, the three ORFs were shown to encode polypeptides involved in the formation of magnesium-protoporphyrin monomethyl ester, chlorophyllide, and protochlorophyllide respectively. From these results, as well as those of Zsebo and Hearst [204], it was proposed that *f108* was *bchL* and *f1025* corresponded to *bchM*. However, *f0* was shown to encode the carboxyl-terminal portion of the *bchH* gene. The open reading frames were found to be part of a large 11-kb operon that encodes numerous genes involved in early steps of the bacteriochlorophyll *a* biosynthesis pathway. Furthermore, these were located upstream from the *puhA* gene, which is known to encode the reaction center H (RC-H) polypeptide. As with the *bchC* gene [217] analysis of the polypeptide sequences encoded by *f0, f108* and *f1025* indicated very few hydrophobic residues, and no obvious membrane-spanning regions were evident from hydrophobicity plots. This raises the possibility that the polypeptides are only peripherally associated with the membrane fractions.

The polypeptide encoded by *f108* [219] exhibits no obvious sequence similarity to the light-dependent protochlorophyllide oxido-reductase gene product from barley [220] even though the two enzymes appear to catalyse similar reactions. It is however possible that the two enzymes are unrelated when it is considered that the reaction in higher plants requires a light activation step that is not found in bacteria. This possibility is supported by the observation that *Chlorella sp.* appears to have two separate enzymes catalyzing this reaction, one which is light dependent and another which is light independent [221].

Although the *bch* genes have been cloned and their regulation studied, none of the activities encoded by the cloned genes have been demonstrated in any of the photosynthetic bacteria. It is hoped that the recent progress in the cloning and determination of these gene sequences will facilitate future purification of the encoded enzymes for functional studies of enzymatic activity, as well as of the overall physiology of chlorophyll synthesis and biosynthetic complex biogenesis. Demonstration of the activities of the cloned gene products *in vitro* will be needed in confirming the proposed steps in the bacteriochlorophyll biosynthesis pathway.

3.2. Chlorophyll biosynthesis genes
The study of the genes encoding the enzymes which catalyse the transformation of protoporphyrin IX into chlorophyll *a* and *b* in higher plants has been largely neglected due to the difficulty in growing chlorophylless mutants of higher plants and the lack of knowledge about the enzymes involved.

In higher plants, a cDNA clone for barley, *Hordeum vulgaris* protochlorophyllide reductase has been isolated by Apel et al. [222] and this has been used extensively to demonstrate that light represses the transcription of the reductase gene (see review by Harpster and Apel [223]). The cDNA has now been sequenced [220]. The predicted amino acid (388 residues in length) includes a transit peptide of 74 amino acid. The cDNA has been expressed in *E. coli*. The activity of the protein in bacterial lysate was completely dependent on the presence of NADPH and protochlorophyllide and requires light. More recently Darrah et al. [224] have isolated a putative protochlorophyllide reductase cDNA clone from an etiolated oat, *Avena sativa*, cDNA library and have used it to confirm qualitatively most of the findings of Apel et al. [222] with respect to the light regulation of the synthesis of protochlorophyllide reductase. Furthermore they sequenced the cDNA clone of the reductase. The sequence data revealed an open reading frame (ORF) coding for a 315 amino acid peptide of M_r 33,758. Since the isolated mature oat enzyme has an apparent M_r of 41,000, this does not represent a full copy of the gene. In the absence of any information about the N-terminal residues of the mature protein, the proportion of the mature reductase represented by the gene sequence could not be specified. That the nucleic acid sequence was that of protochlorophyllide reductase was confirmed by comparison of the amino acid composition, calculated from the derived sequence, with that experimentally determined for the purified protein. Furthermore, the derived amino acid sequence was shown to be similar to that determined from sequence analysis of a cyanogen bromide cleavage fragment of the purified reductase. We have compared the gene and predicted amino acid sequence with that reported for barley [220] and found that the reported sequence may represent that of the mature oat protochlorphyllide reductase. Furthermore, the predicted amino acid sequence of the mature oat and barley protochlorophyllide reductases showed almost complete similarity.

Investigations on the early stages of chlorophyll biosynthesis in barley have resulted in the isolation of glutamyl-tRNA synthase, gluyamyl-tRNA Reductase and glutamate semialdehyde aminotransferase [225]. The purification and sequencing of the aminotransferase enzyme has yielded sequence information permitting the isolation and cloning of the gene [87]. Comparision between the nucleotide sequences and the derived protein sequences with that for *hemL* from *S. typhimurium* [30] indicate that the aminotransferases of plants and Enterobacteriaceae are structurally similar and thus are likely to have ancient evolutionary origins.

References

1. Lascelles, J. (1975) Annals N.Y. Acad. Sci. 244, 334–347.
2. Lennox, E.S. (1955) Virology 1, 190–206.
3. Lederberg, E.M. (1947) Genetics 32, 505–525.
4. Ivanovics, G. and Koczka, S. (1952) Acta. Physiol. Acad. Sci. Hung. 3, 441–457.
5. Sasarman, A., Surdeanu, M., Szegli, G., Horodniceanu, T. Greceanu, V. and Dumitrescur, A. (1968a) J. Bacteriol. 96, 570–572.
6. Sasarman A., Chartrand, P., Proschek, R., Desrochers, M., Tardif, D. and Lapointe, C. (1975) J. Bacteriol. 124, 1205–1212.
7. Sasarman, A., Sanderson, D.E., Surdeanu, M. and Sonea, S. (1970) J. Bacteriol. 102, 531–536.
8. Beljanski, M. and Beljanski, M. (1957) Ann. Inst. Pasteur Paris 92, 396–412.
9. Lascelles, J. (1968) in: Advances in Microbial physiology (A.H. Rose and J.F. Wilkinson, Eds.) vol. 2, pp. 1–42, Academic Press, New York.
10. Sasarman, A., Surdeanu, M. and Horodniceanu, T. (1968) J. Bacteriol. 96, 1882–1884.
11. Chartrand, P., Tardif, D. and Sasarman, A. (1979) J. Gen. Microbiol. 110, 61–66.
12. Powell, K.A., Cox, R., McConville, M. and Charles, H.P. (1973) Enzyme 16, 65–73.
13. Wulff, D.L. (1967) J. Bacteriol. 93, 1473–1474.
14. Haddock, B.A. and Schairer, W.V. (1973) Eur. J. Biochem. 35, 34–35.
15. Li, J-M, Braithwaite, O., Russell, C.S. and Cosloy, S.D. (1989) J. Bacteriol. 171, 2547–2552.
16. Avissar, Y.J. and Beale, S.I. (1989) J. Bacteriol. 171, 2919–2924.
17. McConville, M.L. and Charles, H.P. (1979) J. Gen. Microbiol. 111, 193–200.
18. Drolet, M.,Peloquin, L., Echelard, Y., Cousineau, L. and Sasarman, A. (1989) Mol. Gen. Genet. 216, 347–352.
19. Oh-hama, T., Seto, H. and Miyachi, S. (1986) Arch. Biochem. Biophys. 246, 192–198.
20. Avissar, Y.J., Ormerod, J.G. and Beale, S.I. (1989) Arch. Microbiol. 151, 513–519.
21. Elliott, T. and Roth, J.R. (1989) Mol. Gen. Genet. 216, 303–314.
22. Thomas, S.D. and Jordan, P.M. (1986) Nucleic Acids Res. 14, 6215–6226.
23. Thomas, S.D. (1986) PhD. Thesis Southampton Univ.
24. Jordan, P.M., Mgbeje, B.I.A., Thomas, S.D. and Alwan, A.F. (1987) Nucleic Acid Res. 15, 10583.
25. Sasarman, A., Nepveu, A., Echelard, Y., Dymetryszyn, J., Drolet, M. and Goyer, C. (1987) J. Bacteriol. 169, 4257–4262.
26. Bachmann, B.J. (1983) Microbiol. Rev. 47, 180–200.
27. Bachmann, B.J. (1990) Microbiol. Rev. 54, 130–197.
28. Alefounder, P.R., Abell, C. and Battersby, A.R. (1988) Nucleic Acids Res. 16, 9871
29. Sasarman, A. and Desrochers, M. (1976) J. Bacteriol. 128, 717–721.
30. Elliott, T., Avissar, Y.J. Rhie, G-E., Beale, S.I. (1990) J. Bacteriol. 172, 7071–7084.
31. Tien, W. and White, D.C. (1968) Proc. Natl. Acad. Sci. USA 61, 1392–1398.
32. Kiss, I., Berek, I., Ivanovics, G. (1971) J. Gen. Microbiol. 66, 153–159.

33 Berek, I., Miczak, A. and Ivanovics, G. (1974) Mol. Gen. Genet. 132, 233–239.
34 Miczak, A., Berek, I. and Ivanovics, G. (1976) Mol. Gen. Genet. 146, 85–87.
35 Hartford, N. (1975) J. Bacteriol. 121, 835–847.
36 Bard, M., Wood, R.A. and Haslam, J.M. (1974) Biochem. Biophys. Res. Comm. 56, 324–330.
37 Gollub, E.G., Liu, K-P., Dayan, J., Adlersberg, M. and Springon, D.B. (1977) J. Biol. Chem. 252, 2846–2854.
38 Urban-Grimal, D. and Labbe-Bois, R. (1981) Mol. Gen. Genet. 183, 85–92.
39 Sasarman, A., Chartrand, P., Lavoie, M., Tardif, D., Proschek, R. and Lapointe, C. (1979) J. Gen. Microbiol. 133, 297–303.
40 Zagorec, M., Buhler, J-M., Treich, I., Keng, T., Guarente, L. and Labbe-Bois, R. (1988) J. Biol. Chem. 263, 9718–9724.
41 Kurlanzka, A., Zaladek, T., Rytka, J., Labbe-Bois, R. and Urban-Grimal, D. (1988) Biochem. J. 253, 109–116.
42 Meisler, M., Wanner, L., Eddy, R.E. and Shows, T.B. (1980) Biochem. Biophys. Res. Comm. 95, 170–176.
43 Wang, A-L., Arredondo-Vega, F.X., Giampietro, P.F., Smith, M., Anderson, W.F. and Desnick, R.J. (1981) Proc. Natl. Acad. Sci. USA 78, 5734–5738.
44 Sutherland, G.R., Baker, E., Callen, D.F., Hyland, V.J., May, B.K., Bawden, M.J., Healey, H.M. and Borthwick, I.A. (1988) Am. J. Hum. Genet. 43, 331–335.
45 Potluri, V.R., Astrin, K.H., Wetmur, J.G., Bishop, D.F. and Desnick, R.J. (1987) Hum. Genet. 76, 236–239.
46 Mattei, M.G., Dubart, A., Beaupain, D., Goossens, M., Mattei, J.F. (1985) Cytogenet. Cell Genet. 40, 692.
47 Grandchamp, B., Weil, D., Nordmann, Y., Van Cong, N., de Verneuil, H., Foubert, C. and Gross, M.S. (1983) Hum. Genet. 64, 180–183.
48 Desnick R.J. et al. (in press).
49 Nadeau, J.H., Berger, F.G., Kelly, K.A., Pitha, P.M., Sidman, C.L. and Worral, N. (1986) Genetics 114, 1239–1255.
50 Moore, M.R. (1990) in: Biosynthesis of Heme and Chlorophylls (H.A. Dailey Ed.) pp. 1–54, H.A. McGraw-Hill Inc.
51 Nordmann, Y. and Deybach, J-C. (1990) in: Biosynthesis of Heme and Chlorophylls (H.A. Dailey Ed.) pp. 491–542, H.A. McGraw-Hill Inc.
52 Bagust, J., Jordan, P.M., Kelley, Kerkut (1985) Neuroscience Letters Suppl. 21, 584.
53 Meissner P.N., Adams, P. and Kirsch, R.E. (1989) 'A Century of Porphyria' Univ. of Glasgow, 28. Abstract.
54 Grandchamp, B. and Nordmann, Y. (1988) Semin. Hematol. 25, 303–311.
55 Leong, S.A., Williams, P.H. and Ditta, G.S. (1985) Nucleic Acid Res. 13, 5965–5976.
56 Stanley, J., Dowling, D.N. and Broughton, W.J. (1985) Mol. Gen. Genet. 215, 32–37.
57 Guerinot, M.L. and Chelm, B.K. (1986) Proc. Natl. Acad. Sci. USA 83, 1837–1841.
58 Tai, T-N., Moore, M.D. and Kaplan, S. (1988) Gene 70, 139–151.
59 Arresse, M., Carvajal, E., Robinson, S., Sambunaris, A., Panek, A. and Mattoon, J.M. (1983) Curr. Genet. 7, 175–184.
60 Urban-Grimal, D., Ribes, V. and Labbe-Bois, R. (1984) Curr. Genet. 8, 327–331.
61 Keng, T., Alani, E. and Guarente, L. (1986) Mol. Cell. Biol. 6, 355–364.
62 Borthwick, A., Srivastava, G., Day, A.R., Pirola, B.A., Snoswell, M.A., May, B.K. and Elliott, W.H. (1985) Eur. J. Biochem. 150, 481–487.
63 Yamamoto, M., Yew, N.S., Federspiel, M., Dodgson, J.B., Hayashi, N. and Engel, J.D. (1985) Proc. Natl. Acad. Sci. USA 82, 3702–3706.
64 Maguire, D.J., Day, A.R., Borthwick, I.A., Srivastava, G., Wigley, P.L., May, B.K. and Elliot, W.H. (1986) Nucleic Acid Res. 14, 1379–1391.

65 Schoenhaut, D.S. and Curtis, P.J. (1986) Gene 48, 55–63.
66 Yamamoto, M., Kure, S., Engel, J.D. and Hiraga, K. (1988) J. Biol. Chem. 263, 15973–15979.
67 Bawden, M.J., Borthwick, I.A., Healy, H.M., Morris, C.P., May, B.K. and Elliot, W.H. (1987) Nucleic Acids Res. 15, 8563.
68 Urban-Grimal, D., Volland, C., Garnier, T., Dehoux, P. and Labbe-Bois, R. (1986) Eur. J. Biochem. 156, 511–519.
69 Robertson-McClung, C., Somerville, J.E., Guerinot, M.L. and Chelm, B.K. (1987) Gene 54, 133–139.
70 Suwanto, A. and Kaplan, S. (1989) J. Bacteriol. 171, 5850–5859.
71 Neidle, E. and Kaplan, S. (in preparation).
72 Tuboi, S., Kim, H.J. and Kikuchi, G. (1970) Arch. Biochem. Biophys. 138, 155–159.
73 Michalski, W.P. and Nicholas, D.J.D. (1987) J. Bacteriol. 169, 4651–4659.
74 Watanabe, N., Hayashi, H. and Kikuchi, G. (1983) Biochem. Biophys. Res. Commun. 113, 377–383.
75 Elferink, C.J., Srivastava, G., Maguire, D.J., Borthwick, I.A., May, B.K. and Elliot, W.H. (1987) J. Biol. Chem. 262, 3988–3992.
76 Srivastava, G., Borthwick, I.A., Maguire, D.J., Elferink, C.J., Bawden, M.J., Mercer, J.F.B. and May, B.K. (1988) J. Biol. Chem. 263, 5202–5209.
77 Elferink, C.J., Sassa, S. and May, B.K. (1988) J. Biol. Chem. 263, 13012–13016.
78 Riddle, R.D., Yamamoto, M. and Engel, J.D. (1989) Proc. Natl. Acad. Sci. USA 86, 792–796.
79 Loveridge, J.A., Borthwick, I.A., May, B.K. and Elliot, W.H. (1988) Biochim. Biophy. Acta 951, 166–174.
80 Li, J-M., Russell, C.S. and Cosloy, S.D. (1988) J. Cell Biol. 107, 617a.
81 Li, J-M, Russell, C.S. and Cosloy, S.D. (1989) Gene 82, 209–217.
82 Verkamp, E. and Chelm, B.K. (1989) J. Bacteriol. 171, 4728–4735.
83 Elliott, T. (1989) J. Bacteriol. 171, 3948–3960.
84 Petricek, M., Rutberg, L., Schroder, I. and Hederstedt, L. (1990) J. Bacteriol. 172, 2250–2258.
85 O'Neill, G.P., Chen, M-W. and Soll, D. (1989) FEMS. Microbiol. Lett. 60, 255–260.
86 Avissar, Y.J. and Beale, S.I. (1990) J. Bacteriol. 172, 1656–1659.
87 Grimm, B. (1990) Proc. Natl. Acad. Sci. USA 87, 4169–4173.
88 Schneegurt, M.A. and Beale, S.I. (1988) Plant Physiol. 86, 497–504.
89 O'Neill, G.P. and Soll, D. (1990) Biofactors 2, 227–234.
90 O'Neill, G.P. and Soll, D. (1990) J. Bacteriol. 172, 6363–6371.
91 Echelard, Y., Dymetryszyn, J. Drolet, M. and Sasarman, A. (1988) Gen. Genet. 214, 503–508.
92 Li, J-M., Umanoff, H., Proenca, R., Russel, C.S. and Cosloy, S.D. (1988) J. Bacteriol. 170, 1021–1025.
93 Li, J-M., Russell, C.S. and Cosloy, S.D. (1989) Gene 75, 177–184.
94 Myers, A.M., Crivellone, M.D., Koerner, J.J. and Tzagoloff A. (1987) J. Biol. Chem. 262, 16822–16829.
95 Bishop, J.R., Cohen, P.J., Boyer, S.H., Noyes, A.N. and Frelin, L.P. (1986) Proc. Natl. Acad. Sci. USA 83, 5568–5572.
96 Wetmur, J.G., Bishop, D.F., Ostasiewicz, L. and Desnick, R.J. (1986) Gene 43, 123–130.
97 Wetmur, J.G., Bishop, D.F., Cantelmo, C. and Desnick, R.J. (1986) Proc. Natl. Acad. Sci. USA 83, 7703–7707.
98 Barnard, G.F., Itoh, R., Hohbergr, L.H., Shemin, D. (1977) J. Biol. Chem. 252, 8965–8974.
99 Tsukamoto, I., Yoshinaga, T. and Sano, S. (1979) Biochem. Biophys. Acta 570, 167–168.
100 Gibbs, P.N.B. and Jordan, P.M. (1981) Biochem. Soc. Trans. 9, 232–233.
101 Gibbs, P.N.B. and Jordan, P.M. (1986) Biochem. J. 236, 447–451.
102 Umanoff, H., Russell, C.S. and Cosloy, S.D. (1988) J. Bacteriol. 170, 4969–4971.
103 Jordan, P.M. and Warren, M.J. (1987) FEBS Lett. 225, 87–92.

104 Hart, G.H., Miller, A.D., Leeper, F.G. and Battersby, A.R. (1987) J. Chem. Soc. Chem. Commun. 1762–1765.
105 Warren, M.J. and Jordan, P.M. (1988) Biochem. Soc. Trans. 16, 963.
106 Grandchamp, B., Romeo, P.-H., Dubart, A., Raich, N., Rosa, J., Nordmann, Y. and Goossens, M. (1984) Proc. Natl. Acad. Sci. 81, 5036–5040.
107 Gellefors, P.L., Saltzgaber-Muller, J. and Doulas, M.G. (1986) Biochem. J. 240, 673–677.
108 Romeo, P.-H., Raich, N., Dubart, A., Beaupain, D., Mattei, M.G. and Goossens, M. (1986) in: Porphyrins and Porphyria, (Y. Nordmann, Ed.)Colloque INSERM 34, 25–34.
109 Grandchamp, B., de Verneuil, H., Beaumont, C., Chretien, S., Walter, O. and Nordmann, Y. (1987) Eur. J. Biochem. 162, 105–110.
110 Stubnicer, A-C., Picat, C. and Grandchamp, B. (1988) Nucleic Acids Res. 16, 3102.
111 Sharif, A.L., Smith, A.G. and Abell, C. (1989) Eur. J. Biochem. 184, 353–354.
112 Beaumont, C., Porcher, C., Picot, C., Nordmann, Y. and Grandchamp, B. (1989) J. Biol. Chem. 264, 14829–14838.
113 Aiba, H., Mori, K., Tanaka, M., Ooi, T., Roy, A. and Danchin, A. (1984) Nucleic Acids Res. 12, 9427–9440.
114 Desnick, R.J., Ostasiewicz, L.T., Tishler, P.A. and Mustajoki, P. (1985) J. Clin. Invest. 76, 865–874.
115 Raich, N., Romeo, P.-H., Dubart, A., Beaupin, D., Cohen-Solal, M. and Goossens, M. (1986) Nucleic Acids Res. 14, 5955–5968.
116 Chretien, S., Dubart, A., Beaupan, D., Raich, N., Grandchamp, B., Rosa, J., Goossens, M. and Romeo, P.-H. (1988) Proc. Natl. Acad. Sci. USA 85, 6–10.
117 Raich, N., Mignotte, V., Dubart, A., Beaupain, D., Lebaulch, P., Romano, M., Chabret, C., Charnay, P., Papayannopoulou, T., Goossens, M. and Romeo, P.-H. (1989) J. Biol. Chem. 264, 10186–10192.
118 Mignotte, V., Wall, C., de Boer, E., Grosveld, F. and Romeo, P.-H. (1989) Nucleic Acids Res. 17, 37–54.
119 de Boer, E., Antoniou, M., Mignotte, V., Wall, L. and Grosveld, F. (1988) EMBO. J. 7, 4203–4212.
120 Wall, L., de Boer, E., Grosveld, F. (1988) Genes Dev. 2, 1098–1100.
121 Mignotte, V., Elequet, S.F., Raich, N. and Romeo, P.-H. (1989) Proc. Natl. Acad. Sci. USA 86, 6548–6552.
122 Plumb, M., Frampton, S., Wainwright, H., Walker, M., Macleod, K., Goodwin, G. and Harrison, P. (1989) Nucleic Acids Res. 17, 73–92.
123 Frampton, J., Walker, M., Plumb, M. and Harrison, P.R. (1990) Mol. Cell Biol. 10, 3838–3842.
124 Wilson, J.H.P., de Rooij, F.W.M. and Te Velde, K. (1986) Neth. J. Med. 29, 393–399.
125 Grandchamp, B., Picat, C., Mignotte, V., Wilson, J.H.P., Velde, K., Sandkuyl, L., Romeo, P.-H., Goossens, M. and Nordmann, Y. (1989) Proc. Natl. Acad. Sci. 86, 661–664.
126 Grandchamp, B., Picat, C., Kauppinen, R., Mignotte, V., Peltonen, L., Mustajoki, P., Romeo, P.-H., Goossens, M. and Nordmann, Y. (1989) Eur. J. Clin. Invest. 19, 415–418.
127 Treisman, R., Orkin, S.H. and Maniatis, R.T. (1983) Nature (London) 302, 591–596.
128 Dilella, A.G., Marvik, J., Lindsky, A.S., Guttler, F. and Woo, S.L.C. (1986) Nature (London) 322, 799–803.
129 Grandchamp, B., Picat, C., Rooij, F. De, Beaumont, C., Wilson, P., Deybach, S.C. and Nordmann, Y. (1989) Nucleic Acid. Res. 17, 6637–6649.
130 Llewellyn, D.H., Urquhart, A., Scobie, G., Elder, G.H., Kalsheker, N.A. and Harrison, P.R. (1988) Biochem. Soc. Trans. 16, 799–800.
131 Lee, J-S. and Anvret, M. (1987) Nucleic Acids Res. 15, 6307.
132 Lee, J-S., Anvret, M., Lindsten, J., Lannfelt, L., Gellefors, P., Wetterberg, L., Floderus, Y. and Thunell, S. (1988) Hum. Genet. 79, 379–381.
133 Llewellyn, D.H., Elder, G.H., Kalsheker, N.A., Marsh, O.W.M., Harrison, P.R., Granchamp, B., Picat, C., Nordmann, Y., Romeo, P.-H. and Goossens, M. (1987) Lancet 11, 706–708.

134 Lee, J-S., Lindsten, J. and Anvret, M. (1990) Hum. Genet. 84, 241–243.
135 Burton, G., Fagerness, P.E., Hosozawa, S., Jordan, P.M. and Scott, I.A. (1979) J. Chem. Soc. Chem. Commun. 202–204.
136 Jordan, P.M., Burton, G., Nordlov, H., Schneider, M.M. Pryde, L. and Scott, A.I. (1979) J. Chem. Soc. Chem. Commun. 204–205.
137 Battersby, A.R., Fookes, C.J.R., Gustafson-Potter, K.E., Matcham, G.W.J. and McDonald, E. (1979) J. Chem. Soc. Chem. Commun. 24, 1155–1158.
138 Jordan, P.M. (1982) Enzyme 28, 158–167.
139 Wayne, A.W., Straight, R.C., Wales, E.E. and Englert, J. (1979) Biochem. Biophys. Res. Commun. 113, 377–383.
140 Jordan, P.M., Mgbeje, B.I.A., Thomas, S.D. and Alwan, A.F. (1988) Biochem. J. 249, 613–616.
141 Tsai, S.F., Bishop, D.F. and Desnick, R.J. (1988) Proc. Natl. Acad. Sci. USA 85, 7049–7053.
142 Miczak, A., Pragai, B. and Berek, I. (1979) Mol. Gen. Genet. 174, 293–295.
143 Kappas, A., Sassa, S. and Anderson, K.E. (1983) in: The Metabolic Basis of Inherited Disease, 5th Edn. (J.B. Stanbury, J.B. Wyngaarden, D.S. Fredrickson, J.L. Goldstein and M.S. Brown, Eds.) pp. 1325–1332, McGraw-Hill, New York.
144 Deybach, J-C., de Verneuil, H., Boulechfar, S., Grandchamp, B. and Nordmann, Y. (1990) Blood 75, 1763–1765.
145 Romeo, P-H., Dubart, A., Grandchamp, B., de Verneuil, H., Rosa, J., Nordmann, Y. and Goossens, M. (1984) Proc. Natl. Acad. Sci. USA 81, 3346–3350.
146 Romeo, P-H., Raich, N., Dubart, A., Beaupain, D., Cohen-Solal, M. and Goossens, M. (1986) J. Biol. Chem. 261, 9825–9831.
147 Romana, M., Le Boulch, P. and Romeo, P-H. (1987) Nucleic Acid Res. 15, 7211.
148 de Verneuil, H., Grandchamp, B., Beaumont, C., Picat, C. and Nordmann, Y. (1986) Science 234, 732–734.
149 Romana, M., Dubart, A., Beaupain, D., Chabret, C., Goossens, M. and Romeo, P-H. (1987) Nucleic Acid Res. 14, 7343–7356.
150 Grandchamp, B., Beaumont, C., de Verneuil, H. and Nordmann, Y. (1985) J. Biol. Chem. 260, 9630–9635.
151 Elder, G.H., Roberts, A.G. and de Salamanca, R.E.. (1989) Clin. Biochem. 22, 163–168.
152 Sassa, S. (1990) Int. Cell cloning 8, 10–26.
153 Straka, J.G., Rank, J.M. and Bloomer, J.R. (1990)
154 de Verneuil, H., Hansen, J., Picat, C., Grandchamp, B., Kushner, J., Roberts, A., Elder, G. and Nordmann, Y. (1988) Hum. Genet. 78, 101–102
155 Garey, J.R., Hansen, J.L., Harrison, L.M., Kennedy, J.B. and Kushner, J.P. (1989) Blood 73, 892–895.
156 Smith, A.G. and Francis, J.E. (1986) in: Porphyrins and Porphyrias (Y. Nordmann Ed.) pp. 127–136, Colloque INSERM 134, John Libbey, Paris.
157 Camadro, J.M., Chambon, H., Jolles, J. and Labbe, P. (1986) Eur. J. Biochem. 156, 579–587.
158 Yoshinaga, T. and Sano, S. (1980) J. Biol. Chem. 255, 4722–4726.
159 Zagorec, M. and Labbe-Bois, R. (1986) J. Biol. Chem. 261, 2506–2509.
160 Coomber, S.A., Jones, R.M., Jordan, P.M. and Hunter, C.N. (submitted).
161 Ramsier, T.M., Kaluza, B., Studer, D., Gloudemans, T., Bisseling, T., Jordan, P.M., Jones, R.M., Zuber, M. and Hennecke, H. (1989) Arch. Microbiol. 151, 203–212.
162 Labbe-Bois, R. (1990) J. Biol. Chem. 265, 7278–7283.
163 Gokhman, I. and Zamir, A. (1990) Nucleic Acids Res. 18, 6130.
164 Shigeru, T., Nakahashi, Y., Osumi, T. and Tokunaga, R. (1990) J. Biol. Chem. 265, 19377–19380.
165 Bloomer, J.R., Hill, H.D., Morton, K.O., Anderson-Burnham, L.A. and Straka, J.G. (1987) J. Biol. Chem. 262, 667–671.

166 Babior, B.M. (1975) in: Cobalamin biochemistry and pathophysiology (B. Babior, Ed.) pp. 141–212, John Wiley and Son Inc., New York.
167 Jeter, R.M., Olivera, B.M. and Roth, J.R. (1984) J. Bacteriol. 159, 206–213.
168 Cole, J.A., Newman, B.M. and White, P. (1980) J. Gen. Microbiol. 120, 475–483.
169 Peakman, T., Crouzet, J., Mayaux, J.F., Busby, S., Mohan, S., Harborne, N., Wootton, J., Nicholson, R. and Cole, J.(1990) Eur. J. Biochem. 191, 315–323.
170 Warren, M.J., Roessner, C.A., Santander, P.J. and Scott, A.I. (1990) Biochem. J. 265, 725–729.
171 Jeter, R.M. and Roth, J.R. (1987) J. Bacteriol. 169, 3189–3198.
172 Escalante-Semerena, J.C. and Roth, J.R. (1987) J. Bacteriol. 169, 2251–2258.
173 Andersson, D.I. and Roth, J.R. (1989) J. Bacteriol. 171, 6726–6733.
174 Andersson, D.I. and Roth, J.R. (1989) J. Bacteriol. 171, 6734–39.
175 Escalante-Semerena, J.C., Suh, S-J. and Roth, J.R. (1990) J. Bacteriol. 172, 273–280.
176 Roth, J.R., Grabau, C. and Doak, T.G. (1991) in: 8th IUCCP Symposium on Chemical Aspects of Enzyme Biotechnology: Fundamentals, (A. Martell, Ed.) Texas A and M University, Plenum (in press).
177 Lundrigan, M.D., Deveaux, L.C., Mann, B.J. and Kadner, R.J. (1987) Mol. Gen. Genet. 206, 401–407.
178 Heller, K. and Kadner, R.J. (1985) J. Bacteriol. 161, 904–908.
179 Heller, K., Mann, B.J. and Kadner, R.J. (1985) J. Bacteriol. 161, 896–903.
180 Kadner, R.J. and Liggins, G.L. (1973) J. Bacteriol. 115, 514–521.
181 Lundrigan, M.D. and Kadner, R.J. (1989) J. Bacteriol. 171, 154–161.
182 Wolf, J.B. and Brey, R.N. (1986) J. Bacteriol. 166, 51–58.
183 Brey, R.N., Banner, C.D.B. and Wolf, J.B. (1986) J. Bacteriol. 167, 623–630.
184 Cameron, B., Briggs, K., Pridmore, S., Brefort, G. and Crouzet, J. (1989) J. Bacteriol. 171, 547–553.
185 Crouzet, J., Cauchois, L., Blanche, F., Debussche, L., Thibaut, D., Rouyez, M.C., Rigault, S., Mayaux, J-F. and Cameron, B. (1990) J. Bacteriol. 172, 5968–5979.
186 Crouzet, J., Cameron, B., Cauchois, L., Rigault, S., Rouyez, M.C., Blanche, F., Thibaut, D. and Debussche, L. (1990) J. Bacteriol. 172, 5980–5990.
187 Lascelles, J. (1966) Biochem. J. 100, 175–183.
188 Richards, W.R. and Lascelles, J. (1969) Biochemistry 8, 3473–3482.
189 Pudek, M.R. and Richards, W.R. (1975) Biochemistry 14, 3132–3137.
190 Jones, O.T.G. (1978) in: The Porphyrins, Vol.VI (D. Dolphin, Ed.) pp. 179–232, Academic Press, London.
191 Rebeiz, C. and Lascelles, J. (1982) in: Photosynthesis, vol.1, Energy Conversion by Plants and Bacteria, (Godvindjee, Ed.) pp. 699–788, Academic Press, London.
192 Yen, H.C. and Marrs, B. (1976) J. Bacteriol. 126, 619–629.
193 Marrs, B. (1981) J. Bacteriol. 146, 1003–1012.
194 Youvan, D.C., Hearst, J.E. and Marrs, B.L. (1983) J. Bacteriol. 154, 748–755.
195 Kiley, P.J. and Kaplan, S. (1988) Microbiol. Rev. 52, 50–69
196 Young, D.A., Bauer, C.E., Williams, J.C. and Marrs, B.L. (1989) Mol. Gen. Genet.(1989) 218, 1–12
197 Taylor, D.P., Cohen, S.N., Clark, W.G. and Marrs, B.C. (1983) J. Bacteriol. 154, 380–590
198 Cohen-Bazire, A., Sistrom, W.R. and Staner, R.Y. (1957) J. Cell Comp. Physiol. 49, 25–51
199 Biel, A.J. and Marrs, B.L. (1983) J. Bacteriol. 156, 686–694
200 Clark, W.H., Davidson, E. and Marrs, B.L. (1984) J. Biochem. 157, 945–948
201 Zhu, Y.S. and Hearst, J.E. (1986) Proc. Natl. Acad. Sci. USA 83, 7613–7617
202 Zhu, Y.S., Cook, D.N., Leach, F., Armstrong, G.A., Alberti, M. and Hearst, J.E. (1986) J. Bacteriol. 168, 1180–1188
203 Zhu, Y.S. and Hearst, J.E. (1988) Proc. Natl. Acad. Sci. USA 85, 4209–4213
204 Zsebo, K.M. and Hearst, J.E. (1984) Cell 37, 937–947.

205 Youvan, D.C., Alberti, M., Begusch, H., Bylina, E.J. and Hearst, J.E. (1984) Proc. Natl. Acad. Sci. USA 81, 189–192.
206 Bauer, C.E., Young, D.A. and Marrs, B.L. (1988) J. Biol. Chem. 236, 4820–4827.
207 Bauer, C.E. and Marrs, B.L. (1988) Proc. Natl. Acad. Sci. USA 85, 7074–7078.
208 Adams, C.W., Forrest, M.E., Cohen, S.N. and Beatty, J.T. (1989) J. Bacteriol. 171, 473–482.
209 Lascelles, J. (1968) in: Advances in Microbial Physiology (A.H. Rose and J.F. Wilkinson Eds.) Vol. 2, pp. 1–42, Academic Press, N.Y.
210 Sistrom, W.R., Macalusa, A. and Pledger, R. (1984) Arch. Microbiol. 138, 161–165.
211 Hunter, C.N. and Turner, G. (1988) J. Gen. Microbiol. 143, 1471–1480.
212 Hunter, C.N. and Coomber, S.A. (1988) J. Gen. Microbiol. 134, 1491–1497.
213 Hunter, C.N., Ashby, M.K. and Coomber, S.A. (1987) Biochem. J. 247, 489–492.
214 Dierstein, R. (1983) FEBS Letters 160, 281–286.
215 Coomber, S.A. and Hunter, C.N. (1989) Arch. Microbiol. 151, 454–458.
216 Coomber, S.A., Chaudhri, M., Connor, A., Britton, G. and Hunter, C.N. (1990) Mol. Microbiol. 4, 977–989.
217 Wellington, C.L. and Beatty, J.T. (1989) Gene 83, 251–261.
218 Youvan, D.C., Bylina, E.J., Alberti, M., Begusch, H. and Hearst, J.E. (1984) Cell 37, 949–957.
219 Yang, Z. and Bauer, C.E. (1990) J. Bacteriol. 172, 5001–5010.
220 Schulz, R., Steinmuller, K., Klaas, M., Forreiter, C., Rasmussen, S., Hiller, C. and Apel, K. (1989) Mol. Gen. Genet. 217, 355–361.
221 Griffiths, W.T. and Mapleston, R.E. (1978) in: NADPH-protochlorophyllide oxidoreductase in Chloroplast Development (G. Akoyunoglou and J.H. Argyrbudi-Akoyunoglou, Eds.) pp. 99–105, Elsevier/North-Holland Publishing Co., Amsterdam.
222 Apel, K., Gollmer, I. and Batschauer, A. (1983) J. Cell Biochem. 23, 181–189.
223 Harpster, M. and Apel, K. (1985) Physiol. Plant 64, 147–152.
224 Darrah, P.M., Kay, S.A., Teakle, G.R. and Griffiths, W.F. (1990) Biochem. J. 265, 789–798.
225 Kannangara, G.C., Gough, S.P., Bruyant, P., Hoober, J.K., Kahn, A. and von Wettstein, D (1988) T.I.B.S. 13, 139–143.

INDEX

The spellings of haem and heme and also of 5-aminolaevulinate and 5-aminolevulinate are used interchangeably in the chapters. In the index the English spelling is adopted.

19-Acetyl corrin, 120
Active-site of,
    5-aminolaevulinic acid dehydratase, 24, 25, 29
    porphobilinogen deaminase, 58
Acute intermittent porphyria (AIP), 7, 24, 262-263, 269
    enzymatic abnormalities and genetic defects in, 262-263, 271-272
    porphobilinogen deaminase mutant classes, 271-272
    raised 5-aminolaevulinic acid synthase in, 7
Acute porphyrias (Hepatic porphyrias),
    classification of, 262-263
    enzymic defects of, 262-263
    neuropathy in, 24, 262
Adenosyl cobalamin
    (see Coenzyme B_{12}), 101
S-Adenosyl-methionine,
    methyl donor in vitamin B_{12} biosynthesis, 102, 276,
S-Adenosyl-methionine-Mg-protoporphyrin IX methyltransferase, 182
    kinetic mechanism investigated by affinity columns, 183
    linkage to Mg-chelatase in R. sphaeroides, 182
S-Adenosyl-methionine: precorrin-3 methyl transferase,
    cobI in P. dentrificans, 282
S-Adenosyl-methionine: uroporphyrinogen III methyltransferase (SUMT)(M-1), 128
    expressed from cysG, 132
    in formation of dihydrosirohydrochlorin (precorrin-2), 130
    isolation of, 128
    methylation of uroporphyrinogens I and III by, 128
N-Alkylporphyrinogen, 41
N-Alkylporphyrins, 8
    inhibition of ferrochelatase by, 94-95
2-Amino-3-ketoadipic acid, 3, 11-12
Aminomalonate, 10, 14
5-Amino-6-hydroxytetrahydropyran-2-one (HAT), 17-18, 165
    (see Glutamate 1-semialdehyde)
Aminomethyldipyrromethanes, 35, 36

Aminomethyltripyrranes, 35, 36
Aminomethylbilanes, 35, 36
    pseudo-substrates for deaminase/cosynthase, 37, 38, 49
    rearrangements of ring D, 36
5-Aminolaevulinate dehydratase (Doss) porphyria, 24, 263
5-Aminolaevulinate dehydratase (porphobilinogen synthase), 19-28
    'A' and 'P' sites, 26-29
    active-site histidine, 22, 29
    active-site lysine, 24, 25, 29, 268
    alkylation of sulphydryl groups, 29, 30
    cellular location in plants, 210
    chromosomal location in humans, 262
    cDNA of, 22, 268, 281
    EXAFS study of, 23
    gene encoding (hemB), 22, 259, 260
    half-site reactivity 21, 30
    inhibitors of 24, 29, 30, 168
    lead inhibition of, 23, 24
    magnesium requirement, 22, 167,
    mechanism of, 26-28
    mixed pyrrole, 26
    molecular biology, 22, 259-260, 268-269, 281
    occurrence and properties, 19-21
    order of binding substrates of, 26-27, 168
    plant, 19, 22, 25
    quarternary structure of, 20-21
    regulation in plants, 215
    Schiff base intermediate, 24, 26
    steric course, 28
    yeast, 19, 22, 25
    zinc finger, 22
    zinc requirement of, 22-23
5-Aminolaevulinate synthase (ALAS), 6-17
    cellular location, 5, 7
    comparison of protein sequences, 266
    early experiments, 4
    erythroid, 6, 9, 264-265
    exchange reactions catalysed by, 14, 160
    genes encoding erythroid and ubiquitous, 14, 261-263, 263-266
    haem regulation of, 5, 6-9
    half-life of, 7-8

house-keeping (ubiquitous) form of, 6
induction by drugs, 7-8
inhibitors of, 10, 14
in *R. sphaeroides*, 5-6, 159
mitochodrial import of, 5, 7, 264
molecular biology of, 14-15, 263-266, 281
occurrence and properties of, 4-5, 159-161
protein precursors of, 14-15, 264
reaction mechanism, 10-13, 160
regulation in bacteria, 5, 6,
regulation in eukaryotes, 6-9, 265
reports of in plants, 160
requirement for pyridoxal 5'-phosphate, 10, 13-15
mRNA, 264-265
stereochemical course of reaction mechanism, 13-15, 159
stimulation by drugs, 8
structure of the enzyme, 10, 14, 15
substrate specificity and kinetics, 9, 10
transcriptional control, 265
two forms in bacteria, 6
two forms in eukaryotes, 8, 264-266
5-Aminolaevulinic acid (ALA), -(also 5-Aminolaevulinate)
    biosynthesis of, 4-19, 158
    biosynthesis of tetrapyrroles from, 2, 155
    discovery of, 3
    formation of porphobilinogen from, 19, 167-168
    neurological effects of, 24
    precursor of vitamin B_{12}, 103
    precursor of coenzyme F_{430}, 143
5-Aminolaevulinic acid biosynthesis,
    two pathways in *Euglena*, 19
    from glutamate
    (see Glutamate pathway, 15-19 and 161-167)
    from glycine and succinyl-CoA, 3, 19, 158-161
    phylogenic distribution of the two pathways for, 166-167
    regulation of in plants, 211-215, 216-218
    (See also 5-Aminolaevulinate synthase)
5-Aminolaevulinic acid synthase
    (see 5-Aminolaevulinate synthase)
Aminomethylbilanes
    experiments with, in biosynthesis of uroporphyrinogens, 35-38
    intramolecular rearrangement of, 37
    structure of, related to porphobilinogen, 36
Aminomethyldipyrromethanes and aminomethyltripyrranes
    experiments with, as tetrapyrrole precursors, 35

    structures of, related to porphobilinogen, 36
1-Amino-2-propanol, 132
Anaerobiosis effects on greening, 205
Antenna pigments in photosynthetic bacteria, 239
Apo-deaminase, 50-51
    assembly of dipyrromethane cofactor by, 47-49, 50-51
Apo-phycobiliprotein,
    attachment of bilin chromophores by thioether linkage to, 178-179
Arginine,
    in porphobilinogen deaminase catalytic cleft, 60
Arsenite as ferrochelatase inhibitor, 95

Baclobilin, 140
Bacteriochlorophyll biosynthesis, 237-253
    branch points in, 246, 249
    in bacterial and algal mutants, 242
    in Chlorobiaceae, 239
    (see bacteriochlorophylls *c*, *d* and *e*)
    from glutamate, 240
    from glycine, 240
    in purple bacteria, 238-239
    pathway of, 243 -> 244
    regulation of, 283-285
Bacteriochlorophyll biosynthesis genes (*bch*), 282-287
    lacZ fusions to promoters of, 283
    regulation of, by light and O_2, 283, 284-285
Bacteriochlorophyll a, 83, 157, 243
    biosynthetic pathway for, 240, 243
    esterification mechanism at 7-position with 5-amino-1[$C^{18}O_2H$]-laevulinic acid, 245
    4-ethyl group formation, 244
    glycine C-2 as methoxy carbon source, 240
    in purple bacteria, 238-239
    structure of, 238
Bacteriochlorophyll b,
    biosynthetic pathway for, 238, 243, 245
    in purple bacteria, 238-239
    structure of, 238
Bacteriochlorophyll(s) c, d, and e, (*Chlorobium* chlorophylls), 239
    absolute configuration of ring D in, 239
    determination of structures of, 239
    esterifying alcohols of, 239
    5-ethyl group formation from acetic acid, 249
    glutamate as a tetrapyrrole ring precursor, 246
    glycine as methyl group precursor, 246
    in Chlorobiaceae, 239
    δ-*meso*-substituents, 239
    methylation at 4- and 5-positions, 239, 240-242,

246
  nmr studies, 239
  proposed methylation pathways at 4-position, 247-248
  stereochemistry of the 2-(1-hydroxyethyl) group, 239
  structures of, 240-242
Bacteriochlorophyll f,
  biosynthesis of, 239
  structure of, 242
Bacteriochlorophyll g, 240, 242-243
Bacteriopheophytins, 201, 202
bch genes, 282-287
Bile pigments of plants, 157, 172-178
  chemical structures, 157, 175
Bilin biosynthesis
  (see Phycocyanobilin and Phycoerythrobilin)
  algal haem oxygenase in, 173-174
  biliverdin as phycobilin precursor, 173
  biliverdin reduction to phycocyanobilin, 174-177
  biosynthesis of phycobilins, 172-178
  haem as phycobilin precursor, 172
  ligation of phycobilins to apoproteins, 178-179
  relationship between phytochrome and phycobilin chromophores, 157
  structures of phycobilin chromophores, 175
Biliproteins, 157, 219-222
  (see Phycocyanin and Phycoerythrin)
α-Bromoporphobilinogen, 46
btuB gene of *E. coli*, 280

^{13}C NMR in vitamin B_{12} pathway, 103
C-5 pathway
  (see Glutamate pathway),
Cadmium,
  activation of bovine 5-aminolaevulinate dehydratase by, 29
  as ferrochelatase inhibitor, 94
Carotenoid biosynthesis genes (*crt*), 283-284
Chelation of,
  (see Magnesium chelatase)
  (see Ferrochelatase)
  iron in haem and bilin biosynthesis, 91-96, 171-172, 172-178
  magnesium in chlorophyll biosynthesis, 179-182
Chelating agents,
  effects of iron chelating agents on greening, 205
  effects of EDTA on 5-aminolaevulinate dehydratase, 20-21
Chemicals which induce 5-aminolaevulinate synthase, 7-8

Chlorobium chlorophylls
  (see Bacteriochlorophylls *c*, *d* and *e*)
  changes in composition in dim light, 252
  2-(1-hydroxyethyl) in ring A, *(R)* and *(S)* forms, 239-242
  propionate methylation pathway, 247
  vinyl methylation pathway, 247
Chlorophyll(s),
  Chlorobium, 239
  enzymes catalyzing early stages in biosynthesis of, 158-170
  enzymes catalyzing late stages in biosynthesis of, 179-201
  hydroxymethyl intermediate between chlorophyll *a* and *b*, 195
  model for regulation of biosynthesis of, 217
  reaction centre, 201
  turnover in plants, 206-207
Chlorophyll *a*,
  possible multiple forms of, 199-200
  structure of, 157,195
Chlorophyll *a* formation,
  chelation of magnesium, 179-182
  chlorophyll synthase, 193-195
  esterification of ring D propionate, 193-195
  isocyclic (E) ring formation, 183-187
  β-oxidation model for ring E formation, 187
  possible multiple forms of, 199-200
  protochlorophyllide reduction (light dependent), 188-192
  protochlorophyllide reduction (dark), 192-193
  magnesium (Mg) chelatase, 179-182
  Mg-protoporphyrin methyl transferase, 182-183
  vinyl group reduction, 187-188
Chlorophyll a^1 in reaction centre, 201
Chlorophyll *b*
  dephytylation of, 197
  multiple forms of, 199-200
  possible pathway(s) for biosynthesis of, 196-198
  structure of, 195
Chlorophyll *b* formation, 195-199
  (see Chlorophyll *a* formation)
  from chlorophyllide *a* geranylgeranyl ester, 197
  from chlorophyllide a, 196
  from chlorophyll *a*, 195-196
  in vivo studies on, 195-198
  in vitro studies on, 198-199
  precursors of, 197
  protochlorophyllide as precursor for, 198
  putative hydroxyl intermediate in, 195
  requirement for NADP$^+$ in, 198
Chlorophyll(s) *c*,

structure of, 157,200
structures of chlorophyll c_1, c_2 and c_3, 200
Chlorophyll biosynthesis genes
    barley protochlorophyllide reductase, 240
    regulation by light, 287
Chlorophyll biosynthesis regulation,
    5-aminolaevulinate synthesis levels in, 203
    effect of light on, 202
    effect of iron chelators on, 205-206
    effect of oxygen on, 205-206
    effector-mediated control in, 211-213
    environmental effects on, 202
    hormonal effects on, 204
    in gymnosperms and algae, 204
    induction of enzymes by light, 203
    in mitochondria and chloroplasts, 209
    model for, 217
    phytochrome, role in, 203
Chlorophyll synthase, 193-195
Chlorophyllase, 193
Chloroplasts,
    effects of Fe chelators and anaerobiosis on greening in, 205-206
    possible location of all chlorophyll synthesis enzymes in, 209
cobA gene of *Salmonella typhimurium*,
    in adenosylation of vitamin B_{12} intermediates, 279
cobB gene of *Salmonella typhimurium*, 280
cob gene clusters in *Pseudomonas denitrificans*, 281-282
    group A (*cobF,G,H,I,J,K,L,M*), 281
    group B (*cobB,E*), 281
    group C (*cobE,A,B,C,D*), 281
    group D (cobinamide -> coenzyme B_{12} genes), 281
cob gene clusters in *Salmonella typhimurium*
    cis-dominant mutations of, 279
    CobI region (cobinamide), 278
    CobII region (5,6-dimethylbenzimidazole), 278
    CobIII region (cobalamin), 278
    CobIV region, 279
CobI operon in *Salmonella typhimurium*, 279
Cobalt pathway to vitamin B_{12}, 101,102
Cobester (cobyrinic acid ester), 113
Cobinamide, 102,132,133
    genes encoding biosynthesis of, 278-282
Cobyric acid, 102
    as precursor for cobalamin, 132
Cobyrinic acid,
    transformation into cobalamin, 132
Cobyrinic acid *a, c* diamide synthase,
    *cob*B in *P.dentrificans*, 282

Coenzyme B_{12},
    evolutionary aspects of, 133-134
    genes encoding the biosynthesis of, 278-282
    role of, 102
    structure of, 101
Coenzyme F_{430},
    absorption maxima, 144
    amide groups in, 141
    autoxidation to F_{560}, 146
    biosynthesis of, from glutamate, 143
    discovery of, 139
    epimerisation of, 144
    ESR spectrum of, 146
    in methanogenesis, 139
    in methyl coenzyme M reductase, 139, 147-152
    oxidation state of nickel in, 149
    presence of nickel in, 143
    properties of, 144-146
    redox behaviour of, 144-146
    reduction state of ring system, 141
    relationship to other macrocyclic tetrapyrroles, 140-141
    role as prosthetic group, 139
    structure of, 139-141
Coenzyme F_{430} biosynthesis,
    from ^{13}C labelled 5-aminolaevulinate, 143
    from dihydrosirohydrochlorin, 143
    from glutamate, 143
    stages in biosynthesis of, 143,145
Coenzyme $F_{430}M$, 144
    ESR spectrum of, 146
    reduced form, 146
    uv/vis spectrum of, 147
Coenzyme M (CoM) formation, 148
Congenital erythropoietic porphyria (CEP),
    enzymic abnormalities in, 263,
    genetic lesions in, 273
Coproporphyria, 263
Coproporphyrinogen III,
    enzymic formation of, from uroporphyrinogen III, 67-77
    decarboxylation by coproporphyrinogen III oxidase, 77-84
    structure of, 68
Coproporphyrinogen III oxidase,
    anaerobic and aerobic forms of, 79
    chromosomal location of, 262-263
    cofactor requirements for anaerobic forms of, 83
    cDNA of, 275
    harderoporphyrinogen as substrate, 79
    ß-hydroxypropionate intermediates in aerobic, 81-82

iso-harderoporphyrinogen, 79
intermediates in decarboxylation, 79
mechanism of aerobic form of, 81-83
mechanism of anaerobic form of, 83-84
mechanistic considerations, 81-84
molecular biology, 78, 262, 274-275
order of decarboxylation reactions, 79, 79
occurrence, isolation and properties, 78-79
purification of, 78
reaction mechanism of, 79-84
requirement for oxygen, 81-83
stereochemical course of aerobic, 79-81
stereochemical course of anaerobic, 83-84
substrates for, 77, 79
yeast, 78
Coproporphyrinogenase
(see Coproporphyrinogen III oxidase)
Corrin biosynthesis
(see Vitamin B_{12} biosynthesis)
Corrinoids in prebiotic period, 134
crt genes of photosyntheic bacteria, 282-287
Cutaneous hepatic porphyria
(see Porphyria cutanea tarda)
Cyanocobalamin
(see vitamin B_{12})
Cycloheximide, 214
Cysteine trisulphide, 6
cysG gene in *E. coli*, 132,282
Cytochrome P450:
induction by phenobarbitone, 8
suicide destruction by drugs, 8

Decarboxylation at C-12 in B_{12} biosynthesis, 116
steric course of, 119
2-Desacetyl-2-hydroxyethyl bacteriochlorophyllide *a* dehydrogenase, 285
Deuteroporphyrin, as substrate for ferrochelatase, 93
4,5-Dioxovalerate, 18,19
transaminase, 18,19
4,5-Diaminovalerate, 18
3,5-Dicarbethoxy-1,4-dihydrocollidine (DDC), 8, 94
as ferrochelatase inhibitor via N-alkylporphyrins, 8
effect of, on haem biosynthesis, 8
suicide substrate for cytochrome P_{450}, 8
Dicyanocobinamide, 104
Didehydro-F_{430} (F560), 146
Dihydro-factor II (precorrin-2), 111
Dihydro-factor III (precorrin-3), 111
Dihydrogeranylgeranyl group, 193-195
5,6-Dimethylbenzimidazole (DMBI), 101
5,6-Dimethylbenzimidazole phosphoriboside (DMBI-R-P), 133
Dimethylsulfoxide (DMSO), effects on hepatic and erythroid cells, 7
Dinoflagellate luciferin, 158
4,6-Dioxoheptanoic acid
(see Succinylacetone)
Dipyrromethane cofactor in porphobilinogen deaminases, 45-48
assembly of, 45, 49, 50-51
attachment to enzyme through cysteine, 48
discovery of, 43-44
labelling from 5-aminolaevulinic acid, 44, 47
labelling from porphobilinogen, 43, 48
reaction with α-bromoporphobilinogen, 46
role as primer, 52
structure of, 44, 47, 48
universal occurrence of, 44, 45
cDNA encoding *hem* genes in eukaryotes
coproporphyrinogen III oxidase, 274-275
ferrochelatase, 275
porphobilinogen deaminase, 269-271
uroporphyrinogen III synthase, 272
uroporphyrinogen III decarboxylase, 273
Doss porphyria, 24,263
enzymatic abnormalities and genetic defects, 24, 263
Drugs which induce porphyrias,
phenobarbitone, 8
5ß-steroids, 8
3,5-dicarbethoxy-1,4-dihydrocollidine (DDC), 8

E ring of chlorophylls and bacteriochlorophylls
(see Isocyclic ring formation)
Ehrlich's reaction of dipyrromethane cofactor, 44
Enzyme defects in porphyrias, 263
Enzyme intermediate complexes of porphobilinogen deaminase, 42, 46, 50
Erythroid 5-aminolaevulinate synthase, 264-266
regulatory mechanisms, 7, 9, 264
induction and repression in MEL cells, 7, 9
Erythroid cell(s), regulation of haem biosynthesis in, 7, 9
with leukaemia cells, 9
role of haem pool concentration, 9
Erythropoietic porphyrias,
enzyme defects in, 263
(see Congenital erythropoietic porphyria (CEP); Erythropoietic protoporphyria, EP)
Erythropoietic protoporphyria (EPP),
enzymatic abnormalities, 263
Esterification of ring D propionate in chlorophyll

biosynthesis, 193-195
  non-involvement of chlorophyllase, 193
  esterifying alcohol pyrophosphate derivatives, 194
Esterifying groups in bacteriochlorophyll biosynthesis,
  phytyl ester in bacteriochlorophyll *a* from *R. sphaeroides*
  farnesyl ester in bacteriochlorophylls *c*, *d* and *e*, 239
  farnesyl ester in bacteriochlorophyll *g*, 240
  geranylgeranyl ester in bacteriochlorophyll *a* from *R. rubrum*, 238
  stearyl ester in bacteriochlorophyll *c* from *Chloroflexus aurantiacus*, 239
4-Ethyl group formation in bacteriochlorophyll *a* biosynthesis,
  trans-addition of hydrogen, 244
  two stage mechanism, 244-245
  via ring B ethylidine, 244-245
Ethyl group formation in chlorophyll biosynthesis, 187-188
Euglena 5-aminolaevulinate synthase, 161

F_{430}
  (see Coenzyme F_{430}),
Factor I (precorrin-1), structure of, 108
Factor II (precorrin-2), structure of, 108
Factor III (precorrin-3), structure of, 108-109
Factor F_{430}
  (see Coenzyme F_{430})
Factors S_1-S_4,
  isolation from Co^{++}-free incubations, 123
  properties and structure of, 123
Ferrochelatase, 91-96, 171-172
  activation by lipids, 92
  active site of, 95-96
  branch point regulation of, 171
  catalytic mechanism, 95
  importance of sulphydryl groups, 95
  inhibition by *N*-alkylporphyrins, 94-95, 172
  inhibition by lead and mercury, 94
  in photosynthetic organisms, 171-172
  isolation of, 92, 172
  N-methylprotoporphyrins as inhibitors of, 94-95, 172
  mitochondrial import, 92, 171
  molecular biology of, 96, 275-276
  nucleotide sequence of, 275
  occurrence and cellular location, 92, 95, 171
  point mutation of, 276
  primary structure determination, 275
  properties of, 92, 171-172

  proposed kinetic mechanism for, 95
  purification of, 92
  substrate specificity, 93-94
  two forms in plants, 209
Fischer nomenclature, 237
Free-haem pool, 8

Gabaculine, 17, 164, 165-166
GDP-cobinamide, 133
Genes encoding bacteriochlorophyll biosynthesis enzymes,
  genetic map of, 283
  in *Rhodobacter sphaeroides*, 261, 282-287
  in *Rhodopseudomonas capsulatus*, 282-287
  regulation by oxygen levels, 285
Genes encoding chlorophyll biosynthesis enzymes,
  cDNA for 5-aminolaevulinate dehydratase, 22, 25
  cDNA for protochlorophyllide reductase, 287
Genes/cDNAs encoding haem biosynthesis enzymes specifying,
  5-aminolaevulinate dehydratase, 22, 25, 268-269
  5-aminolaevulinate synthase, 263-266
  coproporphyrinogen oxidase, 274-275
  ferrochelatase, 275-276
  glutamate pathway enzymes, 266-268, 281
  porphobilinogen deaminase, 270
  uroporphyrinogen III decarboxylase, 273-274
  uroporphyrinogen III synthase, 273
Genetic defects in the porphyrias, 262-263
  chromosome location of, 262
Geranylgeranyl pyrophosphate,
  in chlorophyll D ring esterification, 194
Globin synthesis in erythroid cells coordinated with haem synthesis, 9
Glutamate pathway for 5-aminolaevulinate biosynthesis,
  discovery and occurrence of, 15-19, 161-162
  distribution amongst species, 161-162, 163
  enzymes of, 16-19, 162-166
  genes encoding enzymes of the, 266-268, 281
  glutamate 1-semialdehyde, 17-18, 19, 160, 164-165
  glutamate 1-semialdehyde (cyclic form), 17-18, 19, 160, 164-165
  glutamate 1-semialdehyde aminotransferase, 165-166
  glutamyl-tRNA, 16, 268, 163-164
  glutamyl-tRNA synthetase (ligase), 164
  in vitamin B_{12} biosynthesis, 133,
  mechanism of 5-aminolaevulinate formation, 162
  $tRNA_{glu}$, 16, 163-164, 268

structures of intermediates, 19, 160
Glutamate 1-semialdehyde, 17-18, 162-164
   cyclic structure (HAT), 18, 19, 160, 165
Glutamate 1-semialdehyde aminotransferase (1,2-aminomutase), 17, 165-166
   genes encoding, 267-268, 281
   inhibitors, 166
   mechanism, 18, 166
Glutamyl-tRNA synthase (ligase), 16, 164
Glutamyl-tRNA, 16, 163-164, 268
Glutamyl-tRNA reductase (dehydrogenase), 16, 165
   genes encoding, 267-268, 281
Glycine pathway for 5-aminolaevulinate biosynthesis
   discovery of, 1-4,
   enzymology of 4-6,
   in *Euglena* 159-161
Glycine in haem biosynthesis, 10
Gunther's disease
   (see Congenital erythropoietic porphyria)

Half-site reactivity of 5-aminolaevulinate dehydratase, 30
Harderian gland, 79
Harderoporphyrin, 79
Harderoporphyrinogen, 79
iso-Harderoporphyrinogen, 79
HAT (5-Amino-6-hydroxytetrahydropyran-2-one)
   (see Glutamate 1-semialdehyde, cyclic form)
Haematoporphyrin as substrate for ferrochelatase, 93
Haem,
   as phycobilin precursor, 172
   direct feedback regulation on 5-aminolaevulinate synthase, 5
   effect on mitochondrial transport, 7
   control of 5-aminolaevulinate synthase levels by, 6-9, 14
   origin of carbon and nitrogen atoms in, 3
   regulation of chlorophyll biosynthesis by, 211-214
   structure of, 92
   transcriptional regulation by, 7
   turnover in plants, 206-207
Haem biosynthesis genes (*hem* genes),
   in *Bacillus subtilis*, 260-261
   in *Escherichia coli*, 258-259
   in humans, 262-263
   in mammals, 262-263
   in *Rhodobacter sphaeroides*, 261
   in *Salmonella typhimurium*, 259-260
   in *Staphylococcus aureus*, 260
   in yeast, 261-262
Haem biosynthetic pathway,
   early studies, 2
   enzymes (see each individual enzyme entry)
   intermediates of the, 2
   molecular biology of enzymes of the, 263-281
   precursors, 3, 4
   regulation in chloroplasts, 202-218
   regulation in higher animals, 6-9
Haem breakdown, 172-174
   by haem oxygenase, 173
   to bilivedin and other bilins, 172-178
   to *N*-substituted porphyrins, 8, 94
Haem deficient mutants,
   resistance to antibiotics, 258
   use in mapping *hem* genes, 258
Haem degradation, 2, 41, 74, 80
Haem oxygenase from *Cyanidium*, 173
   fractionation of, 174
   ferredoxin containing fraction (I), 174
   haem binding fraction (II), 174
   inhibition by Sn-protoporphyrin
   mesohaem as substrate for, 173
   NADPH binding component (III), 174
   requirement for O_2 and *iso*-ascorbate, 173
Haem pool, regulation of 5-aminolaevulinate synthase through, 5-9
Haem synthase
   (see Ferrochelatase)
Haems, types of, 156
hem genes
   (see also Haem biosynthesis genes)
   hemA, 6,14,258,259
   hemA (glycine path), 263-266
   hemA (glutamate path), 258-260, 266-268
   hemB, 25, 258-260,
   hemC, 34, 259, 269
   hemD, 58, 258-260, 272
   hemE, 259-261
   hemF, 259-260
   hemG, 259-261
   hemL, 267-268, 287
   hemT, 6, 14, 264
   HEM1, 14
   HEM2, 268
   HEM12 (*hemE*), 274
   HEM13, 274
   HEM15, 275
hem operon
   in *E. coli*, 259
   in *Bacillus subtilis*, 260-261, 272
hemC (*hem*) promoter in *E. coli*, 269
Hepatoerythropoietic porphyria (HEP),

genetic lesions in, 263
Heptacarboxylic acid porphyrinogens, 69-72
Hereditary coporporphyria (HC),
    enzymatic abnormalities, 263
Hexacarboxylic acid porphyrinogens, 69-72
Hexachlorobenzene, porphyrinogenic role of, 69
Hydroxymethylbilane synthase
    (see Porphobilinogen deaminase)
Hydride catalytic mechanism for coporphyrinogen oxidase, 82
Hydroxymethyl chlorophyll, putative intermediate, 195
Hydroxymethylbilane isomers,
    structural requirements for ring D inversion 55-56
    substrates for uroporphyrinogen III synthase, 55
Hydroxymethylbilane
    (see Preuroporphyrinogen)
Hydroxymethylbilane synthase
    (see Porphobilinogen deaminase)
Hydroxyporphobilinogen, 49
ß-Hydroxypropionate catalytic mechanism for coproporphyrinogen oxidase, 81-82
ß-Hydroxypropionate intermediates for coproporphyrinogen III oxidase, 81-2, 84
6-ß-Hydroxypropyl protoporphyrin monomethylester, 184
8-Hydroxyquinoline, 184

Inherited defects in human haem synthesis enzymes, 262-263
Iron (Fe) branch of tetrapyrrole biosynthetic pathway,
    ferrochelatase and protohaem formation, 91-96, 171-172
    bilin synthesis from haem, 172-178
Iron (Fe) chelators, effects of, on greening, 184
Isocyclic ring formation in chlorophyll biosynthesis:
    cyclase enzyme system, 186
    inhibitors of, 186-187
    intermediates for, 185, 187
    reconstitution of cyclase for, 186-187
    requirement for O_2, 187
    substrate requirements for, 187
    stimulation by S-adenosyl methionine, 186
Isoharderoporphyrinogen, 79
Isomers of uroporphyrinogens, structure of, 33
IUPAC-IUB nomenclature (International Union of Pure and Applied Chemistry and International Union of Biochemistry nomenclature), 31, 237

6-ß-Ketopropyl protoporphyrin monomethylester,

Laevulinic acid,
    inhibitor of 5-aminolaevulinate dehydratase, 24
    inhibitor of chlorophyll synthesis, 161,214
Lead,
    as ferrochelatase inhibitor, 94
    inhibition of mammalian 5-aminolaevulinate dehydrases by, 23-24
Lead poisoning,
    as measured by 5-aminolaevulinate inhibition, 23
Leukaemia (MEL) cells, studies of erythropoiesis by, 6, 9
Light regulation of chlorophyll and haem synthesis,
    5-aminolaevulinate synthesis, 211-215
    haem as a regulatory device, 212-213
    in organelles, 209-210
    model for, 217
    photoreduction of protochlorophyll(ide), 202-203
    phytochrome directed stimulaton of 5-aminolaevulinate synthesis, 203
    plant hormone involvement, 204
    protochlophyllide reductase, 191, 216-218
Luciferin, dinoflagellate, 158

Macrocyclic tetrapyrroles,
    degree of conjugation and unsaturation in, 141-142
Magnesium (Mg) branch of the tetrapyrrole biosynthetic pathway, 179-199
    bacteriochlorophylls
    (see Bacteriochlorophyll)
    chlorophyll a
    (see Chlorophyll a)
    chlorophyll b
    (see Chlorophyll b)
    chlorophyll c, 200-201
    possible multiple forms of chlorophylls a and b, 199-200
    reaction centre chlorophylls, 201
    reaction centre pheophytins, 201-202
Magnesium (Mg) insertion into protoporphyrin IX, 179-182
Mg-chelatase reaction,
    inhibitors of, 181
    in plants, 179-182
    in *Rhodobacter sphaeroides*, 179, 182
    regulation of chlorophyll synthesis, 215-216
    substrate requirements, 181
Mg-2,4-divinyl phaeoporphyrin a_5, 183-184
    accumulation in *R. sphaeroides*, 184

Mg-divinyl protochlorophyllide, 183
Mg-monovinyl chlorophyllide, 185
Mg-monovinyl protochlorophyllide, 184
Mg-protoporphyrin IX methyltransferase, 182-183
Mg-protoporphyrin IX,
    conversion of, to protochlorophyllide, 179-182
Mg-Protoporphyrin methyl transferase
    (see
    S-Adenosyl-methionine-Mg-protoporphyrinIX methyltransferase)
Map of methyl coenzyme M reductase (mcrA,B,G) genes, 152
Mapping of *hem* genes, 257-263
7-Mercaptoheptanoylthreonine phosphate (also component B; H-S-HTP), 148
    heterodisulphide with methyl coenzyme M, 148
    reaction with methyl coenzyme M, 148
mcrA,B,C,D,G genes, map of in methanogenic bacteria, 152
Mercury as ferrochelatase inhibitor, 94
Mesohaem, 173
Mesoporphyrin as substrate for ferrochelatase, 93
Methane formation,
    from methyl chloride, 146
    from methyl coenzyme M, 148
Methanobacterium thermoautotrophicum, 143, 150-151
Methanogenic bacteria,
    requirement of nickel for growth, 139
Methine protons,
    origin of in vitamin B_{12} biosynthesis, 120
Methyl cobalamin, 101
Methyl coenzyme M,
    analogues of, 149
    in methane formation, 148
    structure of, 149
Methyl coenzyme M reductase (component C), 140
    absorption maxima of, 147
    activation of, 149
    coenzyme M binding subunit, 151
    CoM-S-S-HTP reductase, 148
    derived amino acid sequence of α,β and τ subunits of, 150-151
    function of, 147
    gene cluster, 152
    genes encoding α,β and τ subunits of, 149-152
    inhibitors of, 149
    reduction by 7-mercaptoheptanoylthreonine phosphate (H-S-HTP), 148
    mcr genes encoding, 152
    molecular weights of complex and subunits, 147
    specificity of, 149
    subunit structure of, 147,149
Methyl groups of vitamin B_{12},
    at C-1, 106
    at C-12, 106
    gamma effect, 105
    origin of, 104
    steric course of methylation, 106
Methylation of uroporphyrinogen I, 123-128
Methylation pathways in bacteriochlorophyll biosynthesis, 247-248
Methyltransferases in B_{12} biosynthesis, 115, 128-132
Methyltransferase M-1:
    role in the methylation of uroporphyrinogen III, 132
    similarity to SUMT, 132
N-Methylprotoporphyrin as ferrochelatase inhibitor, 94-95, 172
Molecular biology of haem biosynthesis enzymes, 263-281
    5-aminolaevulinate synthase, 263-266
    coproporphyrinogen III oxidase, 261, 274-275
    ferrochelatase, 275-276
    glutamyl-tRNA reductase, 266-268
    glutamate 1-semialdehyde aminotransferase, 266-272
    porphobilinogen deaminase, 268-272
    uroporphyrinogen III synthase, 272-273
    uroporphyrinogen III decarboxylase, 273
Molecular genetics,
    of haem synthesis, 258-263
    of bacteriochlorophyll synthesis, 282-287
    and porphyrias, 262-263
Multiple genes of 5-aminolaevulinate synthase,
    in photosynthetic bacteria, 6, 14, 263-264
    in mammals, 14-15, 264-266
    sequence comparisons and identification of an enzyme core, 15
Multiple pathways for chlorophyll biosynthesis, 199-200
Murine erythroleukaemia (MEL) cells,
    induction and repression of 5-aminolaevulinate synthase in, 7, 9
Mutants of the bacteriochlorophyll biosynthesis pathway, 286
Mutations of genes of haem biosynthesis in porphyrias,
    porphobilinogen deaminase, 271-272
    uroporphyrinogen decarboxylase, 271
    uroporphyrinogen III synthase, 273
NADPH-protochlorophyllide oxidoreductase
    (see Protochlorophyllide reductase)
Neo-cobinamide, 104

Neuberger, 2,3
Nickel porphinoids
 (see Coenzyme F_{430})
Nomenclature of tetrapyrroles, 31
 Fischer system, 237
 I.U.P.A.C. system, 237
Nonacute porphyrias,
 classification of, 262-263
 enzyme defects in, 262-263
Nucleotide sequencing of haem synthesis genes and cDNA encoding,
 5-aminolaevulinate dehydratase, 268-269
 5-aminolaevulinate synthase, 268-266
 coproporphyrinogen III oxidase, 274-275
 ferrochelatase, 275-276
 glutamyl-tRNA reductase, 267
 glutamate 1-semialdehyde aminotransferase, 267-268
 porphobilinogen deaminase, 269-272
 uroporphyrinogen III decarboxylase, 273
 uroporphyrinogen III synthase, 272-273

Oxygen repression of bacteriochlorophyll biosynthesis, 283-285

P_{700} pigment, 201
Pentacarboxylic acid porphyrinogens, 69, 73
Petroporphyrins, 142
Phenobarbitone,
 effects on ALA synthase and porphyrin biosynthesis, 8
Pheophytins, reaction centre, 201-202
Photosensitivity, skin lesions in porphyrias, 262
Photosynthetic bacteria,
 adaptation to dim light, 250-253
 adaptation to light and oxygen, 282
 bacteriochlorophyll synthesis genes in, 282-287
 genetic map of *puf*, *crt* and *bch* genes in, 284
 photosynthetic membrane assembly in, 285
Photosystem I, 201
Photosystem II, 201
Phriaporphyrin in biosynthesis of coproporphyrinogen III, 69
Phycobilin(s),
 biliverdin as precursor of, 173
 biosynthesis of, 177
 chromophore structures, 175
 haem as precursor of, 172
 hypothetical biosynthetic sequence for, 175
 ligation to apoproteins, 157, 178
 possible biosynthetic relationship of phytochrome chromophore to, 157
 regulation by light, 218-219

responses to nitrogen status, 221
Phycobilin chromophores,
 structure, 175
Phycobiliproteins, 157, 220-222
 biosynthesis of, 218-219, 220
Phycobilisomes, 157,
 pigment changes in light of different wavelengths, 219-220
Phycochromes, 219
Phycocyanin(s), 157, 219
 derived structures of protein subunits, 222
 expression of α- and ß-subunits of in *E. coli*, 221
 genes encoding *a* and *b* subunits, 221
 regulation of apoprotein synthesis, 220
Phycocyanin chromophore, 157
Phycocyanobilin,
 algal, 157
 chemical structure of, 157, 176
 enzyme requirements, 177
 pathway for biosynthesis, 176
 synthesis from ALA in dark, 218
Phycoerythrins, 157, 219
Phycoerythrobilin, chemical structure of, 175
Phytochrome, 157-158
Phytochrome chromophore, 157, 177
 possible biosynthetic relationship to phycobilins, 177-178
Phytyl pyrophosphate in chlorophyll biosynthesis, 194
Pigment changes in response to light wavelength in Chlorobiaceae
 bacteriochlorophyll *d* to *c* switch in dim light, 253
 increased methylation at 4-position in response to dim light, 252
 meso-methylation, 252-253
 red shift of antenna bacteriochlorophylls by methylation, 252-253
 spectral changes in response to light wavelength, 252-253
Pigment composition in response to environmental conditions,
 bacteriochlorophyll regulation by light and O_2, 5-6, 283
 chlorophyll and haem turnover, 206-207
 effects of iron chelators and anaerobiosis on greening, 205-206
 methylation of *Chlorobium* chlorophylls, 252-253
 nonplastid haem synthesis and turnover, 208
 physiology of greening in plants and algae, 202-204

regulation of phycobilin content by light, 218-221
responses to nitrogen status, 221-222
Plastids, development of, 203-204
"Polypyrroles" formed by porphobilinogen deaminase,
'D' and 'P', 35, 36
with nitrogenous bases, 35, 36
Porphobilinogen,
biosynthesis of, 19
discovery of, 3-4
formation of the dipyrromethane cofactor from, 44, 50-51
properties of, 33
structure of, 4, 20
Porphobilinogen deaminase (hydroxymethylbilane synthase), 36-57
active site groups, 43, 57
attachment site for the dipyrromethane cofactor, 48, 269
catalytic cleft, 58
cellular location in plants, 210
chain termination by α-bromoporphobilinogen, 46
conserved protein sequences, 57
cysteine mutagenesis, 48
dipyrromethane cofactor in, 43-48
experiments with aminomethylbilanes in, 35-36
experiments with aminomethyldipyrromethanes in, 35-36
experiments with aminomethyltripyrranes in, 35-36
hydrolysis reaction, 52
inhibition by pyridoxal 5'phosphate, 57, 58
intermediate complexes of, 42, 46, 50
isoforms of, 269-272
lesions in acute intermittent porphyria, 271-272
mechanism of action of, 49-52
molecular biology of, 57, 269-272, 168-169
nature of enzymic group in porphobilinogen covalent binding, 45-48
nucleotide sequences, 57
occurrence and isolation of, 34, 168
order of addition of substrates, 41-42, 169
in plants, 168-169
possible steric course of reaction, 53-55
properties of, 34
protein structure of, 49, 57-58
regulation of erythroid form by NF-E1 and NF-E2, 270
site-directed mutagenesis, 57,58
stereochemical studies, 53, 86-87
structure of dipyrromethane attachment site, 48,49
X-ray studies on, 58
Porphobilinogen synthase
(see 5-Aminolaevulinate dehydratase)
"Porphobilinogenase", 34
Porphyria,
acute intermittent, 263
5-aminolaevulinate (Doss), 263
coproporphyria, 263
cutanea tarda, 263
enzymic lesions in, 262-263
erythropoietic protoporphyria, 263
hepatoerythropoietic, 263
variegate, 263
Porphyria cutanea tarda (PCT),
enzymic lesion, 263
excretion of oxidised intermediates in, 70
genetic lesions in, 274
Porphyrin(s),
(see Tetrapyrroles)
N-alkyl-substituted, 95
biosynthesis pathway, 2
degradation of, 2, 41, 74, 80
enzymes in synthesis of, 1-96
nomenclature, 31, 237
structure of, 92-94, 141-142
Porphyrinogen(s),
structure of uroporphyrinogens, 33
intermediates in the uroporphyrinogen decarboxylase reaction, 68
Porphyrinogenic drugs, 8
Posttranslational regulation by haem, 5-6, 7-9
Precorrin-1 (tetrahydro factor I), 111
Precorrin-2 (dihydro factor II), 111
Precorrin-3 (dihydro factor III), 111
Precorrin 4a, 118, 120
Precorrin 4b, 118, 120
Precorrin-5, 118, 120
Precorrin-6a, 119
Precorrin-6b, 118
Precorrin-6x, 135
Precorrin-7, 118, 119
Precorrin-8, 120
Precorrin-8a, 118, 119
Precorrin-8b, 118, 119
Preuroporphyrinogen, 44, 49, 53, 102
analogues of, 55-56
chemical synthesis, 40
discovery of, 38
formation of, 39
half life, 38
nmr spectra, 38
order of assembly of pyrrole rings of, 41-42

substrate for uroporphyrinogen synthases, 39, 40
Promoters for,
　porphobilinogen deaminase (erythroid), 270
　porphobilinogen deaminase (ubiquitous), 270-271
　uroporphobilinogen decarboxylase, 273-274
Propionobacterium shermanii, in vitamin B_{12} biosynthesis
　(see Chapter 3)
Protochlorophyllide,
　conversion of Mg-protoporphyrin IX to, 180
　stages in formation from Mg-protoporphyrin IX, 184
　photoconversion of, 189
Protochlorophyllide reductase in plants,
　cDNA sequences from barley and oat, 287
　expression of in *E. coli*, 287
　precursor protein, 287
　transit peptide of, 287
Protochlorophyllide reduction, 188-193
　dark reaction pathway, 192-193
　light-requiring pathway, 188-192
　light-driven back reaction, 191
　organisms having two mechanisms for, 192
　photointermediates in, 190-192
　proteolytic degradation in light, 216
　regulation of, 191, 216-218
　spectroscopic studies on, 190-192
Protohaem
　(see Protoporphyrin IX)
Protohaem ferrolyase
　(see Ferrochelatase)
Protoporphyria, erythropoietic
　(see Erythropoietic protoporphyria)
Protoporphyrin IX,
　branch point for chlorophyll biosynthesis, 155
　biosynthesis in plants, 167
　non enzymic formation, 84
　pathway from ALA, 167
Protoporphyrin XIII, as substrate for ferrochelatase, 93
Protoporphyrin IX biosynthesis, enzymes of,
　(see Tetrapyrrole biosynthesis)
　5-aminolaevulinate dehydratase, 19-28, 167-168
　5-aminolaevulinate synthase, 6-17, 161
　coproporphyrinogen oxidase, 77-84, 170
　glutamyl-tRNA synthase, 16, 164
　glutamyl-tRNA reductase, 17-18, 165
　glutamate 1-semialdehyde aminotransferase, 18-19, 165-166
　porphobilinogen deaminase, 36-57, 168-169
　protoporphyrinogen oxidase, 84-91, 170,

uroporphyrinogen decarboxylase, 67-77, 169-170
uroporphyrinogen III synthase, 30-32, 34, 53-59, 169
Protoporphyrinogen IX, biosynthesis of, 77-84, 170, 241
Protoporphyrinogen IX oxidase, 84-91, 170
　activity in plants, 170
　cellular location of, 85
　cofacial oxidation of protoporphyrinogen IX, 91
　coupling to respiratory chain in prokaryotes, 85
　flavin in, 91
　mechanism of, 86, 90-91
　occurrence and isolation of, 85
　possible catalytic mechanism of, 90-91
　properties of, 86
　stereochemical studies on, 86, 89-91
　stoichiometry of oxidation, 89-90
Puf operon in photosynthetic bacteria, 284
　promoter of, 285
　role of, 284
pufQ,
　in regulation of flux of bacteriochlorophyll intermediates, 284
Pulse experiments in the study of methylation in B_{12} synthesis, 120
Purple bacteria, 238
Pyridoxal 5'phosphate,
　aminolaevulinic acid synthase, 12, 13
　glutamate 1 semialdehyde transaminase, 18
　inhibition of porphobilinogen deaminase, 57-58
Pyridoxamine 5'phosphate, 18
Pyrrole, mixed, formation of, 26
Pyrrocorphins in vitamin B_{12} biosynthesis, 118

Regulation of haem biosynthesis
　direct feedback regulation on 5-aminolaevulinate synthase, 5
　effect on mitochondrial transport, 7
　control of 5-aminolaevulinate synthase levels, 6-9, 14
　transcriptional regulation by, 7
　turnover in plants, 206-207
Reaction centre pigments,
　of Chlorobiaceae, 239
　chlorophylls, 201
　pheophytins, 201-202
Reaction centre (RC) polypeptides,
　genes encoding, 282, 286
　RC-H, 286
Rhodobacter sphaeroides (also *Rhodopseudomonas*

sphaeroides),
    adaptation to photosynthesis, 5
    5-aminolaevulinate dehydratase enzyme from, 20,24
    5-aminolaevulinate synthase from, 10,14
    bacteriochlorophylls in, 238
    bacteriochlorophyll biosynthesis genes (*bch*), 282
    carotenoid biosynthesis genes (*crt*), 282
    coproporphyrinogen oxidase in, 83
    ferrochelatase from, 92

Rhodopseudomonas capsulatus
    bacteriochlorophyll biosynthesis genes (*bch*), 282
    carotenoid biosynthesis genes (*crt*), 282
    gene clusters of *bch* and *crt* genes in, 282
    genetic map of *puf*, *crt* and *bch* genes, 284

Rhodospirillum rubrum,
    coproporphyrinogen oxidase in, 83

Ring contraction mechanism (Eschenmoser) in corrins, 120

mRNA and cDNA specifying,
    5-aminolaevulinate dehydratase, 22, 268-269, 281
    5-aminolaevulinate synthase, 6-9, 264-266, 281
    bacteriochlorophyll genes, 285
    coproporphyrinogen III oxidase, 261, 274-275, 281
    ferrochelatase, 275-276, 281
    glutamyl-tRNA reductase (dehydrogenase), 266-268, 281
    glutamate 1-semialdehyde aminotransferase, 266-268, 281
    porphobilinogen deaminase, 269-272, 281
    uroporphyrinogen III decarboxylase, 273, 281
    uroporphyrinogen III synthase, 58, 272-273,

$tRNA_{glu}$,
    role in glutamate pathway, 16, 163

$15,17^3$-Seco-F_{430}-17^3-acid (seco-F_{430}), 143
Shemin, 2,3
Sirohaem, 102
Sirohydrochlorin (factor II, dihydro-precorrin-2), 102, 108
Spiro-mechanism for uroporphyrinogen III synthase, 56
Spiro lactams, 56, 57
Spiro-pyrrolenine, 56
5ß-Steroids,
    induction of 5-aminolaevulinate synthase, 8
    stimulation of haem biosynthesis, 8

Succinylacetone as inhibitor of haem biosynthesis,
    effect on 5-aminolaevulinic acid synthase, 7
    inhibitor of 5-aminolaevulinic acid dehydratase, 24
Succincyl-CoA, haem precursor, 5, 12, 159
Sulphydryl (SH) groups, importance of in
    5-aminolaevulinate dehydratase, 29-30
    attachment of dipyrromethane cofactor to apo-deaminase, 48, 269
    ferrochelatase, 94
Tetrahydrogeranylgeranyl group, 193-195
Tetrapyrrole(s)
    biosynthesis pathway, of, 2, 156,
    early investigations, 2-4
    numbering system, 237
    structural relationships between, 140-142
Tetrapyrrole biosynthetic pathway,
    outline of, 2, 156
Tetrapyrrole biosynthesis in photosynthetic organisms,
    5-aminolaevulinic acid biosynthesis from glutamate, 15-19, 161-166
    5-aminolaevulinic acid biosynthesis from glycine, 5-6, 159-161
    iron (Fe) branch, 171-172
    magnesium (Mg) branch, 179-202
    protoporphyrin IX biosynthesis from 5-aminolaevulinate, 167-170
Tetrapyrrole regulation,
    of biosynthetic steps, 215,
    branch points, 215-216
    effector-mediated, of ALA-forming activity, 211-213
    expression and turnover of ALA-forming enzymes, 213-215
    of metabolic activity, 210-211
    of protochlorophyllide photoreduction, 188-192, 192-193
    subcellular compartmentation of tetrapyrrole biosynthesis, 209-210
Transcriptional regulation by haem, 7-9
Trimethylpyrrocorphin, 132
Tunichlorin, 142
Turacin, 140

Ubiquitous genes
    (see individual enzymes and genes)
Uroporphyrin
    (see Uroporphyrinogens)
Uroporphyrinogen(s),
    biosynthesis of, 30-59
    discovery of, as intermediates in tetrapyrrole biosynthesis, 67

isomerisation in acid, 33
isomers, 33
properties of, 32, 33
Uroporphyrinogen isomers, structure of, 33
Uroporphyrinogen I, 33
    factors S_1-S_4 derived from, 123, 127
    methylation of in vitamin B_{12} biosynthesis, 123-127
    formation of, 39-40
Uroporphyrinogen I synthase
    (see Porphobilinogen deaminase)
Uroporphyrinogen III
    (see also uro'gen in some texts),
    decarboxylation to coproporphyrinogen III, 67-68
      (See Coproporphyrinogen III, biosynthesis of)
    early studies on biosynthesis of, 31-32
    origin of pyrrole rings of, 41-42, 169
    porphobilinogen deaminase in biosynthesis of, 38-40, 49-52
    from preuroporphyrinogen, mechanism for, 56
    mechanisms for formation, 35, 53-57
    structure of, 33
    as vitamin B_{12} precursor, 102,106-107
Uroporphyrinogen III cosynthase
    (see Uroporphyrinogen III synthase)
Uroporphyrinogen III decarboxylase, 67-77
    catalytic mechanism of, 76
    chromosomal location of gene for, 262
    hexachlorobenzene poisoning, 69
    inhibition of by sulphydryl reagents, 72
    interaction of substrates, 70-71
    intermediates in the decarboxylation, 68-70
    mechanism of, 75-77
    molecular biology of, 273-274
    number of active sites, 73-74
    occurrence and properties of, 72
    ordered sequence of decarboxylation, 69-72
    reduced half life in mutants, 274
    steric course of, 74-75
    structure of, 72, 273-274
    substrate specificity of, 68
    uroporphyrinogens as substrates, 67
Uroporphyrinogen III isomerase
    (see Uroporphyrinogen III synthase)
Uroporphyrinogen III synthase, (also uroporphyrinogen cosynthetase and isomerase)
    early studies on, 31-32
    hydropathy plot, 272
    hydroxymethylbilanes, studies with, 55-56
    mechanisms of action of, 56
    molecular biology of, 58, 272-273
    mutations in, 273
    N-termini, 58

occurrence and isolation, 34,58
over expression of, 58
preuroporphyrinogen as substrate, 38-41
properties of, 34, 58
rapid assay of, 41
steric course of, 86-89
use of synthetic analogues to investigate, 55-56
spiro-mechanism of ring inversion and cyclization, 56
spirolactams as inhibitors, 56
α-bromopreuroporphyrinogen, 46
substrate for 38-41
(see also Preuroporphyrinogen)
wheat, 32, 169

Variegate porphyria (VP), 24
    enzymatic abnormality, 263
Vitamin B_{12} (Cyanocobalamin),
    structure of, 101
Vitamin B_{12} biosynthesis,
    adenosylation in, 279
    attachment of 1-amino-2-propanol, 132
    biosynthesis of the nucleotide loop, 132-133, 108
    biosynthetic pathway, 278
    deacetylation at C-19 during, 120-121
    decarboxylation at C-12, 116-119
    dihydrofactor II, 108
    dihydrofactor III, 108
    Eschenmoser ring contraction model, 120
    evolution of, 133-134
    general pathway for, 102
    genes of, 133, 276-282
    glutamate (C-5 pathway) in, 133
    incorporation of methionine, 112-116, 122
    in *Pseudomonas denitrificans*, 133
    in *Salmonella typhimurium*, 133
    intermediates of
    (see Precorrins)
    in the absence of, 112
    isobacteriochlorins and their relationship to, 107-112
    isotopic shifts during, 120
    methylation sequence using pulsed experiments in, 112-116, 122
    migration of C-20 -> C-1 in, 119
    origin of methine protons in, 120
    proposed timing of cobalt insertion, 120
    proton exchange during, 120-123
    regulation of, 279
    reverse pulse experiments for order of methylation, 115-116
    tetrahydro factor I (precorrin-1), 108
    transport of, 280

Vitamin B$_{12}$ (cobalamin) biosynthesis genes (*cob* genes), 276-282
    catabolite repression in *S. typhimurium* on, 279
    chromosomal location of, 278
    cluster of, in *P. denitrificans*, 282
    cluster of, in *S. typhimurium*, 278
    clusters of, in *Pseudomonas denitrificans*, 281-282
    *cob*A in adenosylation, 279
    genes encoding methyl transferases, 282
    identification of using *cob* mutants, 278-282
    in *Agrobacterium tumefaciens*, 281
    in *Bacillus megaterium*, 281
    in *cob*I operon, 278
    in *Pseudomonas denitrificans*, 278
    in *Pseudomonas putida*, 281
    in *Salmonella typhimurium*, 276
    repression by molecular oxygen, 278
    transcription control, 278

Vitamin B$_{12}$ membrane transport,
    role of *btuB* gene in *E. coli*, 280

Yeast, haem synthesis enzymes,
    5-aminolaevulinate dehydratase, 19, 25
    5-aminolaevulinate synthase, 14-15
    coproporphyrinogen III oxidase, 78
    ferrochelatase, 92, 96
    molecular biology, 261-262
    protoporphyrinogen oxidase, 85

Zinc,
    in 5-aminolaevulinate dehydratases, 22-23
    binding-site sequences in 5-aminolaevulinate dehydratases, 22
    finger in 5-aminolaevulinate dehydratase, 22
Zinc containing corphin, 123